T0305115

MAKING SURVEILLANCE STATES

Transnational Histories

Making Surveillance States opens up new and exciting perspectives on how systems of state surveillance developed over the nineteenth and twentieth centuries. Taking a transnational approach, the book challenges us to rethink the presumed novelty of contemporary surveillance practices, while developing critical analyses of the ways in which state surveillance has profoundly shaped the emergence of contemporary societies.

Contributors engage with a range of surveillance practices, including medical and disease surveillance, systems of documentation and identification, and policing and security. These approaches enable us to understand how surveillance has underpinned the emergence of modern states, sustained systems of state security, enabled practices of colonial rule, perpetuated racist and gendered forms of identification and classification, regulated and policed migration, shaped the eugenically inflected medicalization of disability and sexuality, and contained dissent. While surveillance is thus bound up with complex relations of power, it is also contested. Emerging from the book is a sense of how state actors understood and legitimized their own surveillance practices, as well as how these practices have been implemented in different times and places. At the same time, contributors explore the myriad ways in which these systems of surveillance have been resisted, challenged, and subverted.

ROBERT HEYNEN is an assistant professor in the Department of Communication Studies at York University.

EMILY VAN DER MEULEN is an associate professor in the Department of Criminology at Ryerson University.

Making Surveillance States

Transnational Histories

EDITED BY ROBERT HEYNEN AND
EMILY VAN DER MEULEN

UNIVERSITY OF TORONTO PRESS
Toronto   Buffalo   London

© University of Toronto Press 2019
Toronto Buffalo London
utorontopress.com
Printed in Canada

ISBN 978-1-4875-0315-4 (cloth)   ISBN 978-1-4875-2248-3 (paper)

Library and Archives Canada Cataloguing in Publication

Title: Making surveillance states : transnational histories / edited by Robert Heynen and Emily van der Meulen.

Names: Heynen, Robert, editor. | van der Meulen, Emily, 1977–, editor.

Description: Includes bibliographical references and index.

Identifiers: Canadiana 20190130245 | ISBN 9781487522483 (paper) | ISBN 9781487503154 (cloth)

Subjects: LCSH: Privacy, Right of – History. | LCSH: Intelligence service – History. | LCSH: Public health surveillance – History. | LCSH: Social control. | LCSH: National security.

Classification: LCC JC596 .M35 2019 | DDC 323.44/8 – dc23

University of Toronto Press acknowledges the financial assistance to its publishing program of the Canada Council for the Arts and the Ontario Arts Council, an agency of the Government of Ontario.

Canada Council   Conseil des Arts
for the Arts   du Canada

ONTARIO ARTS COUNCIL
CONSEIL DES ARTS DE L'ONTARIO
an Ontario government agency
un organisme du gouvernement de l'Ontario

Funded by the   Financé par le
Government   gouvernement
of Canada   du Canada

Canada

MIX
Paper from
responsible sources
FSC   FSC® C103567
www.fsc.org

Contents

Illustrations

Foreword

As a child, I spent some years in the then Belgian Congo, the son of a British doctor and nurse on a remote medical mission, in an area where river transport was the most reliable means. But in order for my father to practise medicine there, as a family we first had to go to Antwerp where he was obliged to study, in French, for his doctor of tropical medicine degree. Though I could not have been aware of it at the time (the early 1950s), this was typical of the tight colonial control over medical practices that had evolved from the more inter-imperial regime that had developed in the Congo in the earlier part of the century. At either time, however, the relationship between health surveillance – especially in the fight against sleeping sickness and malaria – and national-state interests was strong, affecting both medical professionals and local African populations.

In the 1970s, I often travelled through Eastern Europe, visiting the labour and concentration camps that were the repugnant remnants of the Nazi regime and also friends living in Soviet satellite countries, especially Romania. Although these were not colonial situations, in each case the role of identification in surveillance and regulation was notable. Again, I did not then know about the ways that early computing technologies were used for identifying and classifying Jews, "gypsies," and homosexuals for grim and gruesome fates, but I could see the ominous ID documents and yellow stars in those haunting museums. Equally, I was shaken to see, behind the Iron Curtain, the extent to which a combination of surveillance tools – such as the *bulletin de identitate*, automobile registrations revealing the city of origin, and, of course, many thousands of informers – enabled the Securitate, the Romanian secret police, to conduct their reign of fear and, sometimes, terror.

A few years ago, I had the opportunity to tour the site of the National Police Headquarters in Guatemala City, where the now notorious surveillance, torture, and assassination operations were based during the years of civil war that ended in 1996. The atrocities carried out *by* the Policía Nacional were meticulously detailed in mouldy dossiers, accidentally discovered in 2005, that were complete with the *cédula* photo ID card, fingerprints, and descriptions of how each victim was sought out, identified, followed, kidnapped, and "disappeared." This evidence of state surveillance and related atrocities is so recent that few Guatemalans do not know of some family member – especially if they were Mayan, in rural areas, or professionals, such as teachers, lawyers, or journalists, in urban ones – who vanished during those fearful years. One family, well known to us for some years, has no idea exactly what befell their journalist husband, father, and grandfather.

I reflect on my own biography, my own experiences, as a reminder that the question of "making surveillance states" is not merely academic or intellectual. Surveillance studies is indeed a fascinating and absorbing area of research, but as my own encounters – and also the chapters of this book – show, they often uncover realities that are viscerally if not fatally experienced. They frequently require analysis that relates not only to privacy or data protection but to urgent questions of rights, indeed, of violence and of life and death. Surveillance is all too often shown to be the means whereby inequitable, unfair, or unjust treatment is meted out and in which certain population groups are rendered acutely vulnerable. These are the consequential "social sorting" processes of surveillance that characterize modern modes of bureaucratic control, ever accented by technologies old – such as the ones mentioned above – or new, which are often described as "data driven."

Many chapters in this book focus on "state surveillance" of extremely unwelcome kinds and this is right. Anyone concerned with surveillance has a duty to be aware of the risks attending the development of any system, however seemingly innocuous. The worst cases stand as warnings about where the establishment of a surveillance infrastructure might lead. They create the conditions of possibility that could, in certain hands or designed in particular ways, facilitate surveillance of very negative kinds. But at the same time, it is important to acknowledge that the conditions of possibility are just that and the historical record shows – as noted in the book – that to install surveillance may also be a means of ensuring that rights are upheld, such as in the case of some registration systems. They are always ambivalent, however, which is why, once again, ethical scrutiny is not a superfluous luxury. Moreover, in all cases, legitimate authorities, the "states" with which this book is concerned, have a duty to ensure that there is adequate independent oversight, just because such systems are so prone to misuse and abuse.

The longer historical perspective may one day conclude that the twentieth century was the era par excellence of *state* surveillance, in all the many guises that this book unstintingly demonstrates. This is not to say that the twenty-first century has seen any decline of state surveillance. Edward Snowden's witness to the massive expansion of "national security" surveillance, especially after 9/11, makes for sobering study. But the development of state surveillance in the twentieth century was, even from early decades, increasingly dependent on the corporate realm. When Snowden was releasing documents about the NSA, the first public outrage was aimed at those telephone and internet companies that had seemingly betrayed their trust by allowing government access to their data. And the supposedly data-driven dimensions of contemporary surveillance I refer to above often originate in the corporate realm. But, to be clear, it is necessary to emphasize that "data" drive nothing. Rather, it is the state or the corporation or a combination of the two that actually "drives" surveillance, typically by fostering surveillance-friendly cultures or what I call "user-generated surveillance."

For all these reasons and more, then, I welcome this book. It takes seriously the historical record of surveillance initiatives, ones predominantly produced by states, showing their variety of pathways and outcomes, none of which was inevitable but each of which was tragically flawed in some important respects. It tells human stories, both in terms of those who operate and those who experience surveillance. One hopes that the enthusiasms of those who promote surveillance as if it were an automatic and unmitigated good might be tempered by these honest appraisals.

David Lyon

Acknowledgments

Many people have supported us intellectually and in countless other ways from the initial conception of this book to its completion; we thank them deeply and collectively. Most important, of course, are the book's authors. They have inspired us to think in new ways about histories of surveillance, and we are grateful to them for responding patiently and thoughtfully at all stages of the process. We also note our special gratitude to David Lyon and Simone Browne. They have been two of the most instrumental people in foregrounding the importance of historical analyses in surveillance studies, and we are indebted both to their work and to their generosity in contributing to this book.

This is the second time we have worked together on an edited collection with the University of Toronto Press, and both experiences have been exceptionally rewarding and fruitful. We are especially appreciative of Doug Hildebrand for his early support and encouragement as the book was first being developed, and of Jodi Lewchuk for expertly ushering it through to completion. They and many others at the Press, including Janice Evans and Charles Stuart, ensured the publishing process was as seamless and successful as possible.

Finally, we thank the three anonymous peer reviewers, whose knowledgeable and probing comments brought greater clarity and rigour to the collection; Stephan Dobson, once again, for his editorial and indexing prowess; and both the Faculty of Liberal Arts & Professional Studies at York University and the Office of the Dean, Faculty of Arts at Ryerson University for generously providing publication support.

INTRODUCTION

1 Unpacking State Surveillance: Histories, Theories, and Global Contexts

EMILY VAN DER MEULEN AND ROBERT HEYNEN

The field of surveillance studies has grown dramatically over the past two or three decades. Located primarily in sociology in its early days, surveillance has moved to the centre of debate in a variety of disciplines, not least in our own areas of criminology and communication studies. This surge in interest is often seen as a reflection of the rise of surveillance itself, a belief that now, unlike in previous periods, we live in a "surveillance society" (Marx, 1985, 2002). In part this novelty is associated with the rise of electronic and digital surveillance, producing a sense that we are all, and always, being watched. At the same time, new approaches in the field stress that while we are all being watched, this homogenizing perspective risks downplaying the extent to which surveillance is also highly differentiated (Gürses, Kundnani, & Van Hoboken, 2016), leading a number of scholars to focus on the dynamics of race (Browne, 2015; Fiske, 1998), sexuality, gender, and gender expression (Beauchamp, 2009; Dubrofsky & Magnet, 2015; van der Meulen & Heynen, 2016), and embodiment and disability (Saltes, 2013). In addition, and reflecting the perspective that surveillance practices are part of "a complex, multi-scalar, interconnected world" (Murakami Wood, 2009, p. 190), there has been increasing attention to how surveillance practices function in different global locations (see, for example, Donovan, Frowd, & Martin, 2016; Firmino, Bruno, & Arteaga Botello, 2012; Svenonius & Björklund, 2018). What remains an open question, though, is how and to what extent this ubiquitous surveillance is radically new.

The aim of this book, then, is to expand further these critical perspectives and analyses, but to do so through a historical lens, asking how we understand, periodize, and conceptualize the history of surveillance practices, and suggesting that there has been a tendency in the field to overemphasize the novelty of contemporary developments. Certainly we are not the first to note the dearth of historical studies of surveillance.

David Lyon (2007), arguably the most prominent figure in the field, has long insisted that studying and theorizing surveillance should "include being historical" (p. 47), and has consistently highlighted the precedents for contemporary surveillance in older practices of, for example, workplace management, military power, and criminal justice. As we discuss at length below, there is also a range of excellent studies of specific surveillance histories, including Simone Browne's (2015) powerful examination of the histories of slavery in the United States, which broadens the scope of what is generally included in the field. Some areas in surveillance studies have generated more comprehensive attention to history, perhaps most notably around transnational histories of state identification and registration (see About, Brown, & Lonergan, 2013; Bennett & Lyon, 2008; Boersma, van Brakel, Fonio, & Wagenaar, 2014; Breckenridge & Szreter, 2012; Caplan & Torpey, 2001). Colonial surveillance is another area where we find important recent historical work that we draw on extensively in what follows. Colonial histories foreground a key dimension of surveillance, namely, its role in "social sorting" (Lyon, 2003a, 2003b), the categorization and differentiation of people and populations. Lyon's (2003b) account of social sorting refers primarily to contemporary computer-based surveillance practices, including the rise of searchable databases and digital information practices that "not only rationalizes but also automates the process" of surveillance (p. 13). However, as the chapters of this book detail, the idea of social sorting can be repurposed in other contexts to conceptualize longer histories of population management that, as Lyon argues as well, are complex and profoundly ambivalent in their implications.

For all the new work in surveillance studies, however, there remain few attempts to develop a more comprehensive understanding of surveillance histories. This is our goal here, with state surveillance our specific focus. An edited collection is especially conducive to offering a wide range of case studies that not only offer insight into an array of surveillance practices, but also foreground global and transnational perspectives. Histories of surveillance, we contend, are inextricable from broader histories of colonialism and imperialism, and the global movement of goods, people, diseases, ideas, and technologies. Such histories reflect the profoundly unequal ways in which surveillance has been applied to different people and populations. Thus, medicine and disability, race and eugenics, gender and sexuality, and the policing of dissent are all key themes in the chapters that follow. The debilitating impacts of surveillance tend to be the focus of these accounts, although we need to keep in mind the deeply ambivalent nature of surveillance, which is potentially (and often simultaneously) both enabling and disabling in its consequences.

In this introductory chapter we outline a number of frameworks through which we might understand the histories of surveillance over the last two or three centuries, highlighting in particular the various threads and themes that are woven through the chapters. We begin with a consideration of periodization and of different ways of conceptualizing surveillance histories and social change. We next turn to historical study itself, and specifically to the archive as both a repository for records of surveillance and the locus of historical research. Thinking about what is included and excluded from these surveillance records and how archives structure information foregrounds methodological issues that, we argue, are necessary in developing a critical and self-reflexive approach to surveillance history. Finally, we consider the role of the state in the elaboration of systems of surveillance, examining how state theory can help us to conceptualize surveillance histories, and vice versa. Our assumption here is that state formation is a global process, but that it is highly differentiated, shaped by the expansion of capitalist social relations, the development of imperialist and colonial systems, and the more recent emergence of nominally socialist or communist states. Along with the chapter authors, whose contributions we introduce at the start of each of the three thematic sections (Medical, Disease, and Health Surveillance; Identification, Regulation, and Colonial Rule; State Security, Policing, and Dissent), we argue that an attention to the complex and diverse histories of surveillance practices offers crucial resources for understanding and contextualizing the dynamics of contemporary state surveillance.

Surveillance Histories and Social Change

Surveillance is a slippery concept, generally defined as some variation of "monitoring people in order to regulate or govern their behaviour" (Gilliom & Monahan, 2013, p. 2). This definition potentially captures a huge range of historical and contemporary practices, allowing us to reconceptualize various social dynamics in terms of surveillance, although its sheer scope also poses analytical problems. Our focus in this book is on the surveillance histories of the last two to three centuries – some chapters reach back into the eighteenth century, although most focus on the nineteenth and/or twentieth. Roughly, this is the "modern" period often distinguished from "traditional" societies in the sociological literature; our use of scare quotes, however, signals our scepticism towards these terms. In much of that literature, in particular that which draws on Max Weber's analysis of bureaucracy, surveillance plays an important role. Anthony Giddens (1990) is a key figure here; he has argued most explicitly that it is precisely systematic surveillance that characterizes modern societies (see also Dandeker, 1990). Or, as Edward Higgs (2004) contends in his important work on English state surveillance, "Modern states are 'Information States,' and without information on citizens they would cease to function" (p. vii).

Higgs's approach dovetails nicely with more recent theorizations of the "information society," but the question remains to what extent these claims of novelty are tenable. Higgs is interesting in this respect as his history goes back to 1500, earlier than most accounts of modernity, and more recently he has gone even further in arguing that systematic surveillance is not uniquely characteristic even of this longer period, suggesting "this sociological model of historical change is unhelpful, or, better still, downright wrong" (Higgs, 2014, p. 18). Indeed, if we look at broader histories, including outside of Europe, we find many examples of extensive state surveillance regimes, most evidently in the constitution of major empires. The Babylonian, Egyptian, or Roman Empires are especially familiar, but there are other examples as well. The Maurya Empire (circa 321–185 BCE), centred in present-day India, for example, regulated extensive trading networks and established a strong administrative unity over a vast territory that exceeded 5 million square kilometres at its peak, while the Inca Empire (1438–1533) developed a centrally planned redistributive economy over a forbidding terrain reaching 2 million square kilometres, all using knotted strings (*quipus*) rather than writing as a record-keeping medium. Keith Breckenridge and Simon Szreter (2012) highlight other exemplary cases in their collection on civil registration, which foregrounds longer, and often non-European, histories. Thus, for example, Richard von Glahn (2012) describes the long history of population registers compiled in China going back to the sixth century BCE. Under the Ming dynasty this entailed decennial household surveys, with an early seventeenth-century report indicating that the state held two million volumes of these "yellow registers" (*huangce*) compiling the results of past censuses that were then used by the government for everything from military conscription to the distribution or redistribution of land and wealth.

Given these histories, it would seem that practices of monitoring and surveilling people to shape their behaviours are not confined to the "modern." Arguments about the radical novelty and uniqueness of modernity and its attendant information and surveillance systems thus need to be resisted, in particular insofar as these ideas reinforce a profoundly Eurocentric and progressivist understanding of history. At the same time, however, if we are attentive to the specificities of surveillance histories and the social processes in which they are embedded, periodization can be useful. Rather than a generic "modernity," we argue that we need to take into account the specific dynamics of capitalism, colonialism, industrialism, and urbanization that, without denying longer continuities as well, have given a particular shape to global histories of the last two or three centuries. There are strong arguments to be made, evident especially in Foucault's work, that this period produced practices of state surveillance more intensive and extensive than in earlier periods, and which are novel in both the continuous nature of data collection and the extent to which they saturate the practices of everyday life in new ways. This is arguably what distinguishes the "information state," the rise of which enabled the nineteenth-century imagining of "society" as a coherent entity understandable through statistics, probabilities, and risk (Hacking, 1990). New technologies and systems of information management, including automation, as Lyon (2003a, 2003b) stresses in relation to social sorting, facilitated this systematic and ongoing surveillance of large populations. More importantly, though, systems of surveillance and their attendant technologies developed in tandem with the rise of capitalism, industrial development, urbanization, and colonial expansion, and, as David Vincent (2016) stresses in his history of privacy in England, were (re)produced at the everyday life level through everything from new media technologies (e.g., print, mail, telephones, TV) to changes in architecture, urban design, and housing. "Modernity" from this perspective is a rather imprecise term, and any periodization needs to be tied to the specific and concrete socio-historical processes in which surveillance practices are embedded.

Briefly examining two nineteenth-century technologies – photography and the passport – can help to illustrate some of the issues at stake. Photography played a central role in knitting together a series of surveillance practices, beginning with the introduction of the daguerreotype in the 1830s and culminating with the development of the portable camera towards the end of that century. As Allan Sekula (1986), John Tagg (1988), and others have famously argued, photography was at the heart of new configurations of evidence, objectivity, and truth especially evident in criminological practices. New standardized evidentiary photographs produce

> a portrait of the product of the disciplinary method: the body made object; divided and studied; enclosed in a cellular structure of space whose architecture is the file-index; made docile and forced to yield up its truth; separated and individuated; subjected and made subject. When accumulated, such images amount to a new representation of society. (Tagg, 1988, p. 76)

Photography's promise of a "mechanical objectivity" (Daston & Galison, 2007) informed a host of scientific practices in the nineteenth and early twentieth centuries, ranging from medicine and psychiatry to biology and physics. In the work of photographers like Étienne-Jules Marey and Eadweard Muybridge the human body was conceptualized in new ways, with their insights informing a range of practices, perhaps most notably in the surveillance regimes governing labour and the extension of capitalist social relations more generally. Industrial psychology, the scientific management of Frederick Winslow Taylor, or the time and motion studies of Lillian Gilbreth and Frank Gilbreth, for example, all deployed photography and film to rationalize mental, physical, and physiological dimensions of labour (Brown, 2008; Rabinbach, 1990), these approaches both drawing on and reshaping medical surveillance as well. Finally, photography (and, later on, film) proved itself a crucial technology of colonial expansion and rule, serving everything from the mapping of colonized spaces to the anthropological construction of racist systems of classification that both enabled colonial exploitation and provided ideological legitimation for colonial projects. Certainly we can find many instances in which people resisted the encroachment of these new forms of surveillant objectivity and rationality – one of the most compelling is W.E.B. Du Bois's counter-archive of photographs of African Americans at the 1900 Paris Exposition (S.M. Smith, 2004) – but the repressive dimensions of surveillance proved exceptionally resilient.

The passport is another important technology illustrating these broader arguments, one that incorporated in turn the identificatory power of the photograph. Modern passports, which emerged in the late nineteenth century (Torpey, 2000), were by no means the first form of identification regulating movement and border crossing, however, they were distinctive in their function. Passports reflected a "modern constitution of identity as an object that could be stabilized and known through practices of bureaucratic rationalization" (Robertson, 2009, p. 331). They were tied directly to the formation of the imagined community of the nation state that, as Benedict Anderson (1991) has famously argued, reflected both "a totalizing classificatory grid" and "the assumption that the world was made up of replicable plurals" (p. 184). John Torpcy (2000) suggests that in this process "the west has been dominant, and exported their practices" (p. 3). The language of these accounts is somewhat bloodless, however. As Browne (2015) shows, an important precedent of the passport were the passes used in the United States to track and restrict the movement of slaves. Slave passes also prefigured the pass systems or internal passports that reappeared everywhere from Canada to South Africa to the Soviet Union to Israel/Palestine. Describing the passport as merely "exported" from Europe also downplays its crucial role as an active technology of European colonial hegemony and violent occupation, grounding differentiated systems of racialized recognition, tied eventually to bodily and biometric markers, that enabled some bodies to cross borders with considerably greater ease than others. The passport thus regulated global flows of labour that, as Radhika Mongia (2003) argues in discussing Indian indentured labour migration, "demonstrates that the passport not only is a technology *reflecting* certain understandings of race, nation/nationality, and state, but was also central to *organizing* and *securing* the modern definitions of these categories" (p. 196, emphasis in original).

These accounts foreground the important role played by the emergence of new surveillance technologies in state formation, but also that these technologies were inextricable from the socio-histoᵣical contexts in which they emerge. This perspective is captured in one of the more influential recent theories of surveillance: Kevin Haggerty and Richard Ericson's (2000) notion of the "surveillant assemblage." Photography and the passport, we can say, were key actors or nodes in a broader assemblage that stretched across the social field, encompassing institutions, ideas (e.g., about bodies or race), capitalist logics, new information systems, practices of risk management, scientific developments, patterns of migration, and labour regimes, to cite only some of the possible links. Both also developed and changed over the twentieth century, as technologies and in how they were deployed, in particular as they were increasingly digitized. Technological developments clearly played an important role in the changing dynamics of state surveillance, but the idea of the surveillant assemblage also avoids granting an intrinsic and overinflated significance to technological shifts, as often happens, for example, in accounts that claim the radical novelty of electronic and digital forms of surveillance. As the chapters in this book suggest, where surveillance history has perhaps its greatest analytical significance is in embedding the technological in broader social, political, and cultural dynamics.

Archives, History, and Surveillance

The premise of this book is that historical and transnational perspectives are necessary for understanding contemporary contexts of state surveillance; thus, historians and historical research are of central importance. At the same time, we might say that historians themselves are agents of surveillance. They are detectives of a sort, using common surveillance techniques as part of their craft: interviewing key actors, poring over documents, and mining a range of data sources as a way of teasing out the events, patterns, and meanings of the past. Karen A. Fields and Barbara

J. Fields (2012) capture this well in drily noting that the oral historian needs to approach subjects with a "methodological distrust," deploying similar techniques to "police and special agents" (p. 174). When it comes to archival research, however, which is in many ways the lifeblood of history writing, this detective work is better understood as a kind of second-order surveillance. Archives carry the material traces of the surveillance work of others, most notably state actors that include actual police and special agents, as well as preserving data collected by a broad range of state institutions, from health and welfare systems to those managing labour and migration. Corporations, non-governmental organizations, and even individuals all produce archives as well, which can be of great significance for historical research. The term "archive" is also often used more broadly, including in this introduction, to designate a more diffuse body of data that is not institutionally located, for example in Allan Sekula's (1986) evocation of an "archive" encompassing at its broadest all nineteenth-century bourgeois photography. Historians generally are interested in institutional archives, however, and thus looking at what historians do, their methodological approaches, debates, and how they deal with these surveillance records offers intriguing insights into the logic of surveillance itself. It also introduces an awareness of complicity, suggesting that the historian must remain acutely attuned to the ethical and political challenges of working with the documents of state surveillance: how does one use the archival surveillance record without replicating the subject position of the authority that assembled it, with all of its potentially repressive and sometimes violent implications?

Archiving as a practice is as old as human society, but it took on distinctive features in Europe in the past two centuries. The founding of the Archives Nationales in revolutionary France in 1790 was a key event in this respect, establishing a set of protocols and practices that developed more systematically over subsequent decades and centuries. This "archive fever," as Jacques Derrida (1995) designated it, was part of the intensification of systems of surveillance and the growing collection, systematization, and rationalization of knowledge that we saw with photography. Archives themselves can be read as markers of accountability, with Derrida emphasizing that the degree of openness accorded to public archives is itself an index of the state of democracy; in authoritarian states, archival records are largely closed, while a key founding principle of the Archives Nationales was free public access. The link to democratic accountability is directly related to the enabling power of identification; as David Lyon (2009) stresses, systems of identification are bound up with practices of citizenship. However, the state practices recorded in archives, including those of liberal democracies, are in fact profoundly ambivalent, both enabling (in particular in terms of citizenship) and disabling. A counterpoint to the French Revolution's association of archival openness with freedom thus came in the Paris Commune of 1870, when Communards burned pre-1859 city records as an act of liberation (Sekula, 1986). Record-keeping, this response suggested, was recognized by the revolutionaries as a key technology of the bourgeois state. Centralized archival collections associated with supranational, national, or subnational administrations, or the archives associated with the various institutions and bureaucracies comprising the state (e.g., police, security services, medical institutions, welfare bureaucracies, statistical bureaux, transportation departments, property registries, and a host of others), occupy a crucial and complex role in the political contestations shaping nineteenth- and twentieth-century states.

The ambivalent political implications of archives is reflected in historical methodologies. Public accessibility remains a key site of struggle, with archival records increasingly restricted or closed in many parts of the world, including in ostensibly democratic states. Even when records are available, however, historians are faced with the fact that they reflect the particular interests and orientations of the state actors that produced them. For critical historians, then, archival collections cannot be approached with a naïve faith in their truth-value; the history of the archives themselves, their contested and partial nature, must be subject to investigation. A powerful example of what is at stake is explored by Kirsten Weld (2014), who recounts the discovery in 2005 of the archives of the Guatemalan National Police. Comprising some 75 million pages that had been left mouldering in a former detention and torture centre, the collection had been found largely by accident. In it, researchers and activists discovered evidence of the instrumental role played by police in the brutal repression of a host of perceived opponents of the United States-backed Guatemalan regime, ranging from trade unionists to teachers. Rural Maya were the primary victims of an often-genocidal violence, but the recovered archival documents revealed a murderous urban warfare campaign, previously largely undocumented, that also decimated activist and left-wing communities and movements. State officials had long denied the existence of such police records, with the fortuitous discovery of this massive surveillance archive opening a new window onto this brutal history, and exposing both the role of many sitting officials in murder and torture and that of the United States in enabling this repressive violence.

While the discovery of the records appears to be a story of archival justice, Weld stresses that we cannot lose sight of the fact that the documents "were tools of counterinsurgent state formation, rendering legible those sections of society deemed to be enemies of the state in order to enable their elimination" (2014, p. 14). The documents represent what Verne Harris (2002) has called the "archival sliver," the very limited, tendentious, and state-centric information that resides in official collections. Ann Laura Stoler (2006) thus urges scholars to focus on archives "not as sites of knowledge retrieval, but of knowledge production, as monuments of states as well as sites of state ethnography" (p. 268), a challenge that Weld takes up in exemplary fashion. This goes beyond recognizing simple "bias"; the goal is "to identify the conditions of possibility that shaped what could be written, what warranted repetition, what competencies were rewarded in archival writing, what stories could not be told, and what could not be said" (Stoler, 2006, p. 269). "Truth," in other words, is not to be found in unmediated fashion in the archives.

Such historiographical challenges have been met in numerous ways, which offer important lessons for surveillance studies. Many innovations in historical research over the last half century have involved efforts to trouble the certainties of the archive. The field of social history, for example, grew out of a desire to, in E.P. Thompson's (1963) oft-cited phrase, recover the experiences of ordinary people from "the enormous condescension of posterity" (p. 12). That condescension was in part a function of what is included in the archival record itself, the "sliver" that represents the work of the powerful while leaving blank much of the vast canvas of everyday life. For early social historians like Thompson, it was the experiences of the peasantry and the emerging working class during and after the transition to capitalism that were the focus, a perspective developed especially powerfully in Peter Linebaugh and Marcus Rediker's (2000) "hidden history" of the Atlantic world. In other instances, for example in the micro-histories of Carlo Ginzburg (1982) and Natalie Zemon Davis (1983), the lives of those rare ordinary people who had their experiences documented in archives, most commonly criminal justice archives, served as exemplary figures through which the social and cultural dynamics of a particular place and time could be deciphered.

Social historians, feminist historians, historians looking at queer, trans, disabled, and/or racialized communities, historians of everyday life, and many others have long confronted the challenges of working with archival sources and the documents of state surveillance. As we see in many of the chapters in this book, this involves recovering lost voices, but also seeking to write history when such recovery is not possible. These dilemmas are perhaps most strongly evident in the colonial context, where the "imperial archive" or "colonial archive" (Richards, 1993; Stoler, 2002, 2006) functioned quite differently than in European contexts. This was tied in part to different forms of accumulation. From the enclosure of the commons to colonial conquest, capitalism developed through what Karl Marx (1867/1976) called "primitive accumulation," a logic of "accumulation by dispossession" (Harvey, 2003, 2004) that gave rise to new forms of surveillance, not least in suppressing resistance to these processes. Where in Europe enclosures were tied to proletarianization and to the integration of workers into increasingly dense networks of identification and surveillance, colonial states tended to be much more systematically violent and repressive, and also, perhaps paradoxically, *less* extensive in terms of their surveillance apparatuses.

The colonial drive, in particular in the case of settler colonies, was to produce the landscape as empty, ready to be exploited and settled. The impulse was thus to erase rather than record the presence of colonized peoples. This was especially evident in the physical and conceptual erasure of an Indigenous presence in the Americas and Australia, as well as in parts of Africa. Writing on Australia, Terry Smith (2003) argues that practices of calibration, the surveillance and mapping of land and peoples, enabled obliteration, the erasure of the physical and symbolic presence of Indigenous people. In Canada, Glen Coulthard (2014) emphasizes, the genocidal and assimilationist practices of settler colonialism towards Indigenous peoples entailed their "*elimination*, if not physically, then as cultural, political, and legal *peoples* distinguishable from the rest of Canadian society" (p. 4, emphasis in original; see also Henderson, 2018). Surveillance was not entirely absent in these contexts. Indigenous presence was recorded in minimalist ways in Canada, for example, through a pass system similar to those mentioned earlier (K. D. Smith, 2009) and in nutrition experiments enacted in Residential Schools (Mosby, 2013), but was generally in the service of the erasure of an Indigenous presence.

This limited surveillance means that archival records of colonized peoples are sparse, an absence producing the methodological and political challenges identified earlier. The history of American chattel slavery and the transatlantic slave trade provides particularly stark examples of this broader colonial tendency. As Simone Browne (2015) has argued, surveillance practices guaranteed the property relations at the heart of American chattel slavery. Such practices included everything from paper record-keeping to the early biometric technology of branding, while tracking runaway slaves produced extensive networks of surveillance, buttressed by "lantern laws" that required slaves to carry a light after dark and that enjoined all whites to participate in surveilling Black bodies. At the same time, though, beyond maintaining property rights and absolute bodily control, slave owners and especially the state had little need for or interest in maintaining detailed records of the characteristics or everyday life of slaves; indeed, even at the level of naming, so central to state systems of identification, slave owners exercised complete control. The exclusion of slaves and other Black Americans from civic life meant that they left few traces in the more participatory forms of surveillance and record-keeping that enabled voting, education, social welfare, or forms of mobility, even while the pass system and other forms of biometric control also foreshadowed later forms of repressive surveillance of Black Americans (see also Parenti, 2003).

Browne and other historians argue that understanding slave experiences, including what she calls the "dark sousveillance" by which slaves themselves challenged that institution, requires an attention to non-archival sources like oral histories to supplement the profound archival silences. Those silences are emblematic of what Orlando Patterson (1982) has called the "social death" of slavery, a "secular excommunication" (p. 5) that entailed a radical symbolic and material exclusion from community. Insofar as surveillance enables belonging through processes of recording, registration, and documentation of the self, social death is built on documentary erasure. Simon Gikandi (2015) notes that prior to the nineteenth century there were two archives:

> on one hand, we have the library left behind by powerful masters who deployed modern knowledge to secure the racial ideology of enslavement; on the other hand, there is the repressed archive of the enslaved, deprived of subjectivity and voice in order for white supremacy to be heard. (pp. 91–2)

Silence in this sense is itself a record, but one that signifies the slave experience as what Gikandi calls a "place of pure negativity," or alternately an "Atlantic crypt" (2015, p. 92). The question he poses is, "how does one read this archive without African voices, without African documents, without an African historical a priori?" (p. 86).

The historiographical and political dilemmas identified here are crucial in that they reveal something essential about the contradictory dynamics of surveillance; depending on context, being watched can inhibit or enhance life chances, but so can *not* being watched. Writing critical histories entails reading the archival record against the grain, and doing so in ways that brings these enabling and disabling implications of surveillance to the surface. It also brings new archival possibilities into view. A powerful example of what is at stake came in Canada in 1991 in the precedent-setting Indigenous land claims case *Delgamuukw v. British Columbia*. The claim, brought by the Gitxsan and Wet'suwet'en peoples, was rejected by Chief Justice Allan McEachern of the Supreme Court of British Columbia on the grounds that, as Adele Perry (2005) notes, the Gitxsan and Wet'suwet'en "have a substantial history of contact with and resistance to European encroachment, but little of it is documented with the kind of legal archives that have regulated dispossession elsewhere in northern North America and the antipodes" (p. 327). Oral histories were entered into evidence, but Justice McEachern accorded them little weight, thus finding that any land claims had been extinguished. Still, by being presented in court the oral histories were written down, the case producing a massive collection of evidence akin to a counter-archive by virtue of its liminal and subversive role in the judicial process. Crucially, on appeal the ruling was partially overturned by the Supreme Court of Canada, which accepted oral history as legitimate evidence. That ruling recognized, if in a limited way, the extent to which historical forms of state surveillance, preserved in turn in archives, worked to shape the contours of the social and political landscape over decades and centuries to the extreme detriment of Indigenous peoples; it also opened a window, however limited and partial, to imagining worlds beyond the often suffocating power of state surveillance and its archival traces.

Theorizing Surveillance States

What these archival tales demonstrate is the extent to which nation states are fundamentally built on systems of surveillance. Archives not only contain the proliferating traces of that surveillance, but also allow them to persist in ways that continue to shape later surveillance regimes, opening up both emancipatory and repressive new potentials. Archival collections reflect the workings of states, making visible what Bob Jessop (2016) describes as the state's central characteristic, namely, in an account that echoes the definition of surveillance given at the start of this chapter, their "efforts to establish, exercise, and consolidate political power over the population of a specific territory" (p. 19). James C. Scott's (1998) influential *Seeing Like a State* describes this "as a state's attempt to make a society legible" (p. 2). State theorists debate at length the precise scope and significance of these efforts, and how they fit in the broader social and political landscape. Whether we theorize states in the limited sense of holding a monopoly on violence or in more substantive and extensive ways, surveillance is central, but states also vary in how they function. Ian Hacking (1990) captures this variation well in describing the information turn that shaped state surveillance practices over the last two centuries: "Every state, happy or unhappy, was statistical in its own way" (p. 16). While a comprehensive engagement with debates in state theory is beyond the scope of this introduction, our goal in what remains is to outline some frameworks and questions through which to approach the book's chapters as well as the problematic of state

surveillance more broadly. In particular, we highlight how state formation proceeded in highly differentiated ways, most systematically in terms
of whether the states in question were liberal democratic and capitalist,
colonial, or state socialist, but also how states were interconnected in
global and transnational networks.

Surveillance and State Theory

In many respects, the field of surveillance studies has been founded
on Foucault's (1977) panopticon, the prison initially proposed by Jeremy Bentham that, for Foucault, was emblematic of modern disciplinary forms of power. The guards in this circular prison remained hidden
from the prisoners, leading them to self-surveil and internalize the
gaze of authority. More recently, Foucault's work on governmentality
and biopower has moved to the fore, with many scholars finding those
approaches more conducive to understanding contemporary surveillance. For Foucault, however, while disciplinary power emerged over the
course of the 1700s and biopower only at the end of that century, these
different modalities of power overlap and mutually sustain each other.
The former involves an "anatomo-politics of the human body," the latter "a 'biopolitics' of the human race" (Foucault, 2003, p. 243). These
new forms of power had major implications for the state, with Foucault
(1991) arguing that as power diffuses through the social field, the state
becomes perhaps "no more than a composite reality and a mythicized
abstraction, whose importance is a lot more limited than many of us
think" (p. 103). While the institutional state is downgraded, a new "science of the state" (1991, p. 96) emerges, namely, statistics. This is the
realm of biopower, of what he calls the "apparatus (*dispositif*) of security"
(Foucault, 2007, p. 6), which marks the transition from a conception of
government as a form of family management to one in which the welfare
of the population becomes the object of concern.

Security in Foucault's sense has a broad meaning, one closely linked to
practices of "normalization" that pose the "abnormal," whether in terms
of sexual deviance, criminality, illness, or madness, among others, as a
threat to security. Medicine was one key area in which these forms of
regulation and surveillance developed. It was the late nineteenth- and
early twentieth-century work of the eugenic theorists Francis Galton and
Karl Pearson on population- and bio-statistics, for example, that enabled
the development of risk management approaches on which much subsequent information-oriented surveillance has been based (Louçã, 2009).
Eugenics represents a particularly influential development in surveillance histories, even though it is rarely thought of as such. It had an

especially wide reach, of course in Germany under Nazi rule (Weindling, 1989), but also in the United States (whose policies influenced those of the Nazis), where it shaped everything from racist border regimes to the development of the national parks system (Stern, 2005), and in many other countries as well. Eugenics was also crucial to the formation of global colonial systems (see Bashford & Levine, 2010). In many respects, then, the period in which eugenic approaches were in the ascendancy can be seen as a key transitional moment in the development of new and more extensive systems of surveillance central to the "apparatus of security," as noted by Foucault. Medical science is by no means reducible to eugenics, but the development of statistically based and risk-oriented surveillance apparatuses so influential now in medicine, public health, epidemiology, and a range of other fields owes much to those late nineteenth- and early twentieth-century developments. This is not to say that biologizing conceptions of health went unchallenged at the time. Already in 1906, for example, W.E.B. Du Bois rejected the medicalization of tuberculosis as a "Black disease" by arguing that "it is not a racial disease but a social disease" (quoted in Yudell, 2014, p. 252). Here Du Bois highlights the *mis*-recognition so often produced by surveillance regimes.

The role of the state here is complex, with medicine a field that, in straddling public and private realms, captures the porous nature of the modern state. Foucault's later work (as well as Gilles Deleuze's [1992] theorization of the control society) are at the heart of Haggerty and Ericson's (2000) notion of the surveillant assemblage discussed earlier. They too understand the state less in institutional terms, and more as having "its own characteristic set of operations; the tendency to create bounded physical and cognitive spaces, and introduce processes designed to capture flows" (Haggerty & Ericson, 2000, p. 608). Such arguments have been of great importance in rethinking classical state theory, and especially in terms of considering the ways in which state surveillance practices are diffused throughout society. The risk in the turn to biopower, governmentality, or the surveillant assemblage, however, is that states appear simply as coterminous with the social field as a whole. If state institutions and practices are part of larger assemblages, we also need to examine the distinct institutional sites and practices through which states are formed.

This is where historical work becomes so important. Christopher A. Bayly (1996), writing on surveillance in the Indian context, argues: "In early modern societies, the information order was decentralised, consisting of many overlapping groups of knowledge-rich communities. It was not mediated as in contemporary industrial societies by a dominant state or commercial communications sector" (p. 5). In this account, which contrasts dramatically with the Foucaultian view, over the course of the eighteenth and nineteenth centuries the institutionalized state becomes *more* rather than less important in structuring the social field. Similarly, Scott's (1998) contention that states render populations "legible" has affinities to Foucault's theories, in particular his emphasis on statistics, but is oriented very differently. "Legibility," Scott argues, "implies a viewer whose place is central and whose vision is synoptic" (1998, p. 79). The high modernist state that takes this central place is therefore highly defined, unitary, and authoritarian, a concrete actor that engages in concrete projects of social engineering. Certainly there are many difficulties with Scott's view, not the least being his seeming reification of "the state," and his extrapolation from specific examples of authoritarian states and projects for social engineering (e.g., Germany during the First World War, planned cities like Brasilia and Chandigarh, Leninist projects for remaking society, etc.) to states more generally. His emphasis on legibility also discounts the significance of *not* recording people that we addressed earlier; rendering non-legible is also a significant modality of state power. Nevertheless, Scott's approach offers a valuable counterpoint to Foucault's (1991) view of the state as a "mythicized abstraction" (p. 103).

A number of Marxist-inflected approaches to the state have taken up Foucault, but offer different perspectives on state formation that foreground the structuring role played by capital and capitalist social relations, not least in the realm of surveillance. The more fruitful of these approaches build on a Gramscian foundation, seeing the state not merely as a tool of capital, but as retaining a relative and limited autonomy in relation to capital. In this reading, the state performs a mediating function between different class and other social formations, and it is internally differentiated rather than monolithic (see, for example, Jessop, 1990, 2016; Poulantzas, 1980). In the work of Philip Corrigan and Derek Sayer (1985), the Foucaultian emphasis on processes of normalization remains in their notion of "moral regulation," which seeks to analyse the ways in which people are integrated into and subordinated to national communities. Looking specifically at England, they argue that state formation was a form of "cultural revolution" extending *capitalist* relations throughout the social field, while also building on a host of other forms of differentiation. Moral regulation in this view "is coextensive with state formation, and state forms are always animated and legitimated by a particular moral ethos" (1985, p. 4), but what distinguishes Corrigan and Sayer's work is a focus on the ways in which moral regulation can be tied to specific institutionalized sites. While not always articulated as such, we can see forms of moral regulation at work in many of the chapters in this book, for example, in terms of how surveillance practices differentiated populations in ways that enabled unequal forms of labour exploitation, regulations and restrictions on movement, the constitution of public and private realms, the medicalization of social relations, and the suppression of dissent.

State Formation as a Colonial Project

A central limitation to many of the approaches to state formation noted above is that they are rooted in the European experience and take the liberal, democratic, and capitalist state as a normative model. For Achille Mbembe (2003), drawing on Giorgio Agamben's (1998) theorization of "bare life," the Foucaultian notion of biopower is insufficient when faced with colonial history and its legacies of racist praxis, evident especially in United States chattel slavery, South African apartheid, and Israeli state policy towards Palestinians, notably in the occupation of the West Bank and Gaza. This "necropolitics," as Mbembe calls it, involves "the creation of *death-worlds*, new and unique forms of social existence in which vast populations are subjected to conditions of life conferring upon them the status of *living dead*" (2003, p. 40, emphasis in original). This account echoes Patterson's (1982) notion of social death, and, through Agamben, links colonial violence to Nazi genocide in Europe. But even that comparison needs to be qualified. Whether we look to Zygmunt Bauman's (1991) analysis of the bureaucratic rationality of the Holocaust, or at the role played by new information technologies like IBM's punch card systems in facilitating genocide (Aly & Roth, 2004; Black, 2001), it is clear that Nazi victims were already enmeshed in dense networks of state surveillance in quite different ways than was the case with the limited information gathering characteristic of colonial states. Jewish Germans, for example, had been integrated in the state apparatus as citizens, even while they, and to a much greater degree disabled people and Roma, were marked already before 1933 by debilitating forms of social sorting that were rooted in eugenic ideas (Heynen, 2015). Patterson's social death, in that sense, is quite different from what Bauman (1991) calls the "*dehumanization of the objects of bureaucratic operation*" (p. 102; emphasis in original).

This is not to suggest, however, that colonial states were derivative or less developed forms of European states. Rather, as Ann Laura Stoler and Frederick Cooper (1997) have argued, "Europe was made by its imperial projects, as much as colonial encounters were shaped by conflicts within Europe itself" (p. 1), and that the goal should thus be to "treat metropole and colony in a single analytic field" (p. 4). This has been the approach of writers as diverse as Hannah Arendt (1968) and Aimé Césaire (1972), although the former arguably retains a Eurocentric perspective. As we saw with the archive, the distinctiveness of colonial governance turned on what Frederick Cooper (1996), looking at the ways in which the colonial state sought to manage African labour, contends was "the seeming unknowableness of Africa" (p. 334) that confronted colonial officials. Again there was a fundamental *lack* of information, from budget data to the registration of births, marriages, and deaths, thereby "reveal[ing] the colonial inversion of Michel Foucault's vision of surveillance and control in shaping the 'social' in Europe ... [I]n the African colony, the state could not track the individual body or understand the dynamics of the social body" (Cooper, 1996, pp. 334–5). That characteristic was not the product of an "underdeveloped" Africa, however, but of the colonial state itself.

As a way of compensating for the lack of information, in many colonial contexts biometric systems developed instead as a blunt mechanism for tracking labour. Biometric technologies like fingerprinting in fact found their most extensive early use in colonial contexts rather than in Europe, established first in India (Sengoopta, 2003) before migrating through the British Empire and beyond. In South Africa, as Keith Breckenridge (2014) argues in his analysis of identification systems prior to and during the apartheid period, state agencies gathered little information beyond the minimum threshold of identification, in particular on Black workers. This lack of surveillance enabled the super-exploitation of labour, but also served to limit the ability of people to make claims upon the state. This was very different from the situation in the metropole, where state information systems were robust and increasingly comprehensive, while biometric identification and registration was applied only *selectively* to criminalized and otherwise targeted groups (Cole, 2001). In what we now call the Global North, then, "suspect" populations were relatively circumscribed; in the colonial context, as with specific marginalized communities in Europe, North America, and Australia, the population as a whole was potentially suspect.

Surveillance and State Socialism

Considering the colony and metropole as a single analytic field, as Stoler and Cooper (1997) suggest, implies a transnational system that, in the aftermath of direct colonial rule in most parts of the globe, was increasingly conceptualized as one of three worlds. Here, though, we turn not to the "Third World," as the emerging post-colonial states were named, but to the "Second World," the state socialist countries whose development reshaped global dynamics. The Soviet sphere, whether prior to the Second World War or with its dramatic expansion and the onset of the Cold War in its aftermath, poses another set of questions for any account of state theory and state surveillance. How exactly we characterize the Soviet-style state was, and remains, a matter of heated debate. Was it socialist or communist in a meaningful sense, or was it something else, for example state capitalist? Is totalitarianism a useful concept in understanding these states, or does this approach, premised as it is on similarities between Soviet and fascist states, not only efface important differences but also misunderstand their complex internal dynamics? These debates are complicated further if we think of the centrality of socialist and communist politics to anti-colonial movements and postcolonial states.

Totalitarian theory tends to characterize state socialist countries as Orwellian surveillance states, even while surveillance studies scholars generally dismiss the Orwellian model as crude and unhelpful. Many historians criticize totalitarian theory on similar grounds, in part because that approach tends to overplay the extent to which the desire for totalizing control was in fact possible. Thus, for example, Scott's (1998) contention that Lenin subscribed to a version of the high modernist "claim to authority of a small planning elite" (p. 157) is not only a debatable interpretation of Lenin's thought, but also downplays how intractable society proved to be in the face of this planning desire. In fact, the history of the Russian Revolution from 1917 to Lenin's death in 1924 suggests that state surveillance struggled to glean even basic information on much of the population, with officials deeply concerned about this lack. Peasants made up the bulk of the population in the early 1920s and, as in colonial contexts, state actors had little information on them. Peter Holquist (1997, 2001) suggests that early Soviet surveillance was in many respects not so different from surveillance under Tsarism or even in Europe more broadly; urban-oriented Bolsheviks were very interested in gauging the "mood" of the population, of integrating the peasant population, and of expanding the state's information base. Holquist echoes Scott in arguing that this information gathering was in the service of creating a "new man" rather than any liberal democratic notion of public opinion, but stresses that this was not necessarily so different than state projects elsewhere. The goal was to enact a particular cultural revolution (to use Corrigan and Sayer's term), with its own forms of moral regulation. However, Hugh Hudson Jr. (2012) contends that this was not just about top-down social engineering. Police surveillance was genuinely geared to understanding the situation in the countryside, with police reports (*svodki*) reflecting a kind of conversation between peasants and the state. At least up to 1924, and in a more limited sense after, these *svodki* contained a rather unvarnished record of peasant concerns that, mediated through various state apparatuses, shaped state policy in limited but significant ways. Rather than being controlled by a totalizing apparatus of state surveillance, Hudson's account thus suggests that, especially in rural areas, Soviet society was in fact "undergoverned" and "underpoliced" (2012, p. 1). Even under Stalin, surveillance systems remained uneven and incomplete, although what became clear, as was even more evident in colonial contexts, was that escalating levels of violence and repression did not require fine-grained surveillance and information gathering about the population as a whole.

While this brief account of the early Soviet period suggests that common assumptions about the Soviet sphere need to be questioned, blanket claims about the nature of the state in ostensibly communist countries also cannot be sustained. Whether we look at Eastern European countries after the Second World War, post-revolutionary China, Cuba, or Vietnam, or any of the postcolonial states influenced by socialism, from India to Ethiopia, we find widely divergent dynamics of state surveillance. The German Democratic Republic (Deutsche Demokratische Republik, or DDR) is an interesting case, as prior to the emergence of the digital surveillance regimes exposed by Edward Snowden and others, the DDR was often pointed to as the state that approached most closely the Orwellian dystopia of all-seeing surveillance. Relative to its population, the state apparatus of the DDR, most notably the Stasi, had by far the most extensive network of spies, bureaucrats, and informers gathering and analysing data about the population. Despite the density of surveillance mechanisms, however, totalitarian theory remains unpersuasive even in this case. Mary Fulbrook (2005) contends that it makes little sense to dichotomize state and society as we might when analysing liberal democracies. Rather, we need to understand the DDR as a kind of participatory dictatorship. While it was built around a small and powerful elite, "extraordinarily large numbers of people were involved in its functioning, who were implicated in a complex web of micro-relationships of power in every area of life, serving to reproduce and transform the system" (Fulbrook, 2005, p. 235). Fulbrook's reading of the DDR echoes in some respects that of Holquist and Hudson on the early Soviet state, but with the one key difference that here we do not find "underpolicing."

The experience of the DDR foregrounds the centrality of the surveillance of dissent in state socialist countries, but what is interesting is that we see analogous patterns across the First, Second, and Third Worlds of the Cold War era. The suppression of dissent was a central feature of state surveillance across the second half of the twentieth century, a theme taken up in a number of chapters in this book. In the Soviet bloc this was most spectacularly evident in the suppression of uprisings like that in Hungary in 1956 or in the Prague Spring of 1968. The Cultural Revolution in China (1966–1976) presents us with another brutal episode. However, we find extensive surveillance and repression of dissent elsewhere as well. This was at the heart of the brutal Guatemalan conflict cited earlier, for example, and was a central characteristic of the dirty wars that were fought across Latin America, in the mass killings of communists in Indonesia, or in the suppression of various movements, from socialists to Islamists, in the Middle East.

In many of these instances of brutality we find the influence of the United States, belying their rhetoric of the Cold War as a contest between freedom and repression, and also, in ongoing interventions across the Global South, suggesting that colonial legacies remained central to surveillance histories. As Alfred McCoy (2009) argues, the United States' own domestic policing apparatus was fundamentally shaped by the colonial counter-insurgency campaigns they waged as part of their 1898 occupation of the Philippines in the aftermath of the Spanish-American War, an occupation that only ended in 1946. The United States also supported struggles against anti-colonial movements, including within the United States itself. The predominantly Sikh Ghadar Party, to give one prominent example, was an anti-colonial movement that had a major influence along the west coast of North America as well as in India in the early twentieth century. It attracted migrant activists who "came to see themselves not simply as migrants seeking economic opportunity, but as politicized workers who understood the transnational dimensions of racial subjugation" (Sohi, 2014, p. 16). The response to the party was similarly transnational, with British, Canadian, American, and Indian officials working together to contain this emergent anti-colonial and anti-capitalist network, transforming in turn their own policing, border security, and intelligence operations.

These developments were precursors to later intelligence alliances that cemented US hegemony in the post-Second World War period through what John Bellamy Foster and Robert McChesney (2014) have called "surveillance capitalism." It was in this period that the full suite of Cold War surveillance practices was unleashed, including domestically in programs ranging from 1950s anti-communist monitoring to COINTEL-PRO, which in turn laid the groundwork for more recent NSA bulk surveillance. Given the Cold War histories that we are highlighting here, it is profoundly ironic that today Edward Snowden sits in Moscow, exiled in fear of prosecution by an American state whose surveillance and security apparatus is in many respects far and away more extensive than anything dreamed of by the KGB. If this appears as one bitter legacy of the Cold War, however, the current intensification in the surveillance and policing of migrants in the United States, documented and especially undocumented, points more directly to colonial histories.

Conclusion

The migration of surveillance practices between colony and metropole and the interconnected forms of surveillance that characterized the Cold War capture the complex global interactions that have shaped state surveillance, although here we have only begun to explore the myriad ways in which state surveillance has operated historically. The approaches we have highlighted above help to contextualize many of the chapters in this book. As case studies, those chapters bring a historical depth and specificity that helps us gain an understanding of the concrete institutional politics of particular state actors, and the forms of surveillance that have emerged over the last two or three centuries. There is no unitary or singular "surveillance state" in this sense, but rather interlocking sets of institutions, practices, and norms that approach populations and peoples via a surveillant gaze. States are made up of a wide range of actors, some national in scope (e.g., ministries of health, security, or education), others subnational and local, for example, the hospitals, police forces, or schools that are, for many people, the most visible face of "the state." The character of those different state bodies can vary considerably depending on the nature of the state (e.g., liberal democratic, state socialist, fascist, authoritarian, colonial), although common elements are also evident across these different forms. Who one is and how one is located in relation to the state is likewise crucial in shaping the nature and impact of surveillance. Finally, states are not autonomous, self-contained, or unchanging, but rather emerge through complex historical and transnational relations whose dynamics change over time.

If a comprehensive account of all of these factors is well beyond the scope of this introduction, the following chapters provide a rich and varied set of explorations that help us begin to map these surveillance histories. The authors demonstrate the need to pluralize our accounts, not only in relation to states themselves, but also of those subject to, and often socially sorted through, surveillance regimes. Together, the chapters comprise a transnational history, linking nationally based studies with others that foreground the provisional, porous, and shifting nature of the nation state, in particular how surveillance practices move across national borders even while constituting them. The chapter authors offer diverse approaches that highlight the ways in which surveillance states are sustained by global inequities, systems of exploitation, and power relations. This book is thus organized into three thematic sections that demonstrate these transnational dynamics, namely, Medical, Disease, and Health Surveillance; Identification, Regulation, and Colonial Rule; State Security, Policing, and Dissent. Each section begins with a short preface briefly summarizing and contextualizing the chapters, and reinforcing the links to the themes developed in this introduction. The historical analyses that are presented in the following pages are crucial in understanding the deeper dynamics of state surveillance and in tracing the roots of contemporary surveillance practices, roots that, in the quest for legitimacy, are often obscured by state actors themselves. In that sense, it is the dissidents, those combing the silences of the archives, in short, those engaging in the fraught and challenging work of counter-surveillance, who are our guides.

REFERENCES

About, I., Brown, J., & Lonergan, G. (Eds.). (2013). *Identification and registration practices in transnational perspective: People, papers and practices.* Houndmills, UK: Palgrave Macmillan.

Agamben, G. (1998). *Homo sacer: Sovereign power and bare life* (D. Heller-Roazen, Trans.). Stanford, CA: Stanford University Press.

Aly, G., & Roth, K. H. (2004). *The Nazi census: Identification and control in the Third Reich* (E. Black, Trans.). Philadelphia, PA: Temple University Press.

Anderson, B. (1991). *Imagined communities: Reflections on the origin and spread of nationalism* (Rev. ed.). London, UK: Verso.

Arendt, H. (1968). *The origins of totalitarianism.* New York, NY: Harcourt, Brace & World.

Bashford, A., & Levine, P. (Eds.). (2010). *The Oxford handbook of the history of eugenics.* New York, NY: Oxford University Press.

Bauman, Z. (1991). *Modernity and the Holocaust.* Ithaca, NY: Cornell University Press.

Bayly, C. A. (1996). *Empire and information: Intelligence gathering and social communication in India, 1780–1870.* Cambridge, UK: Cambridge University Press.

Beauchamp, T. (2009). Artful concealment and strategic visibility: Transgender bodies and US state surveillance after 9/11. *Surveillance & Society, 6*(4), 356–66.

Bennett, C. J., & Lyon, D. (Eds.). (2008). *Playing the identity card: Surveillance, security and identification in global perspective.* London, UK: Routledge.

Black, E. (2001). *IBM and the Holocaust: The strategic alliance between Nazi Germany and America's most powerful corporation.* New York, NY: Crown Publications.

Boersma, K., Van Brakel, R., Fonio, C., & Wagenaar, P. (Eds.). (2014). *Histories of state surveillance in Europe and beyond.* London, UK: Routledge.

Breckenridge, K. (2014). *Biometric state: The global politics of identification and surveillance in South Africa, 1850 to the present.* Cambridge, UK: Cambridge University Press.

Breckenridge, K., & Szreter, S. (Eds.). (2012). *Registration and recognition: Documenting the person in world history.* Oxford, UK: Oxford University Press.

Brown, E. H. (2008). *The corporate eye: Photography and the rationalization of American commercial culture.* Baltimore, MD: Johns Hopkins University Press.

Browne, S. (2015). *Dark matters: On the surveillance of Blackness.* Durham, NC: Duke University Press.

Caplan, J., & Torpey, J. (Eds.). (2001). *Documenting individual identity: The development of state practices in the modern world.* Princeton, NJ: Princeton University Press.

Césaire, A. (1972). *Discourse on colonialism* (J. Pinkham, Trans.). New York, NY: Monthly Review Press.

Cole, S. (2001). *Suspect identities. A history of fingerprinting and criminal identification.* Cambridge, MA: Harvard University Press.

Cooper, F. (1996). *Decolonization and African society: The labor question in French and British Africa.* Cambridge, UK: Cambridge University Press.

Corrigan, P., & Sayer, D. (1985). *The great arch: English state formation as cultural revolution.* Oxford, UK: Basil Blackwell.

Coulthard, G. S. (2014). *Red skin, white masks: Rejecting the colonial politics of recognition.* Minneapolis, MA: University of Minnesota Press.

Dandeker, C. (1990). *Surveillance, power and modernity: Bureaucracy and discipline from 1700 to the present day.* Cambridge, UK: Polity Press.

Daston, L., & Galison, P. (2007). *Objectivity.* New York, NY: Zone Books.

Davis, N. Z. (1983). *The return of Martin Guerre.* Cambridge, MA: Harvard University Press.

Deleuze, G. (1992). Postscript on the societies of control. *October, 59,* 3–7.

Derrida, J. (1995). *Archive fever: A Freudian impression* (E. Prenowitz, Trans.). Chicago, IL: University of Chicago Press.

Donovan, K. P., Frowd, P. M., & Martin, A. K. (2016). Introduction. *African Studies Review* (special issue – Surveillance in Africa: Politics, Histories, Techniques), *59*(2), 31–7.

Dubrofsky, R. E., & Magnet, S. A. (2015). *Feminist surveillance studies.* Durham, NC: Duke University Press.

Fields, K. E., & Fields, B. J. (2012). *Racecraft: The soul of inequality in American life.* London, UK: Verso.

Firmino, R. J., Bruno, F., & Arteaga Botello, N. (2012). Understanding the sociotechnical networks of surveillance practices in Latin America. *Surveillance & Society* (special issue – Surveillance in Latin America), *10*(1), 1–4.

Fiske, J. (1998). Surveilling the city: Whiteness, the black man, and democratic totalitarianism. *Theory, Culture and Society, 15*(2), 67–88.

Foster, J. B., & McChesney, R. (2014). Surveillance capitalism: Monopoly-finance capital, the military-industrial complex, and the digital age. *Monthly Review, 66*(3), 1–31.

Foucault, M. (1977). *Discipline and punish: The birth of the prison* (A. Sheridan, Trans.). New York, NY: Vintage Books.

Foucault, M. (1991). Governmentality. In G. Burchell, C. Gordon, & P. Miller (Eds.), *Studies in governmentality: Two lectures by and an interview with Michel Foucault* (pp. 87–104). Chicago, IL: The University of Chicago Press.

Foucault, M. (2003). *"Society must be defended": Lectures at the Collège de France, 1975–1976* (D. Macey, Trans.). New York, NY: Picador.

Foucault, M. (2007). *Security, territory, population: Lectures at the Collège de France, 1977–1978* (G. Burchell, Trans.). Houndmills, UK: Palgrave Macmillan.

Fulbrook, M. (2005). *The people's state: East German society from Hitler to Honecker.* New Haven, CT: Yale University Press.

Giddens, A. (1990). *The consequences of modernity.* Cambridge, UK: Polity Press.

Gikandi, S. (2015). Rethinking the archive of enslavement. *Early American Literature, 50*(1), 81–102.

Gilliom, J., & Monahan, T. (2013). *SuperVision: An introduction to the surveillance society.* Chicago, IL: The University of Chicago Press.

Ginzburg, C. (1982). *The cheese and the worms: The cosmos of a sixteenth-century miller* (J. Tedeschi & A. Tedeschi, Trans.). New York, NY: Penguin Books.

Gürses, S., Kundnani, A., & Van Hoboken, J. (2016). Crypto and empire: The contradictions of counter-surveillance advocacy. *Media, Culture & Society, 38*(4), 576–90.

Hacking, I. (1990). *The taming of chance.* Cambridge, UK: Cambridge University Press.

Haggerty, K. D., & Ericson, R. V. (2000). The surveillant assemblage. *British Journal of Sociology, 51*(4), 605–22.

Harris, V. (2002). The archival sliver: Power, memory, and archives in South Africa. *Archival Science, 2,* 63–86.

Harvey, D. (2003). *The new imperialism.* Oxford, UK: Oxford University Press.

Harvey, D. (2004). The "new" imperialism: Accumulation by dispossession. *Socialist Register, 40,* 63–87.

Henderson, J. (2018). "From one part of the empire to another": Promoting a settler-colonial future in late nineteenth-century Canadian immigration handbooks. In E. van der Meulen (Ed.), *From suffragette to homesteader: Exploring British and Canadian colonial histories and women's politics through memoir* (pp. 110–25). Black Point, NS: Fernwood Publishing.

Heynen, R. (2015). *Degeneration and revolution: Radical cultural politics and the body in Weimar Germany.* Leiden, Netherlands: Brill.

Higgs, E. (2004). *The information state in England: The central collection of information on citizens since 1500*. Houndmills, UK: Palgrave Macmillan.

Higgs, E. (2014). Further thoughts on The Information State in England ... since 1500. In K. Boersma, R. Van Brakel, C. Fonio, & P. Wagenaar (Eds.), *Histories of state surveillance in Europe and beyond* (pp. 17–31). London, UK: Routledge.

Holquist, P. (1997). "Information is the alpha and omega of our work": Bolshevik surveillance in its pan-European context. *The Journal of Modern History, 69*(3), 415–50.

Holquist, P. (2001). To count, to extract, and to exterminate: Population statistics and population politics in late imperial and Soviet Russia. In R. G. Suny & T. Martin (Eds.), *A state of nations: Empire and nation-making in the age of Lenin and Stalin* (pp. 111–44). Oxford, UK: Oxford University Press.

Hudson, H., Jr. (2012). *Peasants, political police, and the early Soviet state: Surveillance and accommodation under the New Economic Policy*. New York, NY: Palgrave Macmillan.

Jessop, B. (1990). *State theory: Putting the capitalist state in its place*. Cambridge, UK: Polity Press.

Jessop, B. (2016). *The state: Past, present, future*. Cambridge, UK: Polity Press.

Linebaugh, P., & Rediker, M. (2000). *The many-headed hydra: Sailors, slaves, commoners, and the hidden history of the revolutionary Atlantic*. Boston, MA: Beacon Press.

Louçã, F. (2009). Emancipation through interaction: How eugenics and statistics converged and diverged. *Journal of the History of Biology, 42*, 649–84.

Lyon, D. (Ed.). (2003a). *Surveillance as social sorting: Privacy, risk, and digital discrimination*. New York, NY: Routledge.

Lyon, D. (2003b). Surveillance as social sorting: Computer codes and mobile bodies. In D. Lyon (Ed.), *Surveillance as social sorting: Privacy, risk, and digital discrimination* (pp. 13–30). New York, NY: Routledge.

Lyon, D. (2007). *Surveillance studies: An overview*. Cambridge, UK: Polity Press.

Lyon, D. (2009). *Identifying citizens: ID cards as surveillance*. Cambridge, UK: Polity Press.

Marx, G. T. (1985). The surveillance society: The threat of 1984-style techniques. *The Futurist, 6*, 21–6.

Marx, G. T. (2002). What's new about the "new surveillance"? Classifying for change and continuity. *Surveillance & Society, 1*(1), 9–29.

Marx, K. (1867/1976). *Capital*, vol. 1. New York, NY: Penguin Books.

Mbembe, A. (2003). Necropolitics. *Public Culture, 15*(1), 11–40.

McCoy, A. W. (2009). *Policing America's empire: The United States, the Philippines, and the rise of the surveillance state*. Madison, WI: University of Wisconsin Press.

Mongia, R. V. (2003). Race, nationality, mobility: A history of the passport. In A. Burton (Ed.), *After the imperial turn: Thinking with and through the nation* (pp. 196–214). Durham, NC: Duke University Press.

Mosby, I. (2013). Administering colonial science: Nutrition research and human biomedical experimentation in Aboriginal communities and residential schools, 1942–1952. *Histoire sociale/Social History, XLVI*(91), 615–42.

Murakami Wood, D. (2009). The "surveillance society": Questions of history, place and culture. *European Journal of Criminology, 6*(2), 179–94.

Parenti, C. (2003). *The soft cage: Surveillance in America from slavery to the war on terror.* New York, NY: Basic Books.

Patterson, O. (1982). *Slavery and social death: A comparative study.* Cambridge, MA: Harvard University Press.

Perry, A. (2005). The colonial archive on trial: Possession, dispossession, and history in Delgamuukw v. British Columbia. In A. Burton (Ed.), *Archive stories: Facts, fictions, and the writing of history* (pp. 325–50). Durham, NC: Duke University Press.

Poulantzas, N. (1980). *State, power, socialism.* London, UK: Verso.

Rabinbach, A. (1990). *The human motor: Energy, fatigue, and the origins of modernity.* New York, NY: Basic Books.

Richards, T. (1993). *The imperial archive: Knowledge and the fantasy of empire.* London, UK: Verso.

Robertson, C. (2009). A documentary regime of verification: The emergence of the US passport and the archival problematization of identity. *Cultural Studies, 23*(3), 329–54.

Saltes, N. (2013). "Abnormal" bodies on the borders of inclusion: Biopolitics and the paradox of disability surveillance. *Surveillance & Society, 11*(1/2), 55–73.

Scott, J. C. (1998). *Seeing like a state: How certain schemes to improve the human condition have failed.* New Haven, CT: Yale University Press.

Sekula, A. (1986). The body and the archive. *October, 39,* 3–64.

Sengoopta, C. (2003). *Imprint of the Raj: How fingerprinting was born in colonial India.* London, UK: Macmillan.

Smith, K. D. (2009). *Liberalism, surveillance, and resistance: Indigenous communities in Western Canada, 1877–1927.* Edmonton, AB: Athabasca University Press.

Smith, S. M. (2004). *Photography on the color line: W.E.B. Du Bois, race, and visual culture.* Durham, NC: Duke University Press.

Smith, T. (2003). Visual regimes of colonization: Aboriginal seeing and European vision in Australia. In N. Mirzoeff (Ed.), *The visual culture reader* (2nd ed., pp. 483–94). London, UK: Routledge.

Sohi, S. (2014). *Echoes of mutiny: Race, surveillance & Indian anticolonialism in North America.* Oxford, UK: Oxford University Press.

Stern, A. M. (2005). *Eugenic nation: Faults and frontiers of better breeding in modern America.* Berkeley, CA: University of California Press.

Stoler, A. L. (2002). Colonial archives and the arts of governance. *Archival Science, 2,* 87–109.

Stoler, A. L. (2006). Colonial archives and the arts of governance: On the content in the form. In F. Blouin & W. Rosenberg (Eds.), *Archives, documentation, and institutions of social memory: Essays from the Sawyer Seminar* (pp. 267–79). Ann Arbor, MI: University of Michigan Press.

Stoler, A. L., & Cooper, F. (1997). Between metropole and colony: Rethinking a research agenda. In F. Cooper & A. L. Stoler (Eds.), *Tensions of empire: Colonial cultures in a bourgeois world* (pp. 1–56). Berkeley, CA: University of California Press.

Svenonius, O., & Björklund, F. (2018). Editorial: Surveillance from a post-communist perspective. *Surveillance & Society* (special issue: Surveillance in Post-Communist Societies), *16*(3), 269–76.

Tagg, J. (1988). *The burden of representation: Essays on photographies and histories.* Minneapolis, MI: University of Minnesota Press.

Thompson, E. P. (1963). *The making of the English working class.* New York, NY: Vintage Books.

Torpey, J. (2000). *The invention of the passport: Surveillance, citizenship, and the state.* Cambridge, UK: Cambridge University Press.

van der Meulen, E., & Heynen, R. (2016). *Expanding the gaze: Gender and the politics of Surveillance.* Toronto, ON: University of Toronto Press.

Vincent, D. (2016). *Privacy: A short history.* Cambridge, UK: Polity Press.

von Glahn, R. (2012). Household registration, property rights, and social obligations in imperial China: Principles and practices. In K. Breckenridge & S. Szreter (Eds.), *Registration and recognition: Documenting the person in world history* (pp. 39–66). Oxford, UK: Oxford University Press.

Weindling, P. (1989). *Health, race and German politics between national unification and Nazism, 1870–1945.* Cambridge, UK: Cambridge University Press.

Weld, K. (2014). *Paper cadavers: The archives of dictatorship in Guatemala.* Durham, NC: Duke University Press.

Yudell, M. (2014). *Race unmasked: Biology and race in the twentieth century.* New York, NY: Columbia University Press.

SECTION ONE

Medical, Disease, and Health Surveillance

The development of state surveillance in the nineteenth and twentieth centuries was driven to a significant degree by public health concerns and the rise of medicine. States increasingly took the lead in governing the health of their subject populations, defining and regulating bodily and social norms, and delineating and policing national borders around notions of "fitness" and hygiene. As Alison Bashford (2004) argues, the languages and practices of hygiene and of state authority were deeply interconnected: "Signalling the constant need for purification from the ever-present contaminating threat over the border, however imagined, hygiene became a primary means of signification by which those borders were maintained, threats were specified, and internal weaknesses managed" (p. 5). At its borders, the hygienic nation state was buttressed by increasingly dense systems of surveillance. Disease surveillance was especially central to the militarization and racialization of borders, for example, and to the emergence of regimes of "bioinsecurity" (Ahuja, 2016). This surveillance underpinned normalizing visions of the nation that, as Douglas Baynton (2016) argues, also targeted disabled people and sought "[t]he exclusion of individuals seen as defective" (p. 2).

In linking anxieties over borders to the management of "internal weaknesses," Bashford highlights the ways in which national populations were simultaneously internally differentiated and categorized. Thus, border threats were tied as well to the internal regulation of labour, intervention into the gendered realm of reproduction, the development of the welfare state, and the remaking of urban space. Here too medical, disease, and disability surveillance were central. Who and what was perceived as a threat is evident in a 1904 comment from Hermann Biggs, one of the main architects of the medical surveillance apparatus in the United States. According to Biggs, anti-tuberculosis surveillance measures should be directed at the "homeless, friendless, dependent, dissipated and vicious consumptives" who were "most likely to be dangerous to the community" (quoted in Fairchild, Bayer, & Colgrove, 2007, p. 45). That community was most often conceived of as white and middle class, with immigrants, racialized and Indigenous communities, disabled people, and other marginalized groups the source of threat. Fears over the proliferating internal dangers Biggs evokes ran through many other public health campaigns over the late nineteenth and twentieth centuries as well, ranging from Hansen's disease (leprosy) to syphilis and HIV/AIDS (Ahuja, 2016). At the same time, of course, and in keeping with the profoundly ambivalent nature of much surveillance, public hygiene led to significant (if highly differentiated) improvements in health outcomes.

The policing of the internal and external borders of the nation state through medical surveillance practices and notions of public hygiene also emerged through dense transnational networks. Between 1851 and 1903, for example, a series of International Sanitary Conferences and other hygiene-oriented gatherings were held in Europe and the United States, helping to establish a global reach for health-related intervention that led to the formation of international groups and associations like the Health Organization of the League of Nations, founded in 1923, predecessor to the World Health Organization established in 1948 (Bashford, 2006). That these systems of cooperation were driven to a significant degree by the demands of colonial governance is especially evident in chapter 2, where Jacob Steere-Williams traces the transnational practices of disinfection in the British Empire, beginning in India and moving to South Africa. Anti-plague measures, most notably "dipping," he argues, highlighted the role of colonial medicine and disease surveillance in the regulation of colonial labour. Analysing the visual rhetoric of these campaigns, as well as the labour of surveillance performed in part by Indian workers, Steere-Williams shows that they produced both a racialized colonial Other and a vision of British imperial masculine health.

The second half of the nineteenth century saw the emergence of eugenics as a primary framework through which hygiene was conceptualized, tying individual bodily health directly to that of the nation or to race. While today we tend to associate eugenics with Nazi-era racial science and genocide, this needs to be qualified in a number of ways. First of all, not only did Nazi eugenics draw on eugenics movements elsewhere, most notably the United States (see Stern, 2005), but interest in eugenics spanned much of the political spectrum, only later coming to be the preserve mainly of right-wing politics. The extent to which eugenic ideas became part of a European and North American common sense is evident in the 1910 entry "Civilization" in the *Encyclopædia Britannica*. It draws on the work of the prominent eugenicist

Francis Galton in claiming that civilization promises that "the average physical mental status of the race will be raised immeasurably through the virtual elimination of that vast company of defectives which to-day constitutes so threatening an obstacle to racial progress" (pp. 408–9). Disability was thus central, as the language of "defect" in the encyclopædia entry suggests (see Baynton, 2016; Davis, 2006), but eugenic developments were also rooted in colonial logics. Galton, for example, developed many of his ideas after travelling to colonial southern Africa, eugenics serving in turn to legitimize and shape colonial rule (Breckenridge, 2014).

Holly Caldwell's examination in chapter 3 of late nineteenth-century Mexican campaigns to "cure" deafness picks up these themes and traces how public hygiene and eugenic surveillance turned on the perceived degenerative threats posed by syphilis and alcoholism. Purported to cause deafness, syphilis and alcoholism were leveraged to regulate reproductive relationships, especially those involving Deaf people. As with the politics of disinfection discussed by Steere-Williams, Caldwell highlights the transnational networks through which eugenic ideas travelled to Mexico, exploring the extent to which eugenic concerns were bound up directly with ideas of national health and with securing the "fitness" of the nation at a time of profound social change. "Fitness" was also central to conceptions of sexuality, as B Camminga's shows in chapter 4, which examines the criminalization of gay and gender non-conforming men before, during, and after apartheid. Camminga links sexology, medicine, and eugenics in arguing that an examination of the South African Disguises Acts opens a window onto the constitution of normative and deviant forms of gendered embodiment that have profound implications both for our understanding of the apartheid surveillance state and for the ways in which surveillance practices shaped the dynamics of twentieth-century South Africa. Tracing the transnational linkages that brought the approaches of European sexology to bear, Camminga joins Steere-Williams and Caldwell in foregrounding the routes along which medical and eugenic ideas and practices of surveillance travelled. Together, they show how notions of "fitness," tied to broader fears around "degeneration" (Heynen, 2015), profoundly shaped the development of medical, disease, and public health surveillance.

REFERENCES

Ahuja, N. (2016). *Bioinsecurities: Disease interventions, empire, and the government of species.* Durham, NC: Duke University Press.

Bashford, A. (2004). *Imperial hygiene: A critical history of colonialism, nationalism and public health*. Houndmills, UK: Palgrave Macmillan.

Bashford, A. (2006). "The age of universal contagion": History, disease and globalization. In A. Bashford (Ed.), *Medicine at the border: Disease, globalization and security, 1850 to the present* (pp. 1–17). Houndmills, UK: Palgrave Macmillan.

Baynton, D. C. (2016). *Defectives in the land: Disability and immigration in the age of eugenics*. Chicago: The University of Chicago Press.

Breckenridge, K. (2014). *Biometric state: The global politics of identification and surveillance in South Africa, 1850 to the present*. Cambridge, UK: Cambridge University Press.

Davis, L. J. (2006). Constructing normalcy: The bell curve, the novel, and the invention of the disabled body in the nineteenth century. In L. J. Davis (Ed.), *The disability studies reader* (pp. 3–16). New York, NY: Routledge.

Encyclopædia Britannica (1910). Civilization (11th ed., pp. 408–9). Cambridge, UK: Cambridge University Press.

Fairchild, A., Bayer, R., & Colgrove, J. (2007). *Searching eyes: Privacy, the state, and disease surveillance in America*. Berkeley, CA: University of California Press.

Heynen, R. (2015). *Degeneration and revolution: Radical cultural politics and the body in Weimar Germany*. Leiden, Netherlands: Brill.

Stern, A. M. (2005). *Eugenic nation: Faults and frontiers of better breeding in modern America*. Berkeley, CA: University of California Press.

2 "Coolie" Control: State Surveillance and the Labour of Disinfection across the Late Victorian British Empire

JACOB STEERE-WILLIAMS

At the height of the Third Plague Pandemic in the late 1890s, at one of its most virulent epidemiological nodes, in Bombay, India, English bacteriologist Dr Ernest Hanbury Hankin received a frantic telephone call. Hankin, born in Hertfordshire, had trained at University College London and St Bartholomew's Hospital, and after working at Charles Roy's Pathological Laboratory in London had spent time in the laboratories of Robert Koch in Berlin and Louis Pasteur in Paris. For the majority of his career, Hankin was Chemical Examiner and Government Analyst in Agra, India. On the other end of Hankin's 1899 phone call was his friend Dr Osvald Valdemar Muller, the Danish-born Professor of History at the University of Bombay, who was assisting the municipal authorities by heading the Relief Fund; Muller's opportunistic title was "Official Plague Hospital Visitor," but he noted that "I played the part of the plague patient's friend" (Muller, 1902, p. 92). His job was to visit plague victims in the Bombay hospitals and provide, when deemed deserving by a medical officer, extra rations or clothing and monetary support for the surviving families. Muller rode his bicycle each morning to the five major hospitals in Bombay; in three months alone in 1899, from January to March, he visited with over 5,000 patients. For his efforts, in 1900 he received the Kaisar-i-Hind medal, an award for distinguished civil service for the British Raj, though he noted, "I was looked upon as very foolhardy ... [M]y friends in Bombay treated me as if I were a leper, and society practically cut me as it was known I was on plague duty" (Muller, 1900, p. 88).

Just days before the phone call to Hankin, Muller had attended a man in hospital, Dattu Koli, whose surname indicated his caste status as an unskilled labourer. Koli had worked as a cotton factory millhand, and died of plague while in hospital. The case found deserving for relief, Muller set out to find Koli's family and assist with funeral expenses.

When he came to the man's residence, the first thing he did was phone Hankin. When Hankin arrived at the address provided, on Cross Lane, Currey Road, in the textile mill district of Lower Parel, he opened the door to find a single-room, 100 square foot Bombay chawl, or tenement house, occupied by a poor Hindu mother and her two daughters. All three were lying prostrate on the earthen cow-dung floor writhing in pain and clearly marked by the signs of bubonic plague. Hankin's first order was to remove the mother and her daughters to the Plague Hospital. His assistants successfully took out the mother and one daughter, but the second, a girl of about eight, died before their eyes.

Hankin stood in the centre of the "ill-ventilated" room, surveying what he called "a great state of disorder" (Hankin, 1901, p. 65). In one corner was an area for cooking with a few brass vessels. A second corner had an open drain, which was oozing human excrement. In a third was an altar with a brass image of Ganesha, the Hindu God of Wisdom. The dead body of the little girl occupied the fourth corner. "Clothes, food, and filth of all sorts covered the floor," Hankin sensorially observed, and the "room was pervaded by an overpowering smell that produced nausea" (Hankin, 1901, p. 65).

Hankin spent an hour at the chawl collecting samples for an experiment he conducted later that day on the infectivity of Indian "things." Historian Robert Peckham (2016) has shown that similar efforts were made by British officials in Hong Kong during the plague pandemic. Back at his laboratory, Hankin subcutaneously injected nearly 100 mice, using what he feared was infected earth from the house's floor, a stain of dried saliva he found on a stool, excrement from the drain, dust from a lamp, and a half-eaten pudding he collected from the floor near the deceased girl. Inanimate Indian objects, in the hands of European bacteriology, were being transformed into the dangerous spreaders of infectious disease.

But before he left the chawl on Cross Lane, Currey Road, to begin laboratory experimentation, Hankin ordered that the house and all of its contents be destroyed and burned by the sanitary inspector and his assistants, a specially appointed "gang" of coolie labourers. "Gang" was the contemporary term British officials used to describe groups of unskilled labourers who practised anti-plague measures in India. Hankin's intrusion into the Currey Road tenement – even his order to demolish the property – was not particularly unique to British India in the 1890s (Harrison, 2012). British officials across the country mobilized massive public health machinery – medical-*cum*-militaristic – to identify plague sufferers, plague corpses, and plague contacts; to expel individual Indians from their houses and into segregation, also called health camps;

and to disinfect, fumigate, cleanse, whitewash, and burn their houses, their possessions, and their bodies.

Hankin's account, then, is largely representative; it dovetails with the kind of frameworks scholars have applied to colonial public health and state formation in the late nineteenth century (Arnold, 1993; Echenberg, 2007; Harrison, 1994). Hankin's European, bacteriologically reinforced gaze as he looked around the Currey Road chawl illustrates the fin-de-siècle Western ethos that Indian domestic spaces were dangerous homes for spreading plague, and that Indian "things" (Henare, Holbraad, & Wastell, 2007) were there-for-the-taking research objects for European colonial scientists. The Hankin anecdote is a brief window to foregrounding two key arguments about the modern surveillance state: that Western hygienic ideals and practices over Indigenous bodies were central to the colonial state, and that colonial surveillance was often inherently violent to the bodies and local environments of Indigenous peoples.

The colonial state used moments of epidemic crisis, as I explore in this chapter, to practise what David Lyon (2003), referring to more recent social phenomena, has called "social sorting," or the insidious and discriminatory surveillance practices of sorting people into culturally assigned categories of worthiness. In this way, and as I demonstrate below, to Hankin and his late nineteenth-century contemporaries, the British surveillance of Indian urban spaces, things, and bodies sought to reinforce cultural and racial difference, and in the process to remake Indigenous peoples and places as unsanitary and infectious. This process dovetails with Alison Bashford's (2004) and Deana Heath's (2010) claims as to the centrality of hygienic ideas and practices in the colonial state, particularly the Western obsession with regulating moral contagion.

Hankin's laboratory research on plague had provided an impetus for colonial surveillance, and placed him at the centre of two controversial debates in the late 1890s: over the aetiology of bubonic plague, particularly Paul Simond's rat-flea theory, and over the effectiveness of Waldemar Haffkine's plague inoculation. Both were met with widespread scepticism across British India. R.B. Stewart, Acting Secretary to the Government of Bombay, for example, noted in 1899: "though inoculation may assist in giving confidence, I would rather trust to evacuation with disinfection" (Stewart, 1900, p. 3). Rather than his theoretical interest in bubonic plague, Hankin's day-to-day activities, like those demonstrated by his intrusion into the Currey Road chawl, exemplify the way in which fear over infectious disease was mobilized by British authorities into a campaign to inspect and disinfect urban spaces and people. Public health, in other words, was a prized tool of colonial rule in British India.

Mridula Ramanna (2002) has called late nineteenth-century British anti-plague measures in India "harshly intrusive" (p. 33), and David Arnold (1993) has made the cogent argument that the British response to plague in the 1890s provided the setting for an "unprecedented assault on the body of the colonized" (p. 203). Less explored has been the agency of epidemic control – not the plague "operations," as W.L. Reade noted above, but the plague "operators." Such an approach, which focuses on labour and is space based and geography induced, reveals a critical aspect of the confluence of race, work, and materiality in the history of colonial regimes of surveillance in the late nineteenth and early twentieth centuries. As I explore in this chapter, surveilling the health of British India – a stand-in metaphor for the health of the broader empire – required, according to British colonial authorities, a multipronged approach that constructed Indian domestic spaces, things, and bodies as dangerous objects in the spread of infectious disease, reifying the centrality of colonial social sorting, even as it demonstrated its weakness and inconsistencies. This deployment of "health" in configuring the relationship between individual bodies, the nation, and the empire was common in the late nineteenth and early twentieth centuries, and was evident as well in the eugenic regulation of deafness in Mexico that Holly Caldwell explores in the next chapter.

In the following, I begin by focusing on the racialized technologies of colonial disinfection practices in British India, particularly the role that labouring coolies played in the imposition of anti-plague measures. According to A.R. Burnett-Hurst (1925), already by the eighteenth century "coolie" meant unskilled labourer, deriving from the fisherman caste "Koli." By the period of the British Raj, the term "coolie" was in widespread use to refer to labourers in diverse settings from agricultural tea or sugar plantations to urban factories. Workers labelled as coolies were often young, able-bodied men who were seen to be able to withstand particularly arduous physical labour (see also Damir-Geilsdorf, Lindner, Muller, Tappe, & Zeuske, 2016; Northrup, 1995). I suggest that both because of and despite their caste status, specially appointed gangs of coolies were trained as experts in the attempt to sanitize Indian domestic spaces, things, and the urban ecology for two conflicting reasons: as a test of imperial citizenship, and as a way for British authorities to closely monitor what they saw as the most dangerous spreaders of infectious disease. Part I of the chapter, then, contributes to a vibrant debate in surveillance studies regarding the extent to which colonial regimes were interested in developing comprehensive forms of knowledge about Indigenous populations.

Part II demonstrates the not-so-subtle irony of coolie labour. Although so-called coolie gangs carried out anti-plague measures, their Indian bodies were themselves intensely scrutinized. In other words, at the same time that they were surveillance experts, they were deemed in need of surveillance. Here I unpack the unexplored practice whereby Indigenous Indians were dipped in specially designed vats of caustic chemicals such as carbolic acid, phenyl, and corrosive sublimate. Critical here are not only the actual practices of disinfecting Indians as an overt act of colonial surveillance, but also the way in which human disinfection was communicated across the empire, particularly back home in Britain. Using contemporary changes in photography and the growing middle-class media, British authorities remade the sanitizing colonial project into a public spectacle that the wider British public took part in supervising.

Part III places British dipping efforts in India in the broader historical context of imperial networks by showing that Indigenous Africans were also dipped by British officials in the late nineteenth century. Interrogating disinfection practices during the South African War, I show that British authorities feared both Africans and contract-labouring Indians as spreaders of disease, necessitating heavy-handed and full-bodily measures like dipping. And while both Africans and Indians were given expertise in disinfecting the dangerous things that white British bodies produced, particularly their excrement, in British South Africa, Indian coolies were imagined as superior to African kaffirs. The dually constructed identities of the coolie and the kaffir reflected racial assumptions. This chapter's central interrogation, then, is the practice of colonial state surveillance, showing how boundary objects – goods, clothes, material technologies, and even human excrement – were inescapably tied to imperial identities of sturdy British masculinity and feebly diseased Indigenous "other." Examining the practices of colonial public health demonstrates a unique dimension of modern forms of biopower as they act as technologies of surveillance.

Part I: Experimentation, Flying Plague Columns, and the Indian Urban Ecology

According to an internal memo between members of the Indian Plague Commission in 1898, "the principal source of infection is to be found in the houses ... plague is essentially a disease of locality" (Hankin, 1901, p. 101). That localization was particularized, as seen in Hankin's example above, on Indian homes, especially the homes of the poor. More broadly than an attack on Indian domestic spaces, such practices were steeped in an ecological discourse that borrowed from the language and rhetoric of neo-Darwinism, known in contemporary bacteriological circles as the saprophytic existence and competition of germs outside of the body.

There was a flurry of commingled laboratory-infused field research in the 1880s and 1890s on the saprophytic life of disease-causing bacilli. At the Plague Research Laboratory in Bombay, Drs. Mackie and Winter, for example, set up an experimental space using the small outbuilding attached to their laboratory. They used the 14 x 7 foot building, "without much light or ventilation," because it was "similar to many a house occupied by the natives of Bombay" (Plague Advisory Committee, 1906, p. 511). Splitting the room in half, they prepared two floors, one made of earth covered with cow dung, the other with chunam, a cement made of sand and lime. They then, in their words, "grossly contaminated" the floors "with pure virulent broth cultures of B. pestis" (Plague Advisory Committee, 1906, p. 511). For each of the next five days they took scrapings of earth, emulsified them in sterile broth, and rubbed it into the shaved abdomens of rats and guinea pigs. Mackie and Winter found that plague bacilli could be found in cow dung earth floors after forty-eight hours, while in cement floors after less than twenty-four hours. In a similar set of experiments in 1904, Dr William Glen Liston, Indian Medical Service and staff of the Plague Research Laboratory in Parel, conducted what he called a more a "natural" experiment, introducing guinea pigs into working-class houses where plague was present; he let the guinea pigs run free for a day or two, then collected and chloroformed them, scraped off the fleas, and finally placed the fleas onto healthy animals (typically white rats imported from England). He then examined the stomach contents of the fleas for plague bacilli. Liston even experimented with fumigating a Bombay dwelling with sulphur dioxide, not dissimilar to the widespread contemporary use at ports of the American-developed Clayton machine, which fumigated ships using sulphur as a disinfecting agent (Weindling, 2000). Even in houses that were fumigated, Liston found, guinea pigs left for a day gathered plague-bacilli-positive fleas. European experimental research on bubonic plague, in other words, sought to both take the Indian urban ecology into the space of the European laboratory and, in the case of Liston, to bring the European laboratory to the Indian urban ecology. Both were integral to the colonizing project of sanitary surveillance.

Experimental research on the saprophytic existence of plague bacilli was braced by studies as to the power of particular disinfectants. In Bombay, Hankin and Wilfred Watkins-Pitchford, who later led British public health efforts in South Africa, were leading advocates for what they termed a complete system of disinfection. Hankin found that the most reliable way of disinfecting tenement houses was to dig up the floor with a pick axe about four inches deep, add a layer of dry grass on the floor, burn the grass, and finally disinfect with a solution 1 in 250 of sulphuric acid or a cocktail of corrosive sublimate and hydrochloric acid. Though Hankin argued that limewashing the exterior of houses, seen in numerous photographs of the time, was unnecessary and of limited antibacterial importance, he maintained that it was at least useful in providing visual evidence, we might even say rhetorical surveillance, as to the power of disinfection technologies. It might not be antibacterial, in other words, but limewashing could serve as window dressing for the actual chemical disinfectants used inside.

European experimental research into the infectivity of the Indian urban ecology set the stage for considerable public health intervention. The actual work was pre-eminently hierarchical: provincial and municipal British medical officers and Indian Medical Service officials at the top, followed by mid-tiered British and Indian hospital staff and police officers, and, at the bottom, the largest group, doing most of the work, were coolie labourers hired contractually for minimum wages. "Plague parties" or "search parties," notably in Karachi and Bombay, were composed of European men who conducted house-to-house searches for plague sufferers and plague contacts. The express goal of their surveillance was to discover who was sick or already dead and force their removal to segregation camps or isolation hospitals, where further supervision and inspection would take place. In Poona, for example, "at one time the number of troops employed on plague duty" was over a thousand and was "met with great opposition" (Hankin, 1901, p. 112), the most well known being the plotted assassination of C.W. Rand, chairman of the Poona Plague Committee, by the Chapekar brothers (Choudhury, 2010).

At this time, British plague labour engaged in surveillance worked at different levels than Indian plague labour, the coolie gangs tasked with implementing British-designed anti-plague measures. Figure 2.1 shows a striking contemporary photograph of a typical, all-male coolie gang from Karachi under the command of Lieutenant Anderson, Karachi's Superintendent of Market and Jail Quarters. Itself a means of colonial supervision, or a way of seeing the Indigenous other, the photograph illustrates the power and performativity of visual representations of colonial public health. As Lukas Engelmann (2017) and Christos Lynteris (2016) have recently suggested, photographs of plague victims, workers, and plague environments were inextricably part of the ecology of an epidemic. In figure 2.1, members of a coolie gang of men and some boys deliberately posed in front of a tenement chawl building, demonstrating

Figure 2.1 "Lt Anderson's Gangs," 1897. Photo taken by R. Jalbhoy. Wellcome Library, icv no. 29771, no. 30102i, V0029295, *Karachi Plague Committee*, Creative Commons.

at least the appearance of an organized labour force. We should take seriously, however, the rhetorical power of the photograph's staging of successful disinfection and destruction. As yet another visual reinforcement of colonial surveillance, the gang went as far as to specially mark the building with the date, a red encircled cross, and to decorate it with three rats. The photograph reveals the gendered dynamics of the colonial bacteriological revolution, as disinfection was deemed a male activity. The photograph is also a lens into the materiality of colonial surveillance, with technologies such as pump sprayers in hand to demonstrate successful anti-plague measures. In critical ways, anti-plague technologies in the hands of coolie labourers who placed marks on urban spaces were an extension of other forms of biometric surveillance outlined by Claire Anderson (2004) and Keith Breckenridge (2014). They mirror especially the tattoos that, as Anderson discusses, were a key biometric technology deployed by the colonial state in India, but in this instance markings were used to delimit urban spaces and render them accessible to state and medical surveillance.

Coolie gangs like the one seen in figure 2.1 were also known by the military metaphor of "Flying Plague Columns." These groups opened roofs, tore down walls, and burned so-called dangerous personal items such as rags and clothes. Inhabitants were either quickly forced to Plague Hospitals or segregation camps, or, if healthy, left outside their homes to see the destruction of their property in real time. In 1897 Bombay authorities issued instructions for preparing and using disinfectants, advocating Hankin's use of 1 in 1,000 corrosive sublimate, but reports across British India indicated that in practice the concentrations, and even the particular disinfectant used, were anything but uniform; perchloride of mercury, izal, carbolic acid, and sulphur fumigation were all widely employed inside homes, on Indian things, and on Indian bodies, as discussed below. Consider also the distribution of labour. Some members of coolie gangs responsible for digging earthen floors and destructing walls carried picks and shovels, as illustrated by the men seated in the photograph. Others carried imported Chinese wooden or brass pumps for spraying disinfectant, though when such technologies were unavailable the men were instructed to throw disinfectant on the walls of houses directly with their hands. In the picture above, for example, many of the men standing outside of the disinfected building hold, almost boastfully, their sprayers, which we can see were stand-in weapons in the fight against infectious disease and urban disorder. Other members of coolie gangs were armed with cloth-wrapped sticks and whitewashing buckets for applying freshly slaked lime. When available, steam disinfectors were employed, typically at ports, hospitals, and cantonments. Massive quantities of disinfectant were used. In Dharwar, for example, Major T.H. Hardy (1900), who oversaw a coolie gang, noted that ten buckets of perchloride of mercury were used, at a 1 in 1,000 solution, and "squirted all over the room ... the stuff runs on the floor and is swept on again ... it is all expended on the house; so it floods the ground" (p. 95).

Though such units were designed to have a British hospital staff or soldier in command, with a middling caste "native overseer," in practice "only in a very small proportion of cases were the disinfecting arrangements directly supervised by anyone, medical or lay, who had any specific scientific training or any previous experience of disinfecting operations" (Hankin, 1901, p. 367). The groups could be as small as 10 or 20, like at Igatpuri, or 100 or 200, as they were in Karachi and Bombay. Bombay in particular seems to have been a hotbed for "training" coolie gangs' disinfection practices. Smaller towns and villages in the late 1890s often hired a small number of trained coolie gangs from Bombay, "who knew the disinfection work," and added local coolies who could apprentice under the Bombay crews and be "put through a rude course of instruction on the spot," as Surgeon-General Harvey noted in 1899 (Harvey, 1900, p. 283).

Anti-plague measures were an attempt to transform unskilled labouring coolies into "trained" experts, as a kind of public health test for what Daniel Gorman (2006) has called "imperial citizenship," a concept that moves beyond legal definitions of belonging and instead relies on shared cultural understandings, in this case on disease and disease management. Coolie control, it seems, was as much about objectifying coolies and their environments as the dangerous spreaders of disease as it was the use of coolie labour in stopping the chains of infection. We should take seriously coolie agency, however, particularly the subversive anti-anti-plague measures against British colonial authority, as it fits with Antoinette Burton's (2016) recent account of the many ways in which British imperialism was challenged. There was widespread lamentation by British authorities as to the incomplete work done by coolie gangs. R.B. Stewart (1900) of Bombay complained in 1899 that "even after months of instruction ... the coolie does not understand" that no corner of a room could be left untouched (p. 3). Some British officials feared that coolie gangs simply whitewashed the exterior of homes and marked them with either U.H.H. (unfit for human habitation), or as seen in figure 2.1, the encircled red cross, the so-called plague mark, without actually disinfecting. We have to take seriously, then, that figure 2.1 might be *either* a visual representation of a successful colonial regime of surveillance and anti-plague disinfection *or* a deliberate act of Indigenous resistance. Coolie gangs arguably also challenged colonial surveillance, enacting a kind of historical counter-surveillance that is a key theme of this book.

What is certain at least is that anti-plague measures in British India conducted by coolie gangs were "not unattended with risk" (Hankin, 1901, p. 35). Hankin himself, who ordered the destruction of the chawl on Currey Road, had made clear that there was a "risk of coolies being burnt by having to handle strong chemicals" (Hankin, 1900, p. 14), and by the 1890s there was extensive research as to the dangers of human exposure to strong caustic chemicals. In addition, members of coolie gangs contracted plague so frequently that many were given experimental inoculations as well as ordered to live in isolated camps.

Examining the practices of disinfecting the urban Indian landscape, as explored in part I, takes us away from European laboratories in colonial settings, though there was coolie labour there too, as laboratory assistants, and into the city, the home, and the bazaar. When we look across these spaces of British regimes of surveillance against infectious disease in the late nineteenth century, we see not just an attack on the Indigenous body or Indian domestic spaces or things, but rather a reconfiguration of the spaces of infectivity and the labour of those who carry out public health interventions. In other words, we see the legitimization of state power through a particular moral ethos of cleanliness and contagion, what Philip Corrigan and Derek Sayer (1985) term "moral regulation." Whereas part I examined the labour, materiality, and potential supervision of colonial public health, part II turns to the transnational history of a virtually unknown practice of full-body dipping of Indigenous Indians as an overt and heavy-handed surveillance of the Indian body.

Part II: Dipping and Disinfection

The violent and coercive nature of regimes of surveillance and sanitation, of so-called progressive public health, carried out under the guise of Western imperialism is illustrated in figure 2.2. It captures the ironies of modernist assumptions of technoscience and germ theory heroics. The photograph was taken in 1898, the same year as figure 2.1, by J.H.C. Kelly, Assistant Supervisor for the Indo-European Telegraph Department, an arm of the British-Indian Government. Kelly's role was to maintain telegraph lines in Persia that linked British India and the Ottoman Empire, but in his spare time he was an avid photographer. This photograph was taken at the Keamari Quarantine Depot, Karachi, in current-day Pakistan. Before British incursion on the northwest coast of the Indian subcontinent, Karachi stood as a small fishing village. By the 1890s it was transformed into a bustling port of the British Raj, with over 100,000 diverse inhabitants. In 1898, when this photo was taken, Karachi and its inhabitants were constructed – pathologized – as dangerous nodes in epidemiological webs of infection that might spread across the globe.

The photograph appeared in popular print in the British illustrated weekly *Black and White*, which promoted itself in every issue as specializing in "good art, good literature, and good printing" (Black and White, 1891). The title suggests an unintended but apt analogy between photography and the production of race, and, appearing in this context, the Karachi photograph fulfilled this pedagogical mission, serving as evidence of the broad reach of colonial public health surveillance. *Black and White* was the same periodical where late Victorians read the serialized short stories of Robert Louis Stevenson, Arthur Conan Doyle, Bram Stoker, and H.G. Wells; it was a vehicle for entertainment and mass persuasion. The materiality and transitory nature of figure 2.2 illustrates the complex and violent process of colonial state surveillance. To J.H.C. Kelly, the photographer, the technological medium was a way to capture

Figure 2.2 "Keamori Quarantine Depot," 1898. Photo taken by J.H.C. Kelly. *Black and White*, August 6, 1898, p. 23.

a brief and fleeting moment of colonial public health. As the photograph moved in time and across space – through what scholars such as Alan Lester (2001) have called "imperial networks" – it became a means of seeing, of knowing, bubonic plague, Indigenous peoples, and the *de rigueur* of British imperialism and modern public health.

Kelly's own obsession with photography, largely outside of his official duties, was fuelled by a major transition in photography with the introduction of gelatin plates, which drastically increased the speed at which photographs like the Karachi one could be taken. By the mid-1890s the long exposure time of the mid-Victorian daguerreotype had been superseded by a more "instant" depiction (Tucker, 2005). In the case of the Karachi photograph, the technological change revealed something striking, a sense of motion, of action, and, perhaps, emotion.

Figure 2.2 depicts, in terms starkly *black* and *white*, the actual practice of disinfecting humans, one of the few extant photographs to have been taken of this historical practice. It centres on a group of naked or near-naked Karachiites being forced into a dipping tank. A local Karachi – who was called the "native warrant officer" – provided what was called in the text the "necessary persuasion," wielding here a stern finger sharply pointed into the murky vat. Two other Indians anxiously wait their turn, while a white British official sees the process through. Clothes and belongings sit at the front of the picture, those too to be disinfected in the smaller buckets near one of the central labourers in the photo. On the right we see a coolie whose job was to mix, fill, and stir the carbolic acid mixture. "All natives who are suspect of having been in contact with sources of contagion," the accompanying text triumphantly noted, "are required to visit one of these tanks," indicating that there were several in Karachi alone, "and to take a dip in the water, which has the property of instantly destroying microbes." The last line reads, "the natives do not, as a rule, take kindly to the process, but it is insisted on, notwithstanding," suggesting a fascinating process of resistance to colonial surveillance akin to the deliberately incomplete work done by coolie gangs in disinfecting Indian homes (Black and White, 1898, p. 23).

The caption below figure 2.2 in *Black and White* further described the "process" of disinfection. The tank was five feet deep, brick lined, and filled with a "strong solution" of crude phenol, or carbolic acid, a dark oily liquid. Contemporary European chemists defined a "strong solution" as that containing at least 1 part pure phenol to 18 of water. In experimental studies of the microbe-destroying properties of carbolic acid, leading European bacteriologists Robert Koch, Edward Klein, and Edgar Crookshank independently agreed, to use Koch's words, that carbolic acid "kills if of considerable strength and acting for a long period of time" (Rideal, 1895, p. 150). By the 1890s the use of diluted carbolic was well known in Victorian surgery, after Joseph Lister's 1867 article in the *Lancet* promoting his brand of antiseptic surgery. Figure 2.2 depicts another, more disquieting side of the use of carbolic acid in the age of the so-called bacteriological revolution. The photograph is a powerful example of a discourse on disinfection, colonial power, and public health in an epistemic moment of the late nineteenth century. Victorian jingoist bravado made disinfection appear a *modern* technology – a by-product of a burgeoning chemical industry, bacteriologically reinforced, its champions posited that it would stop epidemiological nodes in the web of infection. Conducted in public spaces such as railway stations and village city centres, dipping was also a kind of public health spectacle that, by imprinting what British officials believed were strong Western chemicals on feeble and dangerous naked Indians bodies, legitimated and extended colonial regimes of disease surveillance.

The safety of disinfectants was just becoming a matter of real concern in the last two decades of the nineteenth century. The well-known Jeyes Fluid, for example, was advertised in popular periodicals as the "Best, Safest and Cheapest," and as not burning the skin, suggesting that other disinfectants did of course do so. However, many public health officials – those in other words using disinfectants as anti-disease technologies – noted like C.G. Moor and T.H. Pearmain's did in their 1897 *Applied Bacteriology* that "while it was 'indispensable' for any disinfectant to be effective, it was only 'desirable' that it was non-poisonous and non-caustic" (Whyte, 2016, p. 136). Highly concentrated, "strong" solutions of phenol like those used in the Karachi dipping tank were highly corrosive, a point not lost on late Victorian chemists. Samuel Rideal (1895), Public Analyst for Lewisham and Lecturer on Chemistry at St George's Hospital, London, noted in his pioneering text *Disinfection and Disinfectants* that phenol had been inhaled, ingested, and absorbed (through human skin) "with dangerous, and even fatal, effects" (p. 254). But, as the Karachi photograph shows, in times of epidemic crisis, now suffused with both the fear and reality of bubonic plague, the warnings of late Victorian chemists were pushed aside by colonial officials. In practice, even the weak recommendations were ignored, and there was little regard for the specific concentrations of disinfectant used.

As a means of surveilling Indian bodies, colonial disinfection practices reveal a gendered and racial practice of epidemic control, one that at once reified the Western colonial project and its inherent racial hierarchies. But disinfection was also a corporeal technology of the body and a "technology in transit," to borrow and adapt a phrase from James Secord (2004), that served as a site of knowledge transfer, debate, and contestation, over the environment and bodies to be sure, but also the things that bodies produce. Material technologies, the Karachi vats, popular disinfecting machines such as Thresh's, and the actual chemicals like phenol, perchloride of mercury, or izal were vehicles for this larger cultural process. And so too were photographs, statistical charts, and epidemiological maps.

The short history of human disinfection must be contextualized in the vociferous debates about the nature and existence of germs, and seen as a test to Foucault's notions of governmentality and biopower. Historian Nancy Tomes (1998) outlined the process by which the germ theory was domesticated into the minds and everyday practices of Americans in the late nineteenth and early twentieth centuries, taken up in a religious-like fervour. More recently, Graham Mooney (2015) has suggested that germ practices in Britain were part of larger attempts at governmental surveillance into the lives and habits of everyday people. Mooney argues that the pillars of the "new" public health – identifying germs, isolating sick individuals, and disinfecting mediums of disease transmission – were not necessarily "intrusive" interventions into domestic spaces. Examining the uneven practices of disinfecting colonial environments and bodies demonstrates that in practice the germ theory was taken up by colonial regimes as a justification for overt displays of violence and oppression, and as a way to sort the microbe-clean from the microbe-laden. But rather than a straightforward story of public health as a tool of empire, the colonial history of disinfection suggests two conclusions. The first is that colonial disinfection demonstrates the unevenness of early biomedicine. The germ theory provided a new means for obsessing over bodies and what bodies produce, making it perhaps less of an epistemological or theoretical break from earlier theories about the body or imperial control than historians have assumed. The second is that colonial disinfection as a means of Indigenous surveillance and control revealed a particular irony of modern imperialism; Indigenous peoples in India and Africa, racially typecast by Europeans as coolies and kaffirs, were both blamed for spreading infectious diseases, and at the same time put in charge as experts over disease management and control.

The Keamari Quarantine Station was only one site where humans were disinfected with caustic and dangerous chemicals in hopes of stopping the spread of plague. In Karachi itself there were numerous locations of dipping stations, particularly mobile dipping vats at the segregation camps of Trans-Lyari and Gulumsha. At Gulumsha, for example, individuals and their goods were disinfected upon arrival, and "similar arrangements [were made] on leaving" (Hankin, 1901, p. 230). Across British India human dipping was on full display as a performative practice at railway stations, ports, health camps, and hospitals. British officials copiously described the process of what one British official noted as the "disinfection of the person by washing in antiseptic solutions" and "completely douching each person with an antiseptic solution" (Hankin, 1901, p. 374). In the context of heated debates and scepticism over both Waldemar Haffekine's anti-plague vaccine and Paul Louis-Simond's rat-flea theory of plague's epidemiology, "intrusive interventions" (Mooney, 2015) like dipping were, in the words of one British official, "indubitably" effective (Hankin, 1901, p. 374).

Armed with bacteriological and epidemiological specificity, the combination of disinfecting Indian bodies and Indian urban environments was broadly construed as a practice that closely supported the rubric of colonial rule. W.L. Reade, administrative head of the British plague commission, noted in the oft-quoted line, "I consider that plague operations properly undertaken present some of the best opportunities for riveting our rule in India ... [and] for showing the superiority of our Western science and thoroughness" (Harrison, 1994, p. 143). Such hyperbolic and rhetorical statements about "superior" Western science belie both popular protest by Indigenous Indians to intrusive measures like dipping and deep-seated anxieties by British public health officials about their effectiveness. Before the Royal Commission on Plague, one British official was clear: "there is no doubt that the disinfection of the person is one of the most unpopular of plague measures, and naturally so, for the measure not only involves a great deal of personal discomfort, but it may also, when the disinfecting solution is a fairly concentrated one, entail a certain amount of severe smarting or subsequent irritation" (Hankin, 1901, p. 374). Despite these misgivings on the part of some officials, human dipping of colonized peoples moved out of India to other colonial locations, as part III illustrates in the case of British South Africa in the late nineteenth century.

Part III: Excremental Obsessions

Dipping animals in caustic chemicals was a routine veterinary practice by the late nineteenth century, but its transfer to humans was a result of colonial frustrations with epidemic disease. In 1899, before the South African Medical Congress in Johannesburg, British Railway Medical Officer Herbert Caiger spoke reverentially of the veterinary sine qua non medicalized public health practice. He noted that between November of 1896 and March of 1897 – a span of only four months – over 11,000 African migrant workers were dipped at the Bethulie Bridge cordon station. Bethulie was on the direct route from the lucrative Rand mines near Kimberly (to the north) and the Eastern Cape (to the south), the village homelands of many migrant workers. To Caiger, the temporality of an epidemic necessitated the measure. In 1896–1897 a virulent outbreak of rinderpest ravaged South Africa. At the International Rinderpest Conference in Vryburg, several dipping regulations were formalized by the South African states, all of which mandated the dipping of animals, and some of which permitted the dipping of Indigenous Africans. To the white politicians, what prevailed was expediency and opportunism in a moment of crisis. To would-be epidemiologists like Caiger, however, dipping was more than just expediency, and also more than destroying germs. It was a technology of surveillance that, in his words, "afforded an opportunity for estimating the number of sick native passengers more accurately than is possible by ordinary inquiries" (Caiger, 1899, p. 258). "A native may conceal the fact that he is ill," he continued, "if he is merely questioned and left undisturbed in his carriage; but, when all native passengers had to get out of the train, strip naked, and go down into the dipping tank and out again, few cases of real illness could escape notice" (p. 258). At Bethulie, like Karachi, dipping was a technology of state power, where white public health authorities cast medical and moral imperative over Indigenous bodies by way of what those in power substantiated as biomedical necessity in a moment of epidemic crisis. Chandak Sengoopta (2003) has shown that biometric surveillance practices such as fingerprinting were often pioneered in India and exported to other colonial locations, a claim substantiated by Breckenridge (2014). The dipping of humans was part of this broader history of colonial surveillance; like biometric measures of fingerprinting, dipping was not applied to the entire population, but selectively to individuals and groups based on racial ideologies of the period. Dipping also had corollaries in the twentieth century with United States disinfection practices against typhus fever at the United States-Mexico border (Mckiernan-González, 2012), suggesting that modern practices of bodily surveillance developed out of complex transnational colonial histories.

What is curious about Caiger's approval of dipping over 11,000 Africans is that it came in an 1899 address titled "Johannesburg as a Typhoid-Distributing Centre." The link he drew between bubonic plague, rinderpest, and typhoid is an interesting one, and helps connect Caiger's account to disinfection practices in 1890s India. According to contemporary British public health opinion, typhoid was "exceedingly prevalent" in South Africa, as it was in India, for two reasons: the cultural habits of Indigenous peoples and the lack of an efficient sanitary system. The result was a racially charged version of the European "dream of a pipe-bound city" (Hallstrom, 2002, p. 1). "We have a semi-barbarous native population to deal with," Caiger lamented, "and we shall have long to wait before we can hope to find among our coloured brethren an enlightened public opinion, or a due sense of responsibility, in sanitary matters" (Caiger, 1899, p. 258).

To Caiger and others across the British Empire, typhoid was constructed as a preventable disease in South Africa because of the decline of the disease in Britain. Here again, as scholars such as Ian Hacking (1990) have described, state surveillance in the nineteenth century was deeply tied to statistical ways of measuring and sorting populations and perceived risks. By the 1890s, the rhetoric in Britain held that a successful public health infrastructure was dependent upon both hygienic and sanitary practices of individuals, and of centralized oversight by public health officials. As Philip Curtin (1998) has shown, rates of typhoid amongst British soldiers in South Africa stayed relatively low until the Boer War, when in a period of less than five years there was a massive rise, the biggest single incidence in modern history. Some 80,000 Britons suffered from the disease and 15,000 were killed, not to mention thousands of Indigenous people about whom Western archival sources have been deadly silent. Statistical returns for India showed a similar worrying trend, an increase in the disease in the colonies just as it was decreasing at home.

The 1890s, then, was a particularly fascinating moment across the British Empire, lavished with jingoist fears over war, the working classes, and the imperial economy, but also in the grips of what appeared to be an epidemic crisis, namely, a sharp rise in typhoid amongst white soldiers, bubonic plague rustling through British ports, and once again the threat of epidemic cholera. This focus on ports and the movement of people and disease was reflected as well in projects to identify, police, and exclude migrants that developed across the empire over subsequent decades, as explored by Uma Dhupelia-Mesthrie and Margaret Allen in chapter 6. Attempts to control typhoid amongst white British bodies encompassed a convoluted web of Royal Army Medical Corps officers, private consulting surgeons, volunteer female nurses, male orderlies, and Indigenous labourers, who were all inexorably bound up with a bacterium, with soldier's bodies, with makeshift field hospitals, and with material technologies such as disinfectors, chemicals, photographs, and maps. Take the example of the Portland Hospital, a civilian field hospital at Bloemfontein, a site morally spatialized by Rudyard Kipling (1903) as "Bloeming-Typhoid-ein" (p. 166). Called by the war journalist William Burdett-Coutts (1900) a "tented city of pestilence" (p. 172), the entire hospital operation revolved around controlling typhoid. The field hospital's layout featured separate tents used as wards for enteric patients, separate latrines for enteric bedpans, and separate latrines for enteric convalescents. The geography of disease control, of encampment technology, pathologized white Britons suffering from the disease within a localized environment vis-à-vis with what they produced, their linens, their excreta, and their urine. Obsessing over each as pathologized, contagious things, the practices for handling typhoid excreta are particularly revealing of a complex set of material realities, gendered debates, and racially charged rhetoric. As Bashford and Hooker (2001) have argued, "the global distribution of hegemonic practices of western biomedicine ... situates the very notion of 'contagion' inside a history of imperialism and colonial discourse, as well as a history of biomedicine" (p. 6). Whereas Indigenous Indians required full-body dipping, only the excrement of white Europeans necessitated disinfection, a kind of colonial social sorting between colonizer and colonized that more closely approximates Foucault's notion of biopower as diffused from the state.

William Stokes, Consulting Surgeon to the #7 General Hospital, Natal, lavished these "excremental obsessions," a phrase adapted from Warwick Anderson's (2006) "excremental colonialism" (p. 104). His camp procedure noted that enteric stools were removed by a "conservancy man (Indian coolie), who removes it to the special latrine set apart for enteric stools. The contents are placed in a covered receptacle. The bedpan is then washed in a 1 in 1,000 solution of perchloride of mercury, dried, and a small quantity of izal, 1 in 20, poured in" (Stokes, 1900, p. 598). "The same process is carried out with urinals," Stokes noted, "about 50 yards from the enteric stool shed is a large oblong shallow boiler placed on a brick furnace; into this boiler the enteric stools, urine, and disinfectants previously used are poured, sawdust is mixed with the same fluid and the whole boiled until almost dry from evaporation" (Stokes, 1900, p. 598). In a rare glimpse into the labour involved, Stokes reported that "there are 60 conservancy coolies, a certain number being told off to each row of tents, latrines, lavatories, etc." (Stokes, 1900, p. 598).

At the Royal Army Medical Corps General #3 Hospital, typhoid stools likewise took centre stage. After being disinfected with formalin, they were mixed with sawdust, saturated with paraffin, and removed outside of camp and burned. Enteric dressings were burned in the camp incinerator, and enteric linens were soaked in formalin then placed in a Theal disinfector. The practice of identifying typhoid was left to Royal Army Medical Corps officers and staff surgeons, who employed another kind of quasi-biometric technology of surveillance – physical marking on clothes – to identify those suffering from the disease. A general practice, not dissimilar to the marking of plague positive houses in India, was to mark soldiers' bodies coming to hospital with the initials N.Y.D (not yet determined) or E.F. (enteric fever).

Deep beneath, but of course always at the surface, was a British-imposed racial hierarchy. The system described above for the hospital in Natal, where sixty so-called conservancy coolies were employed, was acclaimed in the British medical press as efficient and effective in guarding against typhoid. This was in part due to the attention paid to dangerous typhoid excreta, shepherding it from the time it left a white British body, dropped into a specially constructed bedpan, was witnessed by a white male orderly, then physically handled and supposedly destroyed by a coolie. Elsewhere, as Sir William Church (1901) snarled before the Clinical Society of London, the work was left "to the Kaffirs, who could not be trusted as could the Indian coolie" (p. 772). In this discourse of "excremental obsessions," apparently only morally appropriate Indigenous peoples could handle British excrement. Here also was an ironic inversion: when whites suffered in the moment of epidemic crisis, it was Indigenous Africans or indentured Asians who practised disinfection, entrusted as they were with sanitizing urban environments. That contemporary British public health officials made distinctions between coolie and kaffir, placing the former above the latter in a racial hierarchy of public health work, says a great deal about the significance of dichotomous relationships of white and non-white labour in colonial settings.

Conclusion

As a corporeal technology of colonial public health, disinfection –
encompassing the theories that supported it and the material practices
illustrated in the two Karachi photographs – stood Janus-faced in temporality and in geography. It was both a forward-looking germ theory
practice and a reification of existing British racial theory. Over the
course of the twentieth century, public health techniques like disinfection became routinized, even domesticated as necessary governmental
interventions into everyday life, representing intrusions onto the body
that are highlighted by Lyon's concept of social sorting. But for a brief
historical moment in the late nineteenth and early twentieth centuries,
disinfection stood at the cusp of Western public health as an uncertain
practice. In the context of imperial settings, as I explore in this chapter, the everyday practices of disinfection tested the reach of state and
colonial biopower, and were fundamentally about pathologizing bodies and environments and what they produce. Class and gender were
key frameworks in which the colonial state surveillance sought to make
and remake bodies. Coolie gangs were overwhelmingly male, for example, which provides a glimpse into the perceived masculinity of public health work at this time. Late nineteenth-century British regimes
of colonial surveillance used moments of medical crisis for political
advantage. Disease and the unhealthy Indigenous body were sites of
biopolitical contestation, but the labour and materiality of disinfection practices demonstrates the partial failings of, or at least resistance
to, colonial surveillance. The unevenness of colonial biopower, as it
attempted to supervise and sanitize the Indigenous sick body, provided
through colonial practices of surveillance an opportunity for agency
and resistance.

REFERENCES

Anderson, C. (2004). *Legible bodies: Race, criminality and colonialism in South Asia.*
    Oxford, UK: Berg.
Anderson, W. (2006). *Colonial pathologies: American tropical medicine, race, and
    hygiene in the Philippines.* Durham, NC: Duke University Press.
Arnold, D. (1993). *Colonizing the body: State medicine and epidemic disease in
    nineteenth-century India.* Berkeley, CA: University of California Press.
Bashford, A. (2004). *Imperial hygiene: A critical history of colonialism, nationalism,
    and public health.* Houndmills, UK: Palgrave Macmillan.
Bashford, A., & Hooker, C. (2001). *Contagion: Historical and cultural studies.*
    London, UK: Routledge.

Breckenridge, K. (2014). *Biometric state: The global politics of identification and surveillance in South Africa, 1850 to the present.* Cambridge, UK: Cambridge University Press.

Burdett-Coutts, W. (1900). *The sick and wounded in South Africa.* London, UK: Cassell.

Burnett-Hurst, A. R. (1925). *Labour and housing in Bombay.* London, UK: P. S. King.

Burton, A. (2016). *The trouble with empire: Challenges to modern British imperialism* Oxford, UK: Oxford University Press.

Caiger, H. (1899). Johannesburg as a typhoid-distributing centre. *South African Medical Journal,* 264–7.

Choudhury, D. K. L. (2010). *Telegraphic imperialism: Crisis and panic in the Indian empire, 1830–1920.* London, UK: Palgrave Macmillan.

Church, W. (1901). Notes from the clinical society of London. *British Medical Journal,* 772.

Corrigan, P., & Sayer, D. (1985). *The great arch: English state formation as cultural revolution.* Oxford, UK: Basil Blackwell.

Curtin, P. (1998). *Disease and empire: The health of European troops in the conquest of Africa.* Cambridge, UK: Cambridge University Press.

Damir-Geilsdorf, S., Lindner, U., Muller, G., Tappe, O., & Zeuske, M. (2016). *Bonded labor: Global and comparative perspectives.* Bielefeld, Germany: Verlag.

Echenberg, M. (2007). *Plague ports: The global urban impact of bubonic plague, 1894–1901.* New York, NY: New York University Press.

Engelmann, L. (2017). What are medical photographs of plague? Retrieved from https://remedianetwork.net/2017/01/31/what-are-medical-photographs -of-plague/

Gorman, D. (2006). *Imperial citizenship: Empire and the question of belonging.* Manchester, UK: Manchester University Press.

Hacking, I. (1990). *The taming of chance.* Cambridge, UK: Cambridge University Press.

Hallstrom, J. (2002). *Constructing a pipe-bound city: A history of water supply, sewerage, and excreta removal in Norrkoping and Linkoping, Sweden, 1860–1910.* Linkoping, Sweden: Linkoping University.

Hankin, E. H. (1900). Evidence of E. H. Hankin. In *Proceedings of the Indian Plague Commission,* vol. II. Great Britain, House of Commons. London UK: Spottiswoode and Eyre.

Hankin, E. H. (1901). *Evidence of E. H. Hankin.* In *Report of the Indian Plague Commission.* Great Britain, House of Commons, vol. v. London, UK: Spottiswoode and Eyre.

Hardy, T. H. (1900). Evidence of T. H. Hardy. In *Minutes of Evidence, Indian Plague Commission,* vol. I. Great Britain, House of Commons. London UK: Spottiswoode and Eyre.

Harrison, M. (1994). *Public health in British India: Anglo-Indian preventive medicine, 1859–1914.* Cambridge, UK: Cambridge University Press.

Harrison, M. (2012). *How commerce spreads disease.* New Haven, CT: Yale University Press.

Harvey, W. L. (1900). Evidence of Surgeon-General Harvey. In *Minutes of Evidence, Indian Plague Commission,* vol. I. Great Britain, House of Commons. London, UK: Spottiswoode and Eyre.

Heath, D. (2010). *Purifying empire: Obscenity and the politics of moral regulation in Britain, India, and Australia.* New York, NY: Cambridge University Press.

Henare, A., Holbraad, M., & Wastell, S. (2007). *Thinking through things: Theorising artefacts ethnographically.* London, UK: Routledge.

Kelly, J. H. C. (1898). Keamori Quarantine Depot. *Black and White* (1891–1912). *A weekly illustrated record and review.*

Kipling, R. (1903). The parting of the columns. In *The Five Nations.* New York, NY: Charles Scribner & Sons.

Lester, A. (2001). *Imperial networks: Creating identities in nineteenth-century South Africa and Britain.* London, UK: Routledge.

Liston, G. (1904). Plague, rats, and fleas. *Proceedings of the Bombay Natural History Society,* 24 November.

Lynteris, C. (2016). *Ethnographic plague: Configuring disease on the Chinese-Russian frontier.* London, UK: Palgrave Macmillan.

Lyon, D. (Ed.). (2003). *Surveillance as social sorting: Privacy, risk, and digital discrimination.* New York, NY: Routledge.

Mckiernan-Gonzalez, J. (2012). *Fevered measures: Public health and race at the Texas-Mexico border, 1848–1942.* Durham, NC: Duke University Press.

Mooney, G. (2015). *Intrusive interventions: Public health, domestic space, and infectious disease surveillance in England, 1840–1914.* Rochester, NY: Rochester University Press.

Muller, O. A. (1900). Evidence of Prof. O. A. Muller. In *Minutes of Evidence, India Plague Commission,* vol. III. Great Britain, House of Commons. London, UK: Spottiswoode and Eyre.

Muller, O. A. (1902). *Report of the municipal commissioner on the plague in Bombay.* Bombay, India: Advocate of the India Press.

Northrup, D. (1995). *Indentured labor in the age of imperialism, 1834–1922.* Cambridge, UK: Cambridge University Press.

Peckham, R. (2016). Hong Kong junk: Plague and the economy of Chinese things. *Bulletin of the History of Medicine, 90*(1), 32–60.

Plague Advisory Committee. (1906). Plague number: Plague investigations in India. *The Journal of Hygiene, vi,* 422–537.

Ramanna, M. (2002). *Western medicine and public health in colonial Bombay, 1845–1895.* Delhi, India: Orient Longman.

Rideal, S. (1895). *Disinfection and disinfectants.* London: Charles Griffin and Company.

Secord, J. (2004). Technology in transit. *Isis*, *95*, 654–72.

Sengoopta, C. (2003). *Imprint of the Raj: How fingerprinting was born in colonial India*. London, UK: Palgrave Macmillan.

Stewart, R. B. (1900). Evidence of R. B. Stewart. In *Minutes of Evidence, Indian Plague Commission*, vol. III. Great Britain, House of Commons. London, UK: Spottiswoode and Eyre.

Stokes, W. (1900, September 1). A visit to no. 7 general hospital, Estcourt, Natal. *British Medical Journal*, 598–601.

Tomes, N. (1998). *The gospel of germs: Men, women, and the microbe in American life*. Cambridge, MA: Harvard University Press.

Tucker, J. (2005). *Nature exposed: Photography as eyewitness in Victorian science*. Baltimore, MD: Johns Hopkins University Press.

Weindling, P. (2000). *Epidemics and genocide in Eastern Europe, 1890–1945*. Oxford, UK: Oxford University Press.

Whyte, R. (2016). Public health and public safety: Disinfection, carbolic acid, and the plurality of risk, 1870–1914. In T. Crook & M. Esbester (Eds.), *Governing risks in modern Britain*. London, UK: Palgrave Macmillan.

3 Surveillance, Medicine, and the Misterios de la Naturaleza: Campaigns to "Cure" Deafness in Late Nineteenth-Century Mexico City

HOLLY CALDWELL

In 1863, the Mexican newspaper *La Sociedad* published a study titled "Misterios de la naturaleza," which examined the prevalence of deaf-mutism in families. After surveying the rate of deaf-mutism in one family, the author described the "inexplicable mysteries of heredity" and concluded:

> A deaf-mute had two children, a boy and a girl; one was also born deaf-mute and died young. The girl married a healthy man like her and had three children, two deaf-mute girls like their grandfather, and a healthy boy. The healthy boy married a young hearing and speaking girl like himself, and the marriage resulted in a deaf-mute child, whereas his deaf-mute sister married a deaf-mute man and had a child without this defect. Here is an example of the inexplicable mysteries of heredity. (Rodriguez, 1863, p. 2)

This case study was among the first of several published in Mexican newspapers in the nineteenth century that examined familial lineages in order to understand the "mysteries of nature" that led to the prevalence of deaf-mutism in families. This passage highlighted a localized, yet at the same time national, concern that enveloped Mexican society at the time. Deafness was frequently associated with a morose demeanour, with the afflicted often purportedly suffering from chronic depression and suicidal tendencies. These views implied that the diagnosis of deafness or deaf-mutism in a child not only affected the emotional well-being of the parents, but also the lifelong countenance of the child. Drawing on Yael Berda's (2013) discussion of how colonial powers have historically treated local populations as both dangerous and inferior through systems of surveillance enabling their diagnosis and categorization, this chapter analyses the ways in which elites in postcolonial Mexico sought to control and monitor their own populations – particularly those who were categorized and defined as inferior – as a means to reform and improve the nation as a whole. I argue that deafness and its perceived associated causes represented the manifestation of several undesirable elements plaguing Mexican society that were central concerns.

Deafness was viewed as the by-product of unsavoury consanguineous unions, deaf intermarriage, and immoral behaviours, which led many elite reformers to consider it one of the most significant obstacles to creating a fit and healthy nation. Late nineteenth century elites monitored what they perceived were risky populations, and public health policies were utilized to either limit or entirely prevent this potential threat to the nation (see Agostoni, 2004; Cruz Cruz & Cruz-Aldrete, 2013). From a medical standpoint, the Deaf were stigmatized as sick and socially backward and therefore posed an impediment to the positivist project of nation-building, as they could not hear and were unable to communicate in a typical fashion. In modernizing Latin American states that subscribed to the tenets of positivism, individuals were valued and judged primarily by their economic usefulness to the nation, and those who were categorized as defective or abnormal in some way were often defined as lacking such usefulness. The persistent question of how to incorporate Mexico's young deaf population into the national economy, and how to tap into this potential workforce, had plagued physicians and policymakers for decades.

To emphasize how language concerning the Deaf has transformed over time, this chapter will use two sets of language. Throughout history, Mexico's Deaf were, and to some extent remain, characterized and categorized *by* their "disability" despite the fact that Deaf individuals typically do not view themselves as disabled.[1] This concept has been examined in recent years, as scholars have described deafness as a cultural construction as well as a physical phenomenon (see Baynton, 1996; Chinn, 2006; Kudlick, 2003). In the interest of respecting the values and identity of the Deaf community, I have elected to not employ the Mexican government's recently devised terms *personas con discapacidad* or *personas con capacidades diferentes*, and have opted instead to use Deaf, deaf, or deaf-mute. (The capitalized "Deaf" refers to the culturally Deaf, those who do not view themselves as disabled but rather as a linguistic minority, thus embracing the cultural norms, beliefs, and values of the Deaf community. The lowercase version of the term refers simply to the condition of having a hearing and communicative difference.) For the scope of this chapter, the term *deaf* or *deaf-mute*, or the Spanish translations *sordo* or *sordomudo*, will only be used when referring to the hearing and communicative difference of deafness or in the translation of historical documents in order to preserve the integrity of the original texts and the positions of the writers of such texts. I acknowledge that the term *deaf-mute* is offensive to many in the Deaf community, as it was derived from historical connotations of calling one "deaf and dumb," which later came to be understood as "deaf and stupid." In this chapter, I will use contemporary terminology, such as Deaf and deaf, when offering my own analysis of the historical material (see Zames Fleischer & Zames, 2011).[2]

Ensuring the Fitness of the Nation

Governing authorities in modernizing Latin American states had long used science and medicine as instruments of social control to manage certain sectors of society. By the late nineteenth century, both science and medicine had come to represent power and modernity and, for European and Latin American authorities alike, contributed to their respective nations' "civilizing project." Scholars have noted that medicine was used to create and enforce boundaries between civilians, to protect both international and domestic borders from disease and "filth," and to determine who was acceptable to belong to the body politic (see Arnold, 1996; Bashford, 2004; Cunningham & Andrews, 1997; Meade & Walker, 1991). Drawing on the values and definitions of citizenship, order, and progress, modernizing states in Latin America such as Argentina, Brazil, and Mexico used state-sanctioned measures to target specific populations that were deemed "degenerate members of a sick race," such as the urban poor, vagrants, criminals, and immigrants (Cueto & Palmer, 2015, p. 166, see also Rodriguez, 2006; Stepan, 1991).

As part of the campaign to "revamp notions of nationhood," these liberal governments used medical surveillance in order to address issues concerning sanitation and disease as well as perceived social problems such as prostitution, crime, and alcoholism (Cueto & Palmer, 2015, p. 164; see also Armus, 2003). David Lyon (2007) has contended that the purpose of surveillance in the modern era is rarely totalitarian domination, rather, it functions in subtle forms through influence and control, and, arguably, often with unwitting (or unwilling) participants (see also Richards, 2013). In Latin America, late nineteenth-century Argentine elites embraced Eurocentric traditions and values and used medical surveillance to diagnose, monitor, and resolve social problems that they regarded as the negative by-products of massive immigration and Indigenous populations (Rodriguez, 2006). Likewise, when Brazilian elites attempted to "implant" civilization and progress in Rio de Janeiro at the turn of the century, their strategy, Teresa Meade (1997) argues, "began to center on the debate over public health, an issue that likewise was closely tied to the question of social control" (p. 63). In Mexico, when Porfirio Díaz seized control of the presidency in 1876 he initiated a series of dramatic changes that reverberated throughout the country but impacted especially strongly on the nation's capital.

In order to create a modern, progressive society, the Díaz regime followed the precepts of Comtean positivism, a scientific and social philosophy that had recently entered Mexican discourse. As Mexican policymakers envisioned their modernized society, they drew on European models of surveillance in order to classify, diagnose, and cure their own population of perceived backwardness, practices that have been identified as fundamentally tied to the making of citizenship (Zamindar, 2007). Historically, governments have surveilled both the health and bodies of a given population as a means "to achieve the material, productive, economic goals of the state" (Wiebe, 2008, p. 335). European imperialist powers often employed different levels of surveillance to monitor and control their colonial populations, a process in which nations engaged in the act of both regulating and disciplining individual bodies (Stoler, 1995). Although Mexico was not an imperial power, Díaz's elite group of government officials, prominent citizens, and urban professionals – collectively known as *científicos* – employed similar tools of surveillance as their European counterparts and implemented them as part of their overarching goal to determine those individuals as possessing or lacking worth (see Piccato, 2001; Wiebe, 2008).

Positivism was believed to provide a cure for a disordered society by providing the moral, educational, and social structure needed to make the population orderly (Barreda, 1863). As Díaz implemented a comprehensive and aggressive campaign to modernize the Mexican state, education, public health, and moral hygiene became mechanisms of surveillance designed to discipline and modernize the urban poor. As Marcos Cueto and Stephen Palmer (2015) contend, urban society was a "social body that had to be protected, controlled, and cured by experts," a belief that was linked to "the propagation of ideal and moralizing archetypes of a healthy body defined in terms of stereotypes of gender, patriotism, and citizenship" (p. 69). Mexico's Deaf children who had either been orphaned or abandoned became part and parcel of this overarching state-directed campaign to reform the poor, which included removing wayward children from urban streets, expanding correctional institutions, and reforming and rehabilitating people labelled delinquents. Thus, the foundations of deaf education in Mexico were grounded in broader reforms to prevent the poor who were housed in the capital's Hospicio de Pobres (Poor House) from becoming vagrants or social parasites that would threaten the national order.[3] Mexico's masses, particularly the poor and those defined as "undesirable" such as the nation's Deaf (and blind) populations, were subject to forms of classification and scrutiny under the new positivist model, which sought to implement order, progress, and draw on the economic usefulness of its citizens.

Eugenics and Deafness

It has been commonly understood that the Mexican eugenics movement emerged in the wake of the Mexican Revolution of 1910, which triggered massive population decline due to political dislocation and rampant malnutrition and disease (Suárez y López Guazo, 2005). The term "eugenesia" began circulating in print form in Mexico in 1911 when it first appeared in an issue of the newspaper El Diario. Relatedly, congenital deafness was labelled in 1908 as one of the social maladies along with epilepsy, hydrocephaly, imbecility, idiocy, and hysteria that contributed to the "degeneration of the species" ("La Compaña contra la avería," 1908, p. 8). Despite the fact that most eugenicist policies in Mexico were officially enacted in the two decades following the Revolution of 1910, other policies that similarly directly affected the education, treatment, and view of the Deaf were implemented in the late nineteenth century. As in many nations during this period, Mexican elites categorized the Deaf as abnormal, subjecting them to what scholars have described as "objectification by medical discourse and techniques of intervention" (Saltes, 2013, p. 58). Tackling the "mysteries of nature" as they pertained to the spread of deafness through marriage and disease led many elite policymakers to view the hearing disorder as one of the gravest reproductive consequences of "undesirable" individuals engaging in what they regarded as immoral behaviour and unions. This systematic categorization of bodies, coupled with the positivist ideals of creating a morally upright and ordered populace, laid the foundation for the eugenics movement that would take root in the early twentieth century. As B Camminga discusses in the next chapter, albeit in a very different context, eugenics targeted relations of sexuality and morality in particular, as this was the locus of reproduction and hence the heart of the project for national or social regeneration. By shifting the historical lens to the late nineteenth century and examining how the terms of normality were outlined against the grain of abnormality, scholars can better understand how public health policy and social discourse concerning deaf intermarriage served as precursors to the more draconian forms of eugenics that would take shape in later decades.

Historically, deafness was medically defined, understood, and analysed in terms of deviance and moral retribution. From the seventeenth through the early nineteenth centuries, medical practitioners and social scientists alike commonly viewed the occurrence of deafness as divine judgment for immoral or unseemly behaviour. Until the early eighteenth century, deaf individuals in European society were viewed as isolated deviations from a norm because there was no significant discourse surrounding the deaf and deafness. Such deviations, Lennard Davis (1997) argues, were rarely tolerated, and little was done for those who in some way disrupted social norms. In Mexico, fears of individuals who could potentially disrupt the norms of society that *científicos* were looking to establish created a context in the latter half of the nineteenth century in which individuals labelled mad, mentally disabled, or insane were "increasingly perceived as a threat to social harmony and stability and the nation's progress and prosperity" (Ballenger, 2009, p. 5). Using various fields of study such as criminology and medicine, government officials worked to gain authority in determining what was "abnormal," which they then used to create categories of sickness that were often associated with lower-class citizens, criminals, and the so-called mentally insane (Thomson, 2010). Since deafness was categorized as a sickness and a disorder that could (and should) be cured through medical intervention, the Deaf were categorized as mentally incapable and classified in the same legal category as the insane, which resulted in the denial of rights and privileges (Dublan & Lozano, 1876–1904) and the establishment of the Deaf as legitimate targets for eugenic surveillance apparatuses. *Científicos* drew on both subtle and direct methods of surveillance to study deafness and the Deaf, such as marriage and disease, with the expressed interest to reform these individuals (and society) of this affliction.

Rather than analysing their nation's economic backwardness in structural and historical terms, Mexican elites often blamed it on behavioural deficiencies of the Mexican people (Vaughan, 1982). Armed with the understanding that their populations were backward and unfit, Mexican policymakers sought to improve them. By the late nineteenth century, *científicos* had begun to advance the idea that society should recognize what historians have identified as the three Ds – disease, disability, and defect – in such a way as to favour reproduction of the physically and morally desirable and normal over the non-desirable and abnormal (Tomes, 2005; see also Branson & Miller, 2002; Edwards, 2005; Turner, 2006). As the categories of citizenship were questioned and disrupted from the nineteenth century onward, Douglas Baynton (2001) notes, "disability was called on to clarify and define who deserved, and who was deservedly excluded from, citizenship" (p. 33). The classification, monitoring, and surveillance of certain groups also provided the way for *científicos* to create terms of normality and abnormality, which later inspired fear of degenerate genes infecting the productive gene pool. As the nation's elites sought to reform the body politic into a useful, disciplined, and modern society, they implemented varying degrees of what David Lyon (2007) has described as social disciplining, a power structure which suggests that "those who have the capacity to influence how people are classified and categorized tend to be in the position of greater power than those who do not" (p. 26). Historically, for example, agencies of the state often employed various techniques, such as the census, to gather information on the citizenry. While this process was primarily used to cull personal data *about* citizens, Lyon argues that it also *produced* citizens in the sense that it provided statistical data to distinguish "genuine" citizens from outsiders (pp. 30–31). The modern era ushered in newer and more rationalized ways of monitoring citizens, such as tracking the spread of disease in the interest of public health (Lyon, 2007). These subtle forms of surveillance "give the watcher increased power to persuade," and essentially exercise control over the watched based on motives (and methods) that can range from seemingly benign to insidious (Richards, 2013, p. 1955). Ensuring the fitness and health of a population according to an established system of normative markers represented a key element of creating a civilized and modern nation (Rose, 2007). Using the definition of an ideal citizen as the benchmark, *científicos* classified and categorized individuals and groups according to the degree to which they embodied these characteristics. Although *científicos* professed that these factors contributed to a more modernized society with a shared national identity, the questionable scientific and medical evidence used to support their views caused significant difficulties for Deaf individuals, who were believed to be mentally and physically unable to adhere to those standards.

Though the decade of the 1910s resulted in the overthrow of a political regime that had embraced social Darwinism and Auguste Comte's positivism, it was replaced with a new regime using new tools to diagnose and distinguish the "normal" from the "abnormal." Scholars have underscored how many Western nations used eugenicist policies, which were neither fixed nor static in their application, to forcefully sterilize individuals deemed as unfit in order to control reproduction (Turda, 2010; Volscho, 2010). With the exception of the state of Veracruz – the only Mexican state to enact legislation that permitted forced sterilization – Mexican elites instead embraced a thread of eugenics called neo-Lamarckism. Neo-Lamarckism and the related concepts of "preventative eugenics," a term originally coined by English physician Caleb Saleeby (1910), stressed that changes could be made to one's environment in order to improve the "race" or human lot (see also Stepan, 1991). This presented a departure from Mendelian eugenics, which suggested that one's "germ plasm" passed from one generation to the next, which inherently made some individuals "undesirable breeders" (Stepan, 1991, p. 87; see also Bashford, 2004). To neo-Lamarckian eugenicists, a true nation had a common purpose, a shared language and culture, and a homogeneous population; achieving these goals necessitated both social and physical monitoring and intervention (Stepan, 1991; see also Alemdarğolu, 2006; Bashford & Levine, 2010). By demanding that the state aim for the purification of its body politic through eugenic means, these eugenicists thought of themselves as working towards a progressive and noble goal: the creation of a strong and healthy nation (Rose, 2007; Turda & Gillette, 2014; see also Ceyhan, 2012). Public health thus came to represent the mechanism through which elites could reform the masses and served as a "positive" and subtler approach to surveilling their unfit populations in order to control reproduction. This monitoring was normally carried out through public health policy – and the moral underpinnings that were often affiliated with it – and different nations struck different balances with regard to the degree to which they also enacted "negative" eugenicist policies, namely, policies that through legal mechanisms or medical intervention sought to bar specific groups from reproduction.

The increased interest in deafness coupled with the professionalization of the medical field contributed to a national discourse that suggested deafness was a condition that could be treated and cured through direct medical intervention, or by preventing certain marriages or reproductive unions from taking place. It was feared that those who were deemed defective would introduce into a population's gene pool the potential for a degenerate, heritable strain, and that such defective genetic material could manifest itself in crime, pauperism, and immorality (Thomson, 2010). These fears legitimized surveillance regimes, which often resembled what David Lyon (2002) has referred to as "categorical suspicion," a form of surveillance that calls for the policing of individuals who are perceived threats to society at large (as cited in Ball & Webster, 2003, p. 7). As scholars such as Richard Jenkins (2008) have emphasized, this categorization or classification is fundamental to "social sorting," or "the identification and ordering of individuals in order to 'put them in their place' within local, national, and global 'institutional orders'" (as cited in Jenkins, 2012, p. 160; see also Lyon, 2007). The Deaf were often lumped into one of two categories in that they were either viewed as helpless individuals or potential menaces to society, and *científicos* took measures to ensure that the Deaf would not pose a risk to others in society, including through such measures as publicizing the importance of limiting or preventing Deaf intermarriage altogether. In this regard, many elites in the medical and scientific communities grappled with the questions of not only what was contributing to deafness, but more importantly, how they could prevent its transmission. Such measures did not meet the strict parameters of forced sterilization laws that were enacted in the state of Veracruz in the early twentieth century, but they nonetheless represented an attempt to understand how, in their role as the nation's key policymakers, *científicos* could potentially prevent the spread of deafness by controlling marriage and immoral behaviours associated with deafness within these populations. However, as compared to the degree to which stringent negative eugenic policies such as sterilization or murder were implemented in the United States and Germany, Mexico applied a less strict interpretation of the movement, which explains in part why eugenicist policies concerning the Deaf have not received much scholarly attention.

Consanguineous and Deaf Intermarriages

The decade of the 1880s marked a pivotal moment in the history of eugenics and its relationship to deaf people worldwide. Consanguineous unions and deaf intermarriage were labelled as among the primary contributing factors in congenital deafness, and were thus targeted as a key national concern in Mexico, Europe, and the United States. Histories of deaf education and eugenics in the contexts of both North America and Europe have established that three factors – the international adoption of oralist programs, the removal of deaf (and signing) teachers from deaf institutes, and the admonition against deaf intermarriage – were fuelled by eugenicist policies in the late nineteenth century (Baynton, 1996; Burch, 2002; Edwards, 2012; Kennedy, 2015; Quartararo, 2008). While research on this area of Mexican history remains scant, two seemingly disparate worlds collided during this period when Mexican elites were implementing the nascent stages of what would later become the foundation for the nation's eugenics movement. It was during this time that the Escuela Nacional de Sordomudos (National School for Deaf-Mutes) began sending its instructors abroad on commissioned trips to Europe and the United States to study oralist techniques. Throughout the last two decades of the nineteenth century, Mexican education officials were building its relationship with the Volta Bureau (initially called the Volta Laboratory), a premier and internationally renowned research centre for deaf education in Washington, D.C., established by Alexander Graham Bell.

Due to the international nature of the discourse and debate over deafness, it is appropriate to situate Mexico's role and policymaking concerning the deaf within this broader context. Reports from the *American Annals of the Deaf and Dumb* (1878), a publication that covered developments in deaf education throughout North America, asserted the following concerning the hereditary nature of deafness and warned against deaf intermarriage: "We all know the proverb 'Like begets like' and it is an indisputable fact that, in certain families, there is a great tendency to children being deaf" (p. 11). Likewise, Edward Allen Fay (1878), a long-term advocate of the deaf and sign language who also served as editor of the *American Annals of the Deaf and Dumb,* maintained that "though there are doubtless many cases of congenitally deaf children ... one great step in advance would be gained if the marriages of first cousins were to become less frequent – as they ought to be, and would be ... were it generally known how many idiots, deformed, blind, and deaf come of such unions" (p. 10). While these texts were deeply influential, the words of Alexander Graham Bell would prove to have an even more enduring impact on the deaf community worldwide for the next century. In November of 1883, the year that Francis Galton first published his eugenic ideas, Bell delivered a speech to the National Academy of Sciences, which was later published as *Memoir Upon the Formation of a Deaf Variety of the Human Race.* In this address, Bell (1883) identified the use of sign language and deaf intermarriage as two factors that were closely related to the burgeoning ideologies of social Darwinism and the related field of eugenics (see also Greenwald, 2009).

Mexican policymakers were seemingly enthralled with understanding how deaf-mutism was inherited and drew heavily from their European and North American counterparts. As the rate of deaf-mutism within blood-related families in various ethnic groups and countries was examined worldwide, it was reported in major Mexican newspapers and elite medical journals such as *Gazeta Médica.* Mexico City newspapers such as *El Foro* and *Diario del Hogar* published a series of European studies conducted on deaf marriage and potential cures for deafness beginning in the 1880s (see "Buena Noticia para los Sordos," 1884, p. 2; Corral y Maira, 1890, pp. 1–2; "Sección Médica: La Sordera," 1902, p. 4). In 1884 *El Monitor Republicano,* another major Mexico City newspaper, reprinted the results of a decade-long study conducted in Barcelona. Designed to measure the rate of deafness and deaf-mutism and their possible links to consanguineous unions between their parents, the study concluded that six deaf-mute children were born to parents of blood relation, and that there are "many cases where consanguineous marriages have more than one child affected by deaf-mutism" (Valle y Ronquillo, 1884, p. 3).

Though the sample size was limited, detailed family lineages of students enrolled at Barcelona's institute for the deaf were published in *El Monitor Republicano* to illustrate the degree to which consanguineous unions contributed to the prevalence of deaf-mutism in society:

> Of the six deaf alluded to in the study, there are three, whose names have been omitted and indicated by initials, so as to not mortify their parents:
> R. Ll. and Ll., daughter of first cousins; has a deaf brother.
> J. C. and B.C., daughter of first cousins; has two deaf-mute sisters.
> J.S. and S., daughter of first cousins; has a deaf brother and sister.
> (Valle y Ronquillo, 1884, p. 3)

In addition to noting the purported rate of deaf-mutism, the author placed emphasis on the implications for passing on deafness. He concluded that "although it may seem to some that the proportion is small compared to the importance we place on the matter, one must remember that the misfortune is overwhelming for a family who has a child who is deprived of one of the major senses" (Valle y Ronquillo, 1884, p. 2). A decade later, Trinidad García, who served as director of Mexico's first Escuela Nacional de Sordomudos expressed similar concern in his annual institutional report. In 1895, García described in detail the devastation of a mother who has realized that her child was born deaf:

> When a mother lavishes an exquisite outpouring of tenderness, uttering heartfelt and delicate phrases into the ear of her newborn child, she will be completely happy if the tender object of her intense affection answers, even with only a wail, to her maternal caresses; but how unhappy and miserable it makes an innocent mother if she observes that her infant does not produce any sound; that his gaze is vague and sad, his countenance morose and somber; then the misfortune of the mother is limitless, because she realizes that the child of her womb is deaf and is destined to live in enduring isolation. (p. 8)

These views also drew attention to the lifelong misfortune of the parents, whose economic, social, and emotional struggles would henceforth know no limits. In the discussion that followed, García provided what seemed to be the cure for this disorder and stressed that parents could overcome this misfortune (and the implied burden) with the help of charitable educators who would teach their deaf-mute children how to speak.

Despite the fact that Mexico aimed to establish a school for the deaf since achieving independence in the 1820s, the nation was still in the early stages of developing a deaf education program when the debate

over oralism, the use of sign language, and marriage (and reproduction) of Deaf individuals erupted on the world's stage in 1880 following
the Congress of Milan. The Congress of Milan comprised educators and
policymakers from nations in Europe and North America and effectively
banned the use of sign language and promoted oralism, or the reading
of lips, a controversial decision that would ultimately lead to the near
extinction of sign language in North America. In light of the fact that
Mexican education officials sent their instructors abroad to learn from
schools in Europe and the United States, and they were heavily influenced by their patterns of educating deaf students and treating deafness,
it is curious why García, who served as director of the Escuela Nacional de Sordomudos, would draw a hard line against an internationally
renowned institution with which he and his colleagues were looking to
collaborate. Yet, García was one of several Mexican education officials
who challenged the prevailing notion that deaf-mutism and consanguineous marriage were inextricably linked, and he further disputed that
marriage between two deaf individuals would inevitably produce, as Bell
(1883) had suggested, "a deaf variety of the human race" (n.p.). Like
many of his contemporaries, Bell (1883, 1891) espoused positive eugenics in that he encouraged procreation among those who were deemed
"good in stock" or genetically fit, and he thus discouraged marriage
of deaf couples. While García took into account the high rate of deaf-
mutism produced as a result of consanguineous unions and marriage
among the Deaf, his conclusions were not as rigid as those presented by
either Bell or the US Department of Commerce and Labor. In his 1895
institutional report, García described the intermarriage of deaf-mutes
as a comforting union: "This experience must be very consoling for the
deaf-mutes, because the charms of marriage can sweeten its existence,
turning the home into a lovely domestic asylum to provide them with
isolation from society" (p. 5). His report drew on evidence that couples
of close familial relation often had children who were not born deaf and
grew into healthy and strong adults:

> Nature presents, however, some mysteries that elude the most thorough
> scientific research, and one of them is, that a vast majority of marriages
> between close relatives have children who speak, and are strong and vigor-
> ous and this suggests that the purity and vigor of the race can resist the
> effects of inbreeding for a significant amount of time. (García, 1895, p. 7)

García acknowledged that not all unions between deaf individuals
resulted in deaf or deaf-mute children and referred to data culled from
the National Institute of the Deaf in Paris, France. Despite legislation

that strictly segregated boys and girls in French schools for the Deaf with the explicit purpose of preventing their circulation in society at large and discouraging them from marrying and reproducing, García contended that there were records of many students who had attended that school, married each other, and had speaking children (García, 1898; see also Lane, 2001).

To further crystallize his assertions that deafness was not a permanent condition, García cited the research published by French aural surgeon Dr Jean-Pierre Bonnafont, who, in a study titled "Curación de la Sordera," alleged to have "perfected" a surgery to cure deafness by drilling into the inner surface of the eardrum and injecting it with pulverized ether. Although Dr Bonnafont had later publicly denounced the mental abilities and legal capacity of the deaf-mute population in 1877, he also acknowledged that the medical research supporting the claim that deaf children would be born to deaf parents was, in his estimation, inconclusive. Bonnafont (1877) documented four known examples of this "variety," and cited one where both the husband and wife were deaf and dumb, who had two sons who spoke; in the other three, one of the two spouses had this ailment and the children of both sexes spoke. Though he did not explicitly encourage consanguineous unions, García likewise contended that such marriages were not the primary cause of congenital deafness or other physical defects, and in fact, referred to the evidence supporting such claims as exaggerations: "The influence of heritage, though important, has been exaggerated; as there have been many marriages between deaf-mutes whose children spoke" (García, 1898, p. 5).

Employing similar methods as studies on deaf intermarriage conducted by Bell (1883) and Fay (1878), García drew on his own studies of Mexican families with blood-linked marriages and their likelihood to produce deaf offspring. Rather than "dwell upon the numerous examples that many enlightened writers have illustrated in their works, to demonstrate the deplorable effects of consanguineous unions" (García, 1895, p. 5), he highlighted one remarkable case in the records of the Escuela Nacional de Sordomudos that served as a counter to the pattern of blood-related marriages increasing the rate of deaf-mutism. He noted that Gregorio Espinosa, a student who had attended the school, had two sisters and two brothers. All five children were deaf-mute. Yet, "the five were the legitimate sons of Gregorio Espinosa and Maria del Refugio Peregrin, who were third-degree relatives and neighbors in Autlán, State of Jalisco" (García, 1895, p. 5). García underscored this example in his report to support his view that even with a "legitimate and pure" marriage amongst distant non-deaf relatives, the transmission of deafness might still occur. Despite the dozens of accounts published in prominent newspapers, periodicals, and medical journals that expressed concern over consanguineous unions and deaf intermarriage and their connection to congenital deafness, not all Mexican policymakers adopted the prevailing stance of discouraging deaf people from marrying.

Moral Degeneration and the "Most Unfortunate Disease"

The nineteenth century marked an upsurge in the study of hearing and speech disorders, with doctors and scientists renewing interest in the study of deafness worldwide. This is where public health surveillance came into its own. As Mexican positivist reformers embarked on sweeping modernizing efforts beginning in the late nineteenth century, literacy and education campaigns merged with moral instruction to combat what were perceived as unhealthy and superstitious behaviours. A series of Hygienic Congresses that were held beginning in 1882 underscored the degree to which growing concern over the spread of contagious disease (particularly among poor communities) and the alleged role of morality in that spread represented a source of continual misunderstanding among policymakers and the Mexican population. These congresses addressed ever-growing concerns of elites over the related hereditary consequences of what was labelled as immoral behaviour, and resulted in the development and the implementation of surveillance practices and hygiene policies concerning alcoholism, prostitution, and syphilis. Since eugenic policies and doctrines were connected to notions of inherent faults and defects in the Mexican character, deafness was, in many ways, considered a by-product of these late nineteenth-century concerns. Medical journals such as *La Revista Positiva* also published articles that described physicians' concerns over the rate of chronic mental and physical diseases, such as deafness, hysteria, melancholy, dementia, and epilepsy (see Barreda, 1902). The first census to document deafness and mutism in Mexico was conducted in 1900, and while Mexico's deaf population arguably had not increased, it was becoming increasingly apparent that there was a larger number of deaf individuals – representing close to 2 per cent of Mexico's total population – than was previously estimated (Peñafiel, 1901). The development of population censuses provided a framework for public health surveillance in which data gathered shaped governmental policies in ways that had beneficial or detrimental effects on certain demographics (Ruppert, 2012). The acquisition of such data thus encouraged *científicos* to take a more aggressive approach in preventing deafness.

Proponents of neo-Lamarckism believed that not only alcoholism, but also any consumption of alcohol, caused hereditary defects including deafness, but that these could be managed through eugenics policies. In addition to undesirable marital unions, deafness was linked to sources of immorality in the late nineteenth and early twentieth centuries – alcoholism and syphilis. Criminologists linked vices and behaviours such as ignorance, alcoholism, gambling, prostitution, and idleness to crime, and held that vices such as alcoholism, promiscuity, deaf-mutism, lunacy, or the excitation of passions, all represented characteristics that not only affected individuals but posed a threat to society at large ("Delito y delincuentes," Arueta, 1893; see also Buffington, 2000; Macedo, 1897). Deafness thus acted as a catalyst enabling the growth of eugenic surveillance and of a range of forms of moral regulation (Corrigan & Sayer, 1985) directed at these various populations, even while eugenic ideas also prompted increasing surveillance of Deaf people themselves.

Hygienists adopted eugenic views that these vices represented a major factor in degeneration, a discourse that provided scientists with the tools to define social and medical deviance. Medical practitioners and researchers often utilized the science of reproductive technologies to determine which innate characteristics or behaviours contributed to the spread of certain congenital anomalies or disorders. At the Second Pan-American Medical Congress held in 1896, Dr Francisco Vázquez Gómez presented his research on what he concluded were the most common causes of deaf-mutism. Vázquez Gómez had frequently conducted health inspections on both the school's building and the student body enrolled at the Escuela Nacional de Sordomudos. In his provocative study, he examined the background of thirty-six deaf-mute children and their families who were enrolled in the school. Based on that sample, he concluded that 40 per cent of deaf-mutes were born deaf or had become deaf within the first two years of life, and that men were more likely than women to suffer from deafness. Regarding what he referred to as the "most unfortunate disease," Vázquez Gómez (1896) asserted that deafness was caused by the "morbid predisposition of the parents which begets hereditary or congenital deaf-mutism in their children" (p. 659). Rather than blame deaf intermarriage or consanguineous unions for the transmission of deafness, Vázquez Gómez explicitly linked an individual's morality to a propensity for passing on deafness to their offspring. Chronic alcoholism on the part of the father, he recorded in two observations, was associated with his child's deafness. A diagnosis of epilepsy in the father, combined with "hysteria" of the mother, was also alleged to figure strongly in several cases of deaf-mutism. Moreover, he concluded that the probable cause of deaf-mutism was due to the "pathological state" of the fathers, and he noted that five times out of seven, it was the father who had an illness-inducing influence on his child's propensity toward deafness or deaf-mutism. Mothers and their immoral behaviours, however, did not escape scrutiny. Women who had either been diagnosed with hysteria or accused of having had multiple abortions were also blamed for imparting the physical malady to their children. In this sense, Vázquez Gómez argued, deaf-mutism did not only concern the hearing organs, but was directly associated with a state, more or less marked by a weakness of the brain, which resulted in apathy of the intellectual faculties. The diminished intellectual faculties resulting from deafness could be linked to either a father's or a mother's choice to imbibe alcohol or other immoral behaviour. Shortly after Vázquez Gómez published his findings, other medical researchers began to conduct similar studies. In 1896 the newspaper *Voz de México* conducted a study that examined 215 families affected by alcoholism, and determined that 212 families and 503 members of those families were victims of various "hereditary lesions" (Arcos, 1899, p. 2). Of these, it concluded that twenty-nine families suffered from various factors such as "deformity of the skull, deafness, deaf-mutism, blindness, and partial paralysis" (Arcos, 1899, p. 2). These findings paralleled those of Vázquez Gómez (1896), who had argued that deafness was a physical manifestation of "the sad legacy of the most odious vices, among which is customary drunkenness" (p. 661).

As Mexican elites attempted to contain "unhealthy desires" through science, they published numerous studies that cited the harmful effects of sexual deviance, and especially the impact of prostitution, which was seen as undermining values of the family, economy, and social order. Education, moral superiority, and prescribed behaviours and customs were key factors in creating a social and ideological barrier between the elite and middle classes on the one hand, and the downtrodden, uncivilized masses that had embedded themselves in the fabric of the capital city on the other (Garza, 2007). Those who were deemed more inclined to spread disease were targeted by state-sanctioned surveillance measures such as mandatory health inspections, national moral health campaigns to adjust the masses, and new forms of residential institutions for wayward juveniles. Attempts to regulate prostitution during the late nineteenth century as well as the sharp increase in the production of sexual knowledge in the field of medicine combined to create the scientific and administrative basis to identify, control, and eventually punish sexual behaviours deemed deviant (Rivera-Garza, 2001; see also Bliss, 2001). Vázquez Gómez conducted his research on the relationship between congenital deafness, alcohol consumption, and immorality in the context of legalized prostitution in Mexico City, which also contributed to the government's near obsession with regular health inspections and sanitary reform. He reported a 25 per cent higher rate of deafness in the nation's capital as compared to other Mexican states, ascribing higher rates of deafness to behavioural factors such as alcoholism and syphilis associated with prostitution.

Similarly, Dr Luis Lara y Pardo (1908) described prostitutes (and the syphilis they supposedly carried with them) as "a parasite" that moved from "North to South," presumably to the Distrito Federal (pp. 51–2), and argued that there was a direct connection between the spread of disease, urbanization, and industrialization, and an overall increase in threat and possible harm to national health. Officials implemented state-sanctioned and regimented health inspections because such "deviant" women were viewed as carriers of venereal disease and were thus considered a threat to the health of the national body politic. Syphilis was believed to be the cause of many cases of deaf-mutism in which only the father was infected. In one such case, he produced four deaf-mute children with different women. Despite the fact that these researchers made sweeping conclusions based on very limited data and study samples, in their view such "evidence" undergirded medical assumptions that fathers often carried what Vázquez Gómez (1896) referred to as "the fatal germ" or, more precisely, the spread of syphilis (p. 662). Reference to this fatal germ no doubt connoted congenital defects, such as psychological inferiority, both moral and social, which contributed to various forms of degeneration. Assertions that one's morality was inextricably tied to the propensity of contracting syphilis and thus passing on deafness to one's children were indicative of prevailing health concerns at the time, and these views would influence the Mexican medical community's surveillance practices through the mid-twentieth century.

Conclusion

From the colonial period onward, governing authorities in modernizing Latin American states used medicine and science as tools of social control to reform and restrict their populations. As part of the nationalizing campaign to create ideal citizens, Mexican elites introduced various forms of state-sanctioned medical surveillance to monitor and control unwieldy and "backward" populations. Such discourse ultimately led to the implementation of a "civilizing" agenda that would be carried out on certain sectors of the population. In many ways, this medical surveillance paralleled David Lyon's (2002) definition of "categorical suspicion," in that it introduced draconian rules that applied to certain members of the population. These guidelines included the monitoring and control of physical and moral hygiene, in addition to marriage and reproduction between those individuals believed to possess degenerative germ plasm. This project paralleled what Nancy Leys Stepan (1982) has referred to in the European context as the "modern period for scientific racism," when the goal of science was to "extract regularities from variation in nature, to discover the laws behind variation" (p. 7).

As late nineteenth-century hereditary riddles known as "misterios de la naturaleza" persisted into the early twentieth century, deafness was considered one of the most undesirable reproductive consequences of unfit individuals engaging in what were perceived as immoral acts. Eugenics offered Mexican reformers the tools needed "to reshape their populations and redefine citizenship along racial and social norms" (Stern, 1999, pp. 371–2) and created a system of social and physiological sanitization that called for those deemed unhealthy, diseased, or antisocial to be surveilled and segregated from the healthy majority. Health and sexual education came to be regarded as instruments that would promote important changes in the lives, practices, and behaviour of the Mexican people, in particular, of Indigenous populations and the poor. These prescribed, elite-driven values of progress, moral and civic virtue, and modernization also affected the Deaf in several respects. As Mexican officials began to embrace the neo-Lamarckian notion that alcoholism, tuberculosis, and syphilis had direct hereditary consequences, the government enacted a series of public health campaigns to identify and subsequently target these innate behavioural and mental flaws (Liceaga, 1911; Stepan, 1991). Though Deaf children were not labelled as flawed in the traditional sense, they were subsumed under the larger umbrella of social and public health improvement.

In the years following the Revolution of 1910, the Mexican government called for active engagement and participation by its citizenry. In doing so, it provided a clear benchmark from which "normal" could be effectively distinguished from that which was "abnormal," a method used in the social sorting of Mexico's "backward" populations. The census, while a seemingly benign tool designed to cull personal data on populations, contributed to the efforts to both track and define the parameters of citizenship. The "abnormal" sector of the population, or those labelled "abnormal," logistically struggled to fit into the nationalizing agenda of the new constitution. The Mexican eugenics campaign marked deafness as an important social problem requiring expertise, protection, and reform, and its Deaf community fell victim to the prevalence of eugenicist thinking in the late nineteenth and early twentieth centuries. The implementation of a prescribed set of sexual and moral ideals, in addition to the promotion of a "normalizing" process of communication through the formal implementation of oralism in 1895, reflected the notion that deviating from such standards would lead to the "degeneration of the species" and threaten the nation at large.

NOTES

1 Though Mexico adopted a National Disabilities Law in 1995 under the regulations of the United Nations, it does not have widespread laws protecting the individual rights of *personas con discapacidad* or "people with disabilities." The Persons with Disabilities Act of the State of Mexico uses the phrase "personas con capacidades diferentes" (people with different abilities). The definition states that these are persons "suffering from a loss, impairment or reduction of an organ or physical, sensory or intellectual function, which restricts daily life activities and prevents their individual and social development" (IDRM, 2004, pp. 261–2).
2 In simple terms, scholars have cited two constructions of deafness that now dominate – deafness as a disability or hearing impairment and Deafness as designating a member of a cultural and linguistic minority. For the culturally Deaf, their use of sign language signifies that they share a common communication difference from hearing members of society, but that their barriers as a result of being deaf disappear once they are part of a signing community (Lane, 1997).
3 The Escuela Nacional de Sordomudos was born out of two projects – the Hospicio de Niños (Foundling Home) and the Hospicio de Pobres (Poor House) of Mexico City. The Foundling Home was responsible for providing care and services to abandoned children whereas the nation's Poor House was designed to rid the streets of beggars. The centralization of the Poor House in the mid-nineteenth century led to the creation of the Foundling Home, which over time further narrowed its educational program and addressed the needs of poor (and often orphaned or abandoned) deaf children.

REFERENCES

Agostoni, C. (2004). *Monuments of progress: Modernization and public health in Mexico City, 1876–1910*. Calgary, AB: University of Calgary Press.
Alemdarğolu, A. (2006). Eugenics, modernity, and nationalism. In D. M. Turner & K. Stagg (Eds.), *Social histories of disability and deformity* (pp. 126–41). New York, NY: Routledge.
Arcos, D. (1899, October 17). Alcoholes y alcoholismo. *La Voz de México*.
Armus, D. (2003). Disease in the historiography of modern Latin America. In D. Armus (Ed.), *Disease in the history of modern Latin America: From malaria to AIDS* (pp. 1–24). Durham, NC: Duke University Press.

Arnold, D. (Ed.). (1996). *Warm climates & Western medicine: The emergence of tropical medicine 1500–1900.* Atlanta: Rodopi.

Arueta, J. (1893, April 21). Delito y delincuentes. *El Siglo Diez y Nueve.*

Ball, K., & Webster, F. (2003). *The intensification of surveillance: Crime, terrorism and warfare in the information age.* London: Pluto Press.

Ballenger, S. S. (2009). *Modernizing madness: Doctors, patients and asylums in nineteenth century Mexico City* (Unpublished doctoral dissertation). Berkeley, CA: University of California.

Barreda, G. (1863, May 3). La educación moral. *El Siglo, XIX*(839).

Barreda, G. (1902, May 1). La homeopatia ó juicio crítico sobre este nuevo sistema. *La Revista Positiva,* Tomo II (17).

Bashford, A. (2004). *Imperial hygiene: A critical history of colonialism, nationalism and public health.* New York, NY: Palgrave Macmillan.

Bashford, A., & Levine, P. (Eds.). (2010). *The Oxford handbook of eugenics: The history of eugenics.* New York, NY: Oxford University Press.

Baynton, D. C. (1996). *Forbidden signs: American culture and the campaign against sign language.* Chicago, IL: The University of Chicago Press.

Baynton, D. C. (2001). Disability and the justification of inequality in American history. In P. K. Longmore & L. Umansky (Eds.), *The new disability history: American perspectives* (pp. 33–57). New York, NY: New York University Press.

Bell, A. G. (1883, November 13). Memoirs upon the formation of a Deaf variety of the human race, A paper presented to the National Academy of Sciences at New Haven, November 13, 1883. Gallaudet University Library Deaf Collections and Archives, Washington, DC.

Bell, A. G. (1891). *Marriage: An address to the deaf.* Washington, DC: Gibson Bros, Printers, and Bookbinders.

Berda, Y. (2013). Managing dangerous populations: Colonial legacies of security and surveillance. *Sociological Forum, 28*(3), 627–30.

Bliss, K. E. (2001). *Compromised positions: Prostitution, public health, and gender politics in revolutionary Mexico City.* University Park: Pennsylvania State University Press.

Bonnafont, J. P. (1877). Du degré de responsabilité légale des sourds-muets. *Bulletins de la Société d'anthropologie de Paris, 12*(1).

Branson, J., & Miller, D. (Eds.). (2002). *Damned for their difference: The cultural construction of Deaf people as disabled.* Washington, DC: Gallaudet University Press.

Buena noticia para los sordos. (1884, October 2). *Diario del Hogar.*

Buffington, R. M. (2000). *Criminal and citizen in modern Mexico.* Lincoln, NE: University of Nebraska Press.

Burch, S. (2002). *Signs of resistance: American Deaf cultural history, 1900 to World War II.* New York, NY: New York University Press.

Ceyhan, A. (2012). Surveillance as biopower. In K. Ball, K. D. Haggerty, & D. Lyon (Eds.), *Routledge handbook of surveillance studies* (pp. 38–46). New York, NY: Routledge.

Chinn, S. (2006). Gender, sex, and disability from Helen Keller to Tiny Tim. *Radical History Review, 94*, 240–8.

Corral y Maira, M. (1890, July 10). Los matrimonios consanguineos y la sordomudez congenita. *Diario del Hogar.* Reprinted in (1890, July 22). *El Foro.*

Corrigan, P., & Sayer, D. (1985). *The great arch: English state formation as cultural revolution.* Oxford, UK: Basil Blackwell.

Cruz Cruz, J. C., & Cruz-Aldrete, M. (2013). Integración social del sordo en la Ciudad de México: Enfoques médicos y pedagógicos (1867–1900). *Cuicuilco, 56,* 173–201.

Cueto, M., & Palmer, S. (2015). *Medicine and public health in Latin America: A history.* New York, NY: Cambridge University Press.

Cunningham, A., & Andrews, B. (Eds.). (1997). *Western medicine as contested knowledge.* Manchester, UK: Manchester University Press.

Curación de la sordera! (1874, October 11). *La Iberia* and *Siglo Médico.*

Davis, L. J. (1997). Constructing normalcy: The bell curve, the novel, and the invention of the disabled body in the nineteenth century. In L. J. Davis (Ed.), *The disability studies reader* (pp. 9–28). New York, NY: Routledge.

Dublan, M., & Lozano, J. M. (Eds.). (1876–1904). *Legislación mexicana o colección completa de las disposiciones legislativas expedidas desde la independencia de la Republica,* vol. 11, Chapter 10, Title 9, Article 561. 34 vols. (pp. 201–449). México, DF: Imprenta del Comercio.

Edwards, R. A. R. (2012). *Words made flesh: Nineteenth-century Deaf education and the growth of Deaf culture.* New York, NY: New York University Press.

Edwards, S. D. (2005). *Disability: Definitions, value, and identity.* Oxford, UK: Radcliffe Publishing.

Fay, E. A. (Ed.). (1878). *American annals of the deaf and dumb.* Washington, DC: Gibson Brothers.

García, T. (1895). *Escuela Nacional de Sordomudos.* México, DF: Tipografia "El Lapiz del Aguila."

García, T. (1898). *Discurso pronunciado por el director de la Escuela Nacional de Sordo Mudos.* México, DF: Tipografía "El Lapiz del Aguila."

Garza, J. A. (2007). *The imagined underworld: Sex, crime, and vice in Porfirian Mexico City.* Lincoln, NE: University of Nebraska Press.

Greenwald, B. H. (2009). The real "toll" of A. G. Bell: Lessons about eugenics. *Sign Language Studies, 9*(3), 258–65.

International Disability Rights Monitor (IDRM). (2004). *Regional report on the Americas.* Washington, DC: International Disability Network.

Jenkins, R. (2008). *Social identity* (3rd ed.). New York, NY: Routledge.

Jenkins, R. (2012). Identity, surveillance and modernity: Sorting out who's who. In K. Ball, K. D. Haggerty, & D. Lyon (Eds.), *Routledge handbook of surveillance studies* (pp. 159–66). New York, NY: Routledge.

Kennedy, E. (2015). *Abbé Sicard's deaf education: Empowering the mute, 1785–1820.* New York, NY: Palgrave Macmillan.

Kudlick, C. J. (2003). Disability history: Why we need another "Other." *The American Historical Review, 108*(3), 763–93.

La compaña contra la avería: La degeneración de la especie. (1908, May 28). *El Imparcial.*

Lane, H. (1997). Constructions of deafness. In L. J. Davis (Ed.), *The disability studies reader* (pp. 153–71). New York, NY: Routledge.

Lane, H. (2001). Hearing agenda II: Eradicating the deaf-world. In L. Bragg (Ed.), *Deaf world: A historical reader and primary sourcebook* (pp. 365–78). New York, NY: New York University Press.

Lara y Pardo, L. (1908). *La prostitución en México.* México, DF: Librería de la Vda de Ch. Bouret.

Liceaga, E. D. (1911). *Algunas consideraciones acerca de la higiene social en México.* México, DF: Tip. Vda. De F. Díaz y León.

Lyon, D. (Ed.). (2002). *Surveillance as social sorting: Privacy, risk, and digital discrimination.* New York, NY: Routledge.

Lyon, D. (2007). *Surveillance studies: An overview.* Cambridge, UK: Polity Press.

Macedo, M. S. (1897). La criminalidad en México: medios de combatirla. México: Oficina Tipografía de la Secretaria de Fomento.Meade, T. (1997). *Civilizing Rio: Reform and Resistance in a Brazilian city, 1889–1930.* University Park: Pennsylvania State University Press.

Meade, T., & Walker, M. (Eds.). (1991). *Science, medicine and cultural imperialism.* New York, NY: St Martin's Press.

Peñafiel, A. (1901). *Dirección general de estadística á cargo del Dr Antonio Peñafiel, censo general de la República Mexicana.* Mexico, DF: Oficina Tip. De la Secretaría de Fomento.

Piccato, P. (2001). *City of suspects: Crime in Mexico City, 1900–1931.* Durham, NC: Duke University Press.

Quartararo, A. (2008). *Deaf identity and social images in nineteenth-century France.* Washington, DC: Gallaudet University Press.

Richards, N. M. (2013). The dangers of surveillance. *Harvard Law Review, 126*(7), 1934–1965.

Rivera-Garza, C. (2001). The criminalization of the syphilitic body: Prostitutes, health crimes, and society in Mexico City, 1867–1930. In C. Aguirre & G. M. Joseph (Eds.), *Crime and punishment in Latin America* (pp. 147–80). Durham, NC, and London, UK: Duke University Press.

Rodriguez, F. (1863). Misterios de la naturaleza. *La Sociedad,* 28 November.

Rodriguez, J. (2006). *Civilizing Argentina: Science, medicine and the modern state.* Chapel Hill, NC: University of North Carolina Press.

Rose, N. (2007). *The politics of life itself: Biomedicine, power, and subjectivity in the twenty-first century.* Princeton, NJ: Princeton University Press.

Ruppert, E. (2012). Seeing population: Census and surveillance by numbers. In K. Ball, K. D. Haggerty, & D. Lyon (Eds.), *Routledge handbook of surveillance studies* (pp. 209–16). New York, NY: Routledge.

Saleeby, C. (1910). Racial poisons. *The Eugenics Review, 2*(1), 30–52.

Saltes, N. (2013). "Abnormal" bodies on the borders of inclusion: Biopolitics and the paradox of disability surveillance. *Surveillance & Society, 11*(1/2), 55–73.

Sección médica: La sordera. (1902, October 1). *Diario del Hogar.*

Stepan, N. L. (1982). *The idea of race: Great Britain, 1800–1960.* Hamdon, CT: Archon Books.

Stepan, N. L. (1991). *The hour of eugenics: Race, gender and nation in Latin America.* Ithaca, NY: Cornell University Press.

Stern, A. M. (1999). Responsible mothers and normal children: Eugenics, nationalism, and welfare in post-revolutionary Mexico, 1920–1940. *Journal of Historical Sociology, 12*(4), 369–97.

Stoler, A. L. (1995). *Race and the education of desire: Foucault's history of sexuality and the colonial order of things.* Durham, NC: Duke University Press.

Suárez y López Guazo, L. (2005). *Eugenesia y racismo en México.* México, DF: UNAM.

Thomson, M. (2010). Disability, psychiatry, and eugenics. In A. Bashford & P. Levine (Eds.), *The Oxford handbook of eugenics the history of eugenics* (pp. 90–107). New York, NY: Oxford University Press.

Tomes, N. (2005). Bodies of evidence. *OAH Magazine of History: Medicine and History, 19*(5), 7–11.

Turda, M. (2010). *Modernism and eugenics.* New York, NY: Palgrave Macmillan.

Turda, M., & Gillette, A. (2014). *Latin eugenics in comparative perspective.* London, UK: Bloomsbury Academic.

Turner, D. M. (2006). *Social histories of disability and deformity.* New York, NY: Routledge/SSHM.

Valle y Ronquillo, F. (1884, January 3). Matrimonios consanguineos. *El Monitor Republicano.*

Vaughan, M. K. (1982). *The state, education, and social class in Mexico: 1880–1928.* Dekalb, IL: Northern Illinois University Press.

Vázquez Gómez, F. (1896). Causas más frecuentes de la sordo-mudez en México y resultados de la enseñanza del lenguaje articulado en la escuela nacional de sordomudos. Paper presented at the Memorias Pan American Congress.

Volscho, T. W. (2010). Sterilization racism and pan-ethnic disparities of the past decade: The continued encroachment on reproductive rights. *Wicazo Sa Review, 25*(1), 17–31.

Wiebe, S. (2008). Opinion. Re-Thinking citizenship: (Un)healthy bodies and the Canadian border. *Surveillance & Society, 5*(3), 334–9.

Zames Fleischer, D., & Zames, F. (2011). *The disability rights movement: From charity to confrontation.* Philadelphia, PA: Temple University Press.

Zamindar, V. F. Y. (2007). *The long partition and the making of modern South Asia: Refugees, boundaries, histories.* New York, NY: Columbia University Press.

4 "Masquerading as a Woman": The South African Disguises Acts and the Ghosts of Apartheid Surveillance, 1906–2004

B CAMMINGA

I felt a tap on my shoulder. I turned round, and there was this man dressed in a suit who said, "Excuse me, lady, you're under arrest for masquerading." He said *lady!*

Mark Gevisser (1994, p. 30, emphasis in original)

They call me into a room. They tell me to strip, until I am standing just in a black bra with a red suspender belt holding up the stockings and the cotton wool sticking out of the bra and men's underpants. If you were ever caught wearing female underpants you were charged even more. It was somehow okay to wear men's underpants. I had to pay ten pounds admission of guilt and then I could go [*sic*].

Mark Gevisser & Zethu Matebeni (2012, n.p.)

Michele Bruno was one of nearly 350 partygoers in attendance when police in Forest Town, Johannesburg, raided a so-called gay party in 1966. South Africa of the time was marked by burgeoning concern within apartheid's National Party government over homosexuality, considered an aberration and a direct threat to the state and to Calvinistic morals. Indeed, until this time non-normative expressions of sexuality and gender, often conflated as one and the same, were under "constant but not severe surveillance" in South Africa (Grunkel, 2010, p. 52). As Marc Epprecht (2004) argues, "The military establishment regarded homosexuality as indicative of psychological weakness or unfitness for the coming battle. It also suggested vulnerability to communist blandishments or political opposition to apartheid" (p. 147). It became clear after the party that the police did not have enough legal clout to address what Mark Gevisser (1994) has termed the "queer conspiracy" (p. 31). The raid and those in attendance made front-page news in South Africa and a climate of public and moral outrage followed.

Mia Fischer (2016) notes how bodies considered non-normative or non-conforming, those whose practices fall outside of (white) heterosexual and binary gender norms, "have largely been invisible in surveillance studies ... but have not been invisible to contemporary systems of surveillance" (p. 185). From 1948 to 1994 the apartheid state of South Africa was one characterized by white minority rule, the institutionalization of racial segregation and discrimination, extreme police surveillance and repression, and "the fervent reinforcing of heteronormativity" (Tsampiras, 2008, p. 492). The legislative structure of apartheid functioned as the key means through which the state applied these various levels of surveillance and "social control" (Jespersen, 1976, p. 324). Alongside this legislative structure, the medical surveillance of various population groups was a critical tool in the web of control exerted by the apartheid state. It was used perhaps most effectively to justify discriminating between groups classified as "different," not only with regard to race but also, as this chapter explores, with regard to sex, sexuality, and gender.

At the time of the 1966 raid, laws used to curtail deviant behaviour – such as those targeting sodomy or public indecency – only addressed *public* offences and therefore could not be utilized against partygoers in a *private* residence (Hoad, 2010). Of the 350 people at the party, only nine, including Bruno, were charged under little-known pre-Union[1] legislation, collectively referred to as the Disguises Acts (1891–1910),[2] for "masquerading as women" (Gevisser, 1994, p. 30). To address this perceived legislative gap, the Immorality Amendment Act (1969) was created as an extension of the former morality-oriented legislation, which limited sexual contact between the races. The Immorality Act was initially passed in 1927. In 1950 it was amended in order to prohibit sexual contact between white and black South Africans. The Act of 1927 and the amendment of 1950 were both repealed by the 1957 Immorality Act, which continued to restrict sexual interaction between black and white South Africans. The new 1969 Act would go further and also put restrictions on sexual contact between people of the same sex, enabling new forms of surveillance in the process.

In the same year, the Disguises Acts were amalgamated into a singular document entitled the Prohibition of Disguises Act (1969). Its reach – though it affected some homosexuals, i.e., those who dressed in drag – went far beyond curtailing sexual deviation and extended to implying a kind of gender normativity ostensibly expected by the state. It must be noted that since the state at the time was actively targeting homosexuality, it is arguable that the Act was created with homosexuals in mind, hence this particular wording. Certainly, there may have been heterosexual cross-dressers at the time, but since cross-dressing for the South

African state was indicative of or directly linked to homosexuality, they were essentially rendered invisible to the state.

The Prohibition of Disguises Act (1969) suggested that this normativity, given the law's history, would now be more actively policed and surveilled, restricting the possibility of access to public space and by extension the visibility of particular bodies. One of the few academic articles to mention the Prohibition of Disguises Act (1969) argues that it represented a clear moment in which the apartheid state's surveillance efforts, which had always included race, were now made manifest with regard to gender and sexuality (see Jespersen, 1976). The aim of the 1969 amalgamation was, arguably, to bring legislation in line with the state's fear of homosexuality, which can be understood as a concomitant fear of deviation from dichotomous heterosexually orientated gender roles. Notably, as B Camminga and Louise Vincent (2009) have argued with regard to this conflation, gender practices perceived as non-normative by the state, such as drag or cross-dressing, all functioned as a clear indicator of deviant sexuality – homosexuality.

The gay community and gay organizations had galvanized around the 1969 amendment of the Immorality Act, starting what would become the beginnings of a gay movement in South Africa. Henriette Grunkel (2010) is critical of this turn in South Africa's history, and the single-minded focus on the Immorality Amendment Act, noting the missed opportunity by the nascent gay movement to contextualize and thus politicize the surveillance of non-normative expressions of gender *and* sexuality within South African society more broadly. This may explain why, while there was much public response to the Immorality Act's amendment, the amalgamated Prohibition of Disguises Act passed virtually unimpeded.

It should be noted at this point that the historical record with regard to how many prosecutions happened under the various iterations of the Disguises Acts is unclear. The lack of cases, though, does not suggest that it was an ineffective surveillance tool but rather, as suggested later in this chapter, that it worked as an effective deterrent to public visibility of gender-transgressive behaviour more broadly. Indeed, its resurrection and use in 2004, a full ten years into South Africa's constitutional democracy, against drag queens at the annual Johannesburg Pride, I argue, is indicative of the very impetus and underlying spirit of its intended use as a means of apartheid surveillance and social control. This also involved the deployment of what Mark Andrejevic (2005) calls "lateral surveillance" by the conservative Gay and Lesbian Alliance (GLA), which served to amplify state and police surveillance at the time.

Jasbir Puar (2014), in an interview with Lewis West, describes surveillance as "a way of managing populations, creating identities and claiming the future" (n.p.). In this definition, surveillance is productive in that it not only corrals, identifies, designates, and defines acceptable population groups in a given time period, but also functions as a temporally anticipatory apparatus (Puar & West, 2014). As legislation intended to surveil and curtail deviant expressions of gender, the ability of the Disguises Acts to not only maintain their initial impetus *but to also be* deployed against particular population groups, regardless of legal protections aimed at expressions of gender in the democratic South Africa, points to the nefariously adaptive nature of surveillance apparatus over time. Since their original incarnation in the late 1800s there are four key instances, aside from Forest Town in 1966, of the various Disguises Acts being used: *R v. Lesson* (1906), *R v. Ntokile Zulu* (1947), *S v. Kola* (1966), and at Johannesburg Pride in 2004. I use these four instances, along with a close reading of the *Journal of the Medical Association of South Africa* (later the *South African Medical Journal*), to track the progression and changing response of the South African state to deviance as it relates to the often conflated positionalities of sex, gender, and sexuality, and to the surveillance and penalizing of gender transgression. The journal, first published in 1884, is an especially important resource as it functioned as the official journal of the Medical Association of South Africa (MASA), the state-recognized body representing South African medical professionals. Read together, these sources suggest an evolving relationship between the law, state surveillance and regulation of homosexual men, and medical science – in particular eugenics and sexology – in South Africa, a relationship marked in apartheid by the implicit surveillance and public curtailing of the gender-transgressive body in an effort to ensure the appearance of a decent, morally upright, and heteronormatively aligned state. Though perhaps for the apartheid state it may have been important more broadly that whiteness in particular remain aligned with moral decency, the law with regard to disguise, as will be seen, did not discriminate on racial grounds.

The arguments I present in this chapter are twofold: firstly, I see the prohibition of disguise during apartheid as a means through which the South African state marked gender-transgressive populations for policing in their perceived link to homosexuality, and secondly, I argue that as an example of the futurity and productive nature of laws that the state intends as tools to police certain populations in a given time period, the Prohibition of Disguises Act (1969) has had "remarkable longevity" (Sears, 2015, p. 4), maintaining its meaning and import into the new democratic South Africa. While the contexts and specific cases are quite different, my arguments here buttress the implications of the chapters in the final section of this book, which together suggest the powerful persistence of the surveillance and repression of dissent and transgression, but also that these practices extend to democratic as well as authoritarian states.

First Instance: Gender and Sexuality in *R v. Lesson* (1906)

Internationally, the twentieth century saw the emergence of two distinct yet mutually constitutive fields of "scientific" study focused on the body, which would have a substantial impact on future notions of race, sexuality, gender, and racial segregation in South Africa: eugenics and sexology. Both were deeply implicated in practices of medical surveillance, as other chapters in this volume demonstrate as well. Eugenics, a form of scientific racism, melded science and social policy and argued, at its core, that the best and fittest of the population should reproduce, while the reproduction of the unfit, particularly so-called inferior races, should be curtailed (Dubow, 2010). The eugenics movement had, at the time, gained a substantial foothold around the world, with societies, institutions, and organizations devoted to its study in Britain, the United States, and elsewhere (Dubow, 2010; see also Holly Caldwell's examination of eugenics in Mexico in the previous chapter). In the colonial metropole there were increasing concerns over issues of degeneracy, as racial decline was "linked to apocalyptically phrased fears about Empire and its ruling race" (Chanock, 2001, p. 76).

Marked by a language of deviance, degeneracy, and abnormality, eugenics was an ideological import to South Africa. By the early 1900s, as evidenced in the *Journal of the Medical Association of South Africa*, it was freely in circulation and "segregationist ideologues were quick to absorb eugenic thinking" (Dubow, 2010, p. 277). Moreover, as a former colony on the Southern African subcontinent, South Africa was considered one of the frontiers of this new racial science and several international figures in eugenics, including Francis Galton – one of its pioneers – travelled to the country. South African historian Keith Breckenridge (2014) describes Galton, cousin to Charles Darwin, as the "archetypical imperial intellectual" (p. 21), a man who found inspiration in South Africa, as Breckenridge suggests, for his racial insights and "evidence for the emerging statistical science of eugenics" (p. 21). New theories about the subcontinent began to emerge from a growing "corpus of colonial-based intellectuals" (Dubow, 2010, p. 276). The reproduction of the fittest and best was a project whose production, maintenance, and realization were deeply intertwined with state surveillance. Indeed, it is arguable that one would not be possible without the other. Drawing from Foucault, South African historian Shula Marks (1997) argues that medical discourses, such as those linked to eugenics, have been crucial to state powers of surveillance, and in this South Africa was no exception. It should perhaps be unsurprising then that as new theories circulated emanating from the subcontinent, a belief developed regarding white settlers – British and Dutch – and their "co-mingling" as two nationalities of "similar blood and character" (Dubow, 2010, p. 256) as advantageous to the new nation, while the separation of Africans from the settlers through explicit racial segregation spoke to the growing field of eugenics in the country. This understanding with regard to race and miscegenation was bound to very clear understandings of sex, sexuality, and appropriate gender roles.

Sexology was closely aligned with the study of eugenics, the two fields developing almost simultaneously: "sometimes they overlapped to such a degree that they were virtually indistinguishable" (Somerville, 2000, p. 31). This commingling of eugenics and scientific racism with sexology, both invested with a particular kind of surveillance culture, is visible in their efforts to delineate the ill from the well or the degenerate from the healthy member of a given population (Chauncey, 1983). Implementing particular kinds of legislative apparatus targeting specific population groups often attained this separation. Given that eugenics was fundamentally about reproduction, it is not surprising that the surveillance and regulation of gender and sexuality became a priority. Siobhan B. Somerville (2000) notes that not only did sexology initially circulate within this "pervasive climate of eugenicist and anti-miscegenation sentiment and legislation," but in many ways depended on it (p. 124).

In exploring comparable issues regarding the surveillance of public expressions of gender transgression, Susan Stryker (2008) notes similar legislative developments, ordinances, and laws forbidding disguise to those of South Africa in the United States beginning in the 1800s. Suggestively, as in South Africa's case, American laws were often not about a broad definition of disguise but specifically addressed people who were legally designated or assigned as male (e.g., "assigned male") yet who dressed as women. Prior to the establishment of the Union of South Africa in 1910 there were several such Acts in existence within the country, directly addressing similar concerns over disguise. The most crucial of these was the Criminal Law Amendment Act of Natal (1910), which stated the following would be understood as a crime: "(In the case of a male person) being found dressed as a woman in circumstances indicating a probable intention of availing himself of such a disguise to commit a crime, whether such crime be known or not" (S6 (2)E). Although the majority of the Acts did not go as far as the Natal Act in directly mentioning gender, what is evident is that when they were utilized it was often to arrest people for wearing or presenting in ways considered non-normative.

The first recorded arrest under the legislation was *R v. Lesson* (1906), where the accused was found "dressed in female attire with a bottle of beer in his pocket" (Hoctor, 2013, p. 316). Stephen Lesson, also known as Sam Booza, was arrested after being discovered "disguised as a female" in a "public place." Initially Lesson was sentenced to six weeks hard labour or a fine of £3. On appeal the judge noted that since the accused was found "lying very drunk in Sydenham road" there appeared to be an absence of evidence to prove criminal intent (*R v. Lesson*, 1906 20 EDC, p. 183). The conviction was quashed because at this stage, there was not yet a judicial consensus that criminality could be linked directly to being found in public cross-dressing. At this point in time evidence of a further criminal act was still required.

Part of the reason behind the still-tenuous nature of this link was that neither sexology nor the opinion of medical science more broadly had significant influence over legal matters, despite the fact that eugenics, its better-known travelling companion, enjoyed a fairly broad-based appeal in South African society. Sir Kendal Franks in his 1909 lecture entitled *The Position of the Medical Professional in South Africa* notes that the relationship between the state and the medical profession at the time was somewhat strained. Whereas in the future medical opinion would be called upon in such cases to assist with conviction, in the early 1900s in South Africa the "opinion of the profession ... [was] ... little considered" (Franks, 1909, p. 306). This was particularly true in matters regarding the public, which medical practitioners asserted should have brought individuals into contact "with Government authorities" (Franks, 1909, p. 306). Franks was a widely respected British surgeon, who had served as the vice-president of the Royal College of Surgeons in Ireland before immigrating to South Africa. He is considered to have played a critical role in the establishment of public health services in the Transvaal Colony and later the Union of South Africa. In his address, he opined that the poor relationship between the profession and the state was partly due to the lack of a South African governing body with the power to rectify the "deplorable" and "despised" state of the profession and its lack of a direct link to the state through "a separate and distinct department in the Government of the Union with a medical man at its head" (Franks, 1909, p. 318).

Returning to the Disguises Acts themselves, it is unclear whether these types of laws regarding disguise specifically existed due to public outcry or a growing number of incidences of assigned male people dressing as women, or to both. Claire Sears (2015) suggests that South Africa was not unique in utilizing judicial means to ensure normative behaviour in the 1900s, but rather that this was part of a wider international turn, particularly visible in countries like South Africa with a growing concern over the surveillance and policing of "the boundaries of sex, race, citizenship and city space" (p. 10). These types of laws ensured that people whose gender presentation was perceived as non-normative were excluded from public participation and city life while also defining the limited sphere of acceptable gender identities that would allow access. Essentially there "were only two ways that people with non-normative gender presentation could avoid arrest – either changing their clothing to comply with the law or evading police detection by fully passing" (Sears, 2008, p. 173). Other laws passed within the same time period, such as the 1914 Immigration Act, which explicitly prohibited persons convicted of sodomy or "unnatural offences" from settling in South Africa, can be read as further evidence of this emergent concern over gender and sexuality (Epprecht, 2004).

Second Instance: Science and Sex in *R v. Ntokile Zulu* (1947)

Nearly twenty years after Sir Kendal Franks's 1909 lecture, it would seem that a far more conducive relationship between medical science and the state had begun to unfold, with a clear perception of the importance of medical science to the "enhanced national status" of the Union of South Africa (Cluver, 1928, p. 434). Indeed, the two key pitfalls pointed out by Franks, the denigration of the profession and the lack of a state medical body, seemed to have been overcome. Emergent here and crucial to this burgeoning relationship between medical science and the South African state was the body as a prominent site of surveillance. As David Lyon (2001) notes, this was a key development for nation states of the time, particularly those like South Africa invested in nationalistic notions of the "best and fittest," which wished to subject their citizenry to "disciplinary shaping towards new purposes" (p. 71). It is the deviant body, including the gender non-conforming body, as a visible threat to acceptable and desirable nationhood that begins to emerge here as something that needs to be actively categorized, surveilled, and curtailed.

In a speech read before the South African Medical Congress in 1928, the Assistant Health Officer for the Union of South Africa thus applauded the creation of "a single united medical association," which was touted as indicative of "increased local patriotism," noting:

With this local patriotism has come to individual practitioners a more intense sense of duty with regard to the building up of a healthy and virile nation. The medical-practitioner's share in nation building is no small one. He certainly does not yet take his proportionate place in the councils of nation building ... he has most important contributions to make to the country's legislation: contributions which cannot possibly be made by any legislator not trained in medicine. (Cluver, 1928, p. 434)

In September of the same year, the *Journal of the Medical Association of South Africa* published arguably its first article directly addressing sexuality and gender entitled "Homosexuality" (Stohr, 1928). At the time, homosexuality was understood as a subcategory of the invert – a diagnosis that referred to the assumed inborn reversal of gender traits *and* sexual interest in the same sex – an intertwining of sexuality and gender. Part of the substance of the article addressed the health and virility of the nation, underlining a belief that medical science, in this case sexology, should naturally be taken into account by leaders and legislators. As the article noted, "the mental health specialist cannot help having his own opinion about how the law can best deal with mental abnormality" (Stohr, 1928, p. 459). It then separates homosexuals into two categories: those who are happy, and are thus actually "female inverts," and those who are not happy, and are actually "at the bottom" heterosexual (Stohr, 1928, p. 454). The article suggests that to punish the invert "is laughable ... you can't make him more than he is" and that the best approach would be to treat inverts with "pity, and, if possible, understanding" (Stohr, 1928, p. 459).

As evidenced by the next known case of arrest under the Disguises Acts in 1947, by the 1940s the courts had begun to call upon medical practitioners in seeking to surveil and understand the behaviours, ailments, and perceived abnormalities of the body. The Durban Magistrate's court charged the accused, Ntokile Zulu from the Cato Manor informal settlement, "who is a native male of the estimated age of 36 years," for masquerading in female attire when he was male (*R v. Ntokile Zulu*, 1947, p. 241). At the request of the Reviewing Judge, who noted that Zulu was what the judge considered "sexually abnormal ... possessed of abnormal inclinations," a psychiatrist examined the accused (*R v. Ntokile Zulu*, 1947, p. 241). The following extract was presented from the psychiatrist's report:

I find with respect to his mental and physical condition that his appearance and behavior display all the signs and symptoms of acquired contrary sexual instinct and effemination. The history, although given by the prisoner, is consistent with this condition. He states that his mother reared him as a female child. He always associated with girls and women. He did the work of women, such as dressmaking. During the full period of my examination the prisoner's voice remained soft and effeminate in pitch and tone ... The accused's present state is practically tantamount to that of congenital sexual inversion. (*R v. Ntokile Zulu*, 1947, p. 241)

The idea of sexual abnormality as tied to perceptions of gender non-conformity is crucial here. It is not simply that being assigned male and wearing female attire was considered problematic, but rather that this stated something about a person's sexual proclivities, a suggestion of inherent deviance, which could be diagnosed as "congenital sexual inversion." Suspicion of such deviance also served in this instance as justification for a particularly intimate medicalized surveillance. Unlike with Lesson, who was recused due to lack of evidence, the crime for Zulu became the disguise itself. Rather than something to be committed with the disguise being indicative of intention, as with Lesson, the disguise for Zulu now suggested something about the individual's sexual proclivities in relation to the "health and virility" of the heteronormative nation.

Medical surveillance and intervention, as well as the language used, particularly the term "inversion," can be read not only as indicative of a growing medico-legal relationship but also of broader shifts regarding gender, placing emphasis on "dichotomies and difference" and policing "the boundaries of morality" (Stern, 2010, p. 187). Internationally, by the late 1940s – largely due to the treatment of those of Jewish descent, among others, in Hitler's Germany – there was a distinct discrediting of eugenics and a growing effort to "engineer a scientific consensus that race was less a biological than a social and cultural phenomenon" (Dubow, 2010, p. 283). Saul Dubow (2010) argues that the then recently established apartheid state had, due to the historical implementation of racial segregation, little need to invest explicitly in the biological difference of race but rather relied on "appeals to cultural difference and ethnic nationalism" (p. 283).

This de-emphasizing of eugenics saw a shift in the relationship between racial and sexual science in South Africa, which included a marked increase in the mention of sexology in the *Journal of the Medical Association of South Africa* and growing refinement in medical categories – partly spurred on by the greater availability of technology and knowledge. In turn, this saw expanded medical engagement with not only homosexuality but, in relation to it, notions of "sex change."[3] Within South Africa, sex reassignment procedures, though they had been taking place since at least the 1950s, began to increase from the 1960s, following a timeline similar to that of the United States and the United Kingdom. The social and political landscapes, however, differed considerably. Homosexuality and transsexuality were in many ways portrayed as two sides of the same coin in South Africa (unlike in the United States and Western Europe) in that, drawing on the ongoing understanding of inversion, sex change was understood as the cure to homosexuality. In light of this, control was explicitly reinvested by state powers in the hands of medico-legal gatekeepers (Camminga &

Vincent, 2009). In the journal's first full-page article on the topic, entitled "Transvestism and Transsexualism," Alexander Don (1963) did not
support sex change surgery; in the case studies his article presents, he
suggests electroshock therapy as a far more effective option, but only in
cases where the patient "genuinely desire[d] to stop cross-dressing" (Don,
1963, p. 485). Crucially, what Don's article suggests is a categorical rigid-
ity between those desirous of a heterosexual life facilitated through sex
change and those who would require electroshock therapy as a means to
dissuade undisciplined and ostensibly immoral desires.

It is notable that in "Transvestism and Transsexualism" Don (1963)
points out a growing concern within the medical fraternity over the
legal impediments of the original Disguises Acts (1891–1910) to those
changing sex. Indeed, doctors had begun to provide their patients with
letters in order to "exempt their patients from the law" (Lock Swarr,
2012, p. 59). This providing of documentation by medical gatekeepers
suggests a clear collusion between law and medicine. If one was being
treated, in essence undergoing sex change, then paperwork could sanction the public presence of gender non-conformity, with the understanding that this was a momentary legal lapse. Medical surveillance links up
here with broader systems of registration and documentation cement-
ing particular understandings of gender and sexuality. Those without
documents, given the state's perceptions regarding homosexuality and
transsexuality, would then be considered criminals and their dress would
simply be indicative of this inherent criminality or implied vice. The next
case of arrest under the Disguises Act makes this patently clear.

Third Instance: Masquerading as a Woman in *S v. Kola* (1966)

One of the most widely referenced cases relating to the production
and policing of gender in South Africa is *S v. Kola* (1966), the same
year as the Forest Town party noted above. This case set the precedent
for what would later come to be an amalgamated Prohibition of Disguises Act (1969), drawing principally on both the Transvaal Law on
Masks, False Beards or Other Disguises (1891) and the Criminal Law
Amendment Act of Natal (1910), and was used definitively to surveil
and police gender-transgressive behaviour. Kola – the accused – wearing a dress and make-up was arrested in 1965 in the company of "two
other males, similarly dressed" (*S v. Kola*, 1966, p. 322). This was not
the accused's first arrest, and in both previous instances Kola had been
dressed in women's clothing and also appeared in court dressed as
such. The case itself was pivotal to further understandings of disguise
in relation to gender normativity in that it raised the question of the

exact connection between criminal intent (and what could be considered as such) and Kola's "disguise" as a woman. In a clear shift towards perceived criminality, Kola became one of the first actually to be convicted for being in a disguise. As with *R v. Ntokile Zulu* (1947), medical opinion was sought from the district surgeon: "Although his general physical configuration (e.g. his build and hips) and sexual organs were those of a male, the pitch and tone of his voice and the style of the hair on his head were feminine, that possibly he had a sexually inverted mind, which was congenital, and that he was in consequence a psychological misfit or deviate [*sic*]" (*S v. Kola*, 1966, p. 327). On appeal, it was decided that "liability would be established by proving the intention to conceal identity" and that this satisfied the requirements as established for criminality (Hoctor, 2013, p. 317). This is unlike in the past where, when courts were clearly uneasy with an assigned male person in female attire, *mens rea* had been difficult to establish. In this case, the court argued that once *actus reus* was established, *mens rea* would naturally follow and it was up to the accused to prove otherwise (Stuart, 1967). Unlike in *R v. Lesson* (1906) and *R v. Ntokile Zulu* (1947), the focus in *S v. Kola* (1966) was solely directed at the meaning and intention of concealing sex:

> It is clear that he must have intended to conceal his sex and pass himself off as a woman, and I think that, in the absence of any evidence from him explaining his conduct, it must further be presumed that he thereby intended to conceal his identity, for that would be a reasonable and probable consequence of concealing his sex. (*S v. Kola*, 1966, p. 327)

It may seem obvious that this case provides a specific historical reading of perceptions of gender, sex, and sexuality shaping medical practitioners' reasoning. Kola is read as physically male but manifesting feminine behavioural patterns through expression. The surgeon's conclusion based on this was that Kola's mind was the opposite of what it should be – female instead of male – and thus congenitally inverted. A reading of congenital sexual inversion is a direct suggestion of problematic sexuality – that is, homosexuality.

S v. Kola (1966) provides a clear instance from which to plot sociocultural perceptions of gender normativity and their wider influences in South Africa around the time the Disguises Acts were amalgamated. Critically, and perhaps what sets this case apart, in 1965 the South African publication *Drum Magazine* actually covered the case of the three friends arrested for "masquerading as women" – John (Joan) Kruger, Edward (Edna) Hobles, and Mohamed (Sonia) Kola. According to Dhianaraj

Chetty (1994), while the three were prosecuted, two of their friends –
"Murial and Sharon" – took what Chetty describes as "extreme measures
to avoid persecution, at least from the law: both had 'sex change' operations" (p. 122).

For Puar (2014) surveillance allows the state to outline ideal subjects,
citizens who remain disciplined within a given culture while also disciplining others through their own scrutiny of those around them. Critically, "surveillance does not simply monitor, it enforces certain behaviors
and certain identities, thereby excluding others" (Puar & West, 2014,
n.p.). As the arrest of Kola makes clear, the issue at hand was the suggested moral and sexual deviance, directly linked to homosexuality,
indicated by public cross-dressing. The solution, as Murial and Sharon's
actions suggest, was to transition or to change sex, thus evading or negat-
ing the surveillant gaze. This solution was patently supported by the state
(Camminga & Vincent, 2009), as doing so realigned one's sex in the
eyes of the state – that is, one no longer masqueraded as a woman but
had become one. For those, like Kola, who evidently did not make the
same choice, public visibility of gender non-conformity, understood to
be directly indicative of homosexuality, was unacceptable and punished
accordingly, ensuring removal or exclusion from the public space.

For the next thirty years, in its new post-Forest Town amalgamation,
the Prohibition of Disguises Act (1969) functioned to ensure that those
who cross-dressed, thus signifying homosexuality through their presentation, were actively criminalized. It bears repeating that for the apartheid state the homosexual and the transsexual were in fact one and the
same, with a singular caveat: the former was resistant to the state-offered
"cure" of sex change surgery, which would function as a form of heterosexual realignment, and the latter was not. So invested was the state
in the preservation of heteronormativity and this notion of cure that it
also amended the 1974 Births, Deaths and Marriages Registration Act
to allow those who had undergone realignment to alter the birth sex
on their identity document. Amanda Lock Swarr (2012) argues that this
amendment highlights a "close and mutually-substantiating relationship
between the state and medical institutions, as legal recommendations
followed the medical opinion of the 1970s" (p. 60).

Those who would not be cured, such as Kola, would find no place in
the public realm, and indeed they did not. From the 1970s, surveillance
and harassment by police was a regular aspect of gender-transgressive
life, taking place alongside increasing violence by members of the public. Newspapers reported the presence of "moffie-jagters" or "faggot
hunters" – men who hunted other "men with the habit of wearing wom-
en's clothing" with seeming impunity (Gays in South Africa, 1983, p. 13;

also see van Heerden, 1981, p. 15). The Phoenix Society, established in the late 1970s, was the first organization in South Africa focused on gender-transgressive identities. They, due to the atmosphere of the time and the general surveillance of transgressive gender, functioned largely as a clandestine organization, with communication happening through anonymous letter writing, rental mailboxes, pseudonyms, and obscure adverts taken out in mainstream men's magazine such as *Scope*. The Society warned their members about instances where individuals had been picked up by "the 'MAIN MANNE'[4] [or top dogs] … element in the South African Police force and taken to a deserted spot and beaten up and left there" (FanFare, 1986, p. 20). They often suggested that if they needed to meet, which they did infrequently, members should gather at each other's homes rather than venture out. In 1989, the Special Branch, also known as the Security Police, the most feared arm of the apartheid state police force, discovered the existence of the Society through intercepting some of its mail and requested a full membership list (FanFare, 1989). For fear of infiltration or having their members arrested, the majority of whom did not want sex change surgery, the leadership of the organization at the time burned much of the Society's information, reading materials, and contact details (FanFare, 1989).

Fourth Instance: Apartheid Ghosts and Johannesburg Pride (2004)

On June 17, 1991, the vote that repealed the legal framework of apartheid took place. In a surprising turn of events in 1992, at the "inception of the transition from apartheid to democracy," the Births, Marriages and Deaths Registration Amendment Act no. 51 of 1974 was repealed (Lock Swarr, 2012, p. 62). Lock Swarr (2012) points out that the reason for the repeal was not directly related to sex reassignment but rather a move to streamline registration processes. In an era where it was believed that homosexuality was a distinct threat to the moral standing of apartheid South Africa, it was accepted by the apartheid state that a person's body could be realigned with their gender, thus supposedly reinstating heterosexual desire. In this transitional period towards the new South Africa where homosexuality would eventually become constitutionally protected, the political organizing for which had already begun, there was no longer a need to perceive a static, heterosexually focused gender at play in the wrong body. Rather, the body fell away as something malleable. Lock Swarr argues that this was a fundamental moment, in what can be described as a medico-legal split, indicating clear "differences between medical and legal understandings of gender as alterable or inalterable" (2012, p. 63). The confluence of medicine and law that we saw developing over the course of the twentieth century now began to reverse.

The restructuring of South Africa towards a constitutional democracy meant the dismantling of the apartheid state's legislative framework. One of the most lauded elements of this new democracy was the Equality Clause including the provision prohibiting discrimination on the grounds of sexual orientation, thus providing the basis for several sig- · nificant court cases in the first years of democracy. Key amongst a series of court rulings in the 1990s was the *National Coalition for Gay and Lesbian Equality vs Minister of Justice and Others* (1998), which declared the criminalization of sodomy unconstitutional while also providing the grounds for recognizing myriad other rights for "sexual minorities" in South Africa (de Vos, 2009). Edwin Cameron's (1993) article "Sexual Orientation and the Constitution: A Test Case for Human Rights" was highly influential in both the majority and minority judgments of the Constitutional Court. In arguing that adequate protection within the Constitution for sexual orientation would entail the right to free speech, association, and conduct Cameron notes that this would also include, reflecting on the case of *S v. Kola* (1966), the freedom to "cross-dress (that is, appear in drag)" (1993, p. 471). Following from this, the majority judgment of Justice Ackerman gave a broad reading of sexual orientation, reliant on *S v. Kola*, that extended the prohibition on discrimination on the basis of sexual orientation, as envisioned in the Bill of Rights (1996), to discrimination against transsexual people.

During this period of constitutional shifting there was a significant increase in political violence in South Africa. In an attempt to bring South Africa's jurisprudence in line with international best practice models as swiftly as possible, the Goldstone Commission of Inquiry Regarding the Prevention of Public Violence and Intimidation was established in 1991. The "enquiry's deliberations flowed from a basic principle that the right to assemble, demonstrate, protest and petition is a necessary condition for making good a democratic society's commitment to universal political participation" (Memeza, 2006, p. 12). The Commission sought to ensure that the possibility of the right to gather was broadly available to all South Africans, making a clear break with "draconian apartheid jurisprudence" (Memeza, 2006, p. 12). One of the key outcomes of the Commission was the Goldstone Bill, which became the Regulation of Gatherings Act (1993), officially enacted only after the 1996 election. The Act was "an enabling legislative framework for public activities prior to the holding of the country's first democratic elections" (Memeza, 2006, p. 12) that, according to Mzi Memeza (2006), was meant to be transitory in nature, and it was created in a time of very real need with the intention that something more suitable to post-apartheid South Africa would replace it. Read as an entirely ahistorical and benign piece of legislation, the Prohibition of Disguises Act (1969) found new life within post-apartheid South Africa by being folded into the Regulation of Gatherings Act (1993).[5]

Chetty (1994), in reflecting on the life of Kola and ostensibly all those to have come into contact with the apartheid position on disguise, muses: "What could be so dangerous about clothes meant for another gender?" (p. 122). In 2004, eleven years after the first South African Pride celebrating constitutional democracy was held in Johannesburg, the surveillance ghosts of various Disguises Acts were resurrected. Seemingly unbeknownst to those at the time, the Act was deployed in the manner it had always been intended – against non-conforming bodies in public spaces. Harkening back to an era of extreme state control, the Johannesburg Metro Police announced that those dressed in drag or crossdressing would not be allowed to take part in Pride. Police spokesperson Wayne Minnaar stated that the decision to become involved "was taken to ensure consistency ... [with the law] ... and that no one in disguise or wearing a mask will be allowed to take part in the march" (Wolmarans, 2004a, n.p.) Notably, a fringe conservative group calling themselves the Gay and Lesbian Alliance (GLA) informed the police about the particular section of the now decade old Gatherings Act (1993). Acknowledging the historical roots of the Prohibition of Disguises Act (1969), the GLA vowed "to make sure drag queens in particular are arrested for contravening the apartheid-era act" (Khangale, 2004a, n.p.). The GLA, taking up the mantle of surveillance embedded within the law, went so far as to state that they would assist police on the day in pointing out "crossdressers" in order to have them arrested. As with the very different forms of lateral surveillance discussed by Andrejevic (2005), in this case too we find that "[r]ather than displacing 'top-down' forms of monitoring, such practices emulate and amplify them, fostering the internalization of government strategies" (p. 479).

Returning to the temporally anticipatory nature of laws intended to classify, surveil, medicalize, manage, and exclude, the GLA, unwittingly, dredged up a far more complex history regarding surveillance, bodies, classification, and assumed access to public space and visibility. As Toby Beauchamp (2009) notes, this is unsurprising given that state policies which may at first seem benign, "are in fact deeply rooted in the maintenance and enforcement of normatively gendered bodies, behaviors and identities" (pp. 356–7). Indeed, speaking to the very heart of who the Prohibition of Disguises Act (1969) was envisioned to target and the "maintenance and enforcement of normatively gendered bodies,"

the GLA stated that they regarded "cross-dressers as unwanted elements within our society" (Wolmarans, 2004b, n.p.). For Beauchamp, in
approaching the relationship between gender non-conformity and state
surveillance, it is important to think beyond the medical and legal monitoring of bodies in order to consider the relationship between the ideals
of citizenship and legitimate forms of gender expression. Critically, the
GLA's position that those who are openly gender non-conforming are
"unwanted elements" of South African society led them to work with the
state, and request that the police provide surveillance and remove these
"unwanted elements." It is this that marks the long history of surveillance within the South African public sphere of gender non-conforming
bodies by the state, delimiting the outlines of acceptable South African
citizenship.

In reaction to the threat, several LGBT organizations met with the
police. They issued a collective statement regarding the decision taken
by the police noting that the interpretation of the Act by the police was
"blatantly unconstitutional" (Wolmarans, 2004b, n.p.). At the time, the
police disputed the initial accusation that they were being homophobic
by stating again that "the law was being enforced to provide consistency"
(Drag Controversy, 2004, n.p.). Harkening back to *R v. Lesson* (1906)
and an era marked by the lack of medico-legal interaction, after meeting
with the police it was agreed

> that revellers could wear their masks, lipsticks and even paint themselves.
> However, they could not drink alcohol ... it was agreed that the parade
> should take place without hindrance ... We don't want to interfere with
> what they will be wearing as long as the procession is decent. (Khangale,
> 2004a, n.p.)

It is these dual ideas of decency and consistency that are key here.
Indeed, though the accusation of homophobia may have been real, in
truth, the police were simply ensuring consistency – applying a law that in
its initial incarnation had been constructed as a tool of surveillance and
utilized to ensure the absence of visible gender non-conformity on the
streets of South Africa. Their use of the law, or threat to do so, poignantly
illustrates Puar's (Puar & West, 2014) argument regarding futurity and
the management of identities. The law's initial intention during apartheid had been outright discrimination. In the democratic South Africa
its continued ability to still be deployed maintained its initial intention –
to persecute those who cross-dress and enforce their public absence.
In fact, as acknowledged by members of the organizations protesting
the GLA and police, this was an instance in which "laws can maintain

continual discrimination, very similar to apartheid South Africa" (Khangale, 2004b, n.p.). With regard to ideas of decency suggested by the police, this too can be read as a call to a particular kind of heteronormativity anchored in the initial meaning of the law. As Evert Knoesen, the director of South African LGBT organization Equality Project, noted, "there are those who think it is only OK to be gay if you fit the so-called 'straight acting' image" (Knoesen & Wolmarans, 2006, p. 158): in other words, an image of respectability, decency, and heteronormativity.

Conclusion

Sears (2008) argues that access to public space is a fundamental precondition of democracy. And indeed, many of the chapters later in this book demonstrate the ways in which the surveillance and regulation of public space serves to limit democratic participation, from the multifaceted repression of Palestinians (Ahmad H. Sa'di, chapter 8) to the state surveillance of dissidents in communist Romania (Cristina Plamadeala, chapter 9) as well as in the United States (Kathryn Montalbano, chapter 10; Elisabetta Ferrari & John Remensperger, chapter 11). In the South African context, by restricting access to public space, laws such as the Disguises Acts and their later amalgamation into the Prohibition of Disguises Act (1969) have been used against those who cross-dress to "effectively exclude multiple people with non normative gender from civic participation and the democratic life of the city" (Sears, 2008, p. 173). South Africa is not the only country to contend with these kinds of laws, which, though seemingly benign, continue to target and police the visibility and behaviour of particular populations. In 2014, a Malaysian court ruled that the banning of cross-dressing was "degrading, oppressive and inhumane" and that it forced people to live "in uncertainty, misery and indignity" (Tizmaghz, 2014, n.p.). The law in question, much like South Africa's, specifically prohibited "any male person who in any public place wears a woman's attire or poses as a woman" (Tizmaghz, 2014, n.p.). The Malaysian Court of Appeal noted that those who might find themselves scrutinized by the law would be considered to be committing a crime in the "very moment they leave their homes to attend to the basic needs of life, to earn a living, or to socialize; and be liable to arrest, detention and prosecution ... [It is] discriminatory and oppressive and denies the appellants the equal protection of the law" (Tizmaghz, 2014, n.p.).

Two years later, in 2016, there were moves within South Africa's Parliament to see the Regulation of Gatherings Act (1993), along with several others, repealed. Zakhele Mbele, South Africa's first openly gay Black member of Parliament, argued that the "continued existence of these laws was 'at best laughable incompetence' and at worst, 'keeping apartheid alive'" (Herman, 2016, n.p.). A report by the South African Parliament's Legal Services division identified over 1,850 pieces of apartheid-era legislation passed between 1910 and 1993 still in effect in the country (Mothapo, 2017). Arguably the continued use of the Prohibition of Disguises Act (1969) in the democratic South Africa carries the echoes of its past into the present – "a remarkable longevity" (Sears, 2015, p. 4) allowing it to be deployed, to function as a "temporally anticipatory apparatus" (Puar & West, 2014, n.p.) in its originary sense, regardless of the shift in structural politics. This has arguably ensured that gender non-conforming South Africans in the democratic dispensation have inherited the "long history of being made suspect" through legislative iterations of surveillance (Stanley, 2011, p. 7). It remains to be seen, as current moves to repeal apartheid era legislation unfold, if the Prohibition of Disguises Act will stay on the books, seemingly understood as an innocuous law posing no threat, itself disguised, and perhaps waiting for some future time when it will be considered useful once again.

NOTES

1 South Africa became a Union in 1910; prior to this the country existed as four separate British colonies – the Cape, Natal, Transvaal, and the Orange River Colonies.
2 These Acts included the Transvaal Law on Masks, False Beards or Other Disguises, no. 2 of 1891; Criminal Law Amendment Act of Natal, no. 10 of 1910 S6(2)E; Cape Colony Police Offences Act, no. 27 of 1882 SVIII(2); and Orange River Colony Police Offences Ordinance, no. 21 of 1902 S25(2).
3 Though the term "sex change" is outdated, and may in fact be considered derogatory, I use it in this paper in an effort to continually signpost what would have been the position and perception of the apartheid state.
4 A hypermasculinized phrase used to refer to the bosses or "top dogs" of the police force, in this instance.
5 According to the Regulation of Gatherings Act, no. 205 of 1993 S 7, "no person shall at any gathering or demonstration wear a disguise or mask or any other apparel or item that obscures his facial features and prevents his identification."

REFERENCES

Andrejevic, M. (2005). The work of watching one another: Lateral surveillance, risk, and governance. *Surveillance & Society*, 2(4), 479–97.

Beauchamp, T. (2009). Artful concealment and strategic visibility: Transgender bodies and US state surveillance after 9/11. *Surveillance & Society, 6*(4), 356–66.

Breckenridge, K. (2014). *Biometric state: The global politics of identification and surveillance in South Africa, 1850 to the present.* Cambridge, UK: Cambridge University Press.

Cameron, E. (1993). Sexual orientation and the constitution: A test case for human rights. *South African Law Journal, 110,* 450–72.

Camminga, B., & Vincent, L. (2009). Putting the "T" into South African human rights: Transsexuality in the post-apartheid order. *Sexualities, 12*(6), 678–700.

Chanock, M. (2001). *The making of South African legal culture 1902–1936: Fear, favour and prejudice.* Cambridge, UK: Cambridge University Press.

Chauncey, G. (1983). From sexual inversion to homosexuality: Medicine and the changing conceptualization of female deviance. *Salmagundi, 58/59*(Fall 1982–Winter 1983), 114–46.

Chetty, D. R. (1994). A drag at Madame Costello's: Cape moffie life and the popular press in the 1950s and 1960s. In M. Gevisser & E. Cameron (Eds.), *Defiant desire: Gay and lesbian lives in South Africa* (pp. 115–27). Braamfontein, SA: Ravan Press.

Cluver, E. (1928, August 8). The medical practitioner's place in local government and health administration in South Africa. *Journal of the Medical Association of South Africa, II,* 434–40.

de Vos, P. (2009). From heteronormativity to full sexual citizenship? Equality and sexual freedom in Laurie Ackermann's constitutional jurisprudence. In J. Barnard, D. Cornell, & F. du Bois (Eds.), *Dignity, freedom and the post-apartheid legal order: The critical jurisprudence of Laurie Ackerman* (pp. 254–73). Cape Town, SA: Juta.

Don, A. M. (1963, May 4). Transvestism and Transsexualism. *South African Medical Journal, XXXVII,* 479–85.

Drag Controversy at Joburg Pride. (2004). Retrieved from www.angelfire.com/zine/gayzim/newsletter/102004.html

Dubow, S. (2010). South Africa: Paradoxes in the place of race. In A. Bashford & P. Levine (Eds.), *The Oxford handbook of the history of eugenics* (pp. 274–88). Oxford, UK: Oxford University Press.

Epprecht, M. (2004). *Hungochani: The history of a dissident sexuality in Southern Africa.* Montreal, QC, and Kingston, ON: McGill-Queen's University Press.

FanFare. (1986). *FanFare Magazine 22 May 1986.* (Joy Wellbeloved Collection – GAL0013). Gay and Lesbian Memory in Action Archives (GALA), William Cullen Library University of Witwatersrand, Johannesburg, South Africa.

FanFare. (1989). *FanFare Magazine May 1989.* (Joy Wellbeloved Collection – GAL0013). Gay and Lesbian Memory in Action Archives (GALA), William Cullen Library University of Witwatersrand, Johannesburg, South Africa.

Fischer, M. (2016). Under the ban-optic gaze: Chelsea Manning and the state's surveillance of transgender bodies. In E. van der Meulen & R. Heynen (Eds.), *Expanding the gaze: Gender and the politics of surveillance* (pp. 185–209). Toronto, ON: University of Toronto Press.

Franks, K. (1909). The position of the medical profession in South Africa. *The South African Medical Record, VII*(12 December), 306–18.

Gays in South Africa: Cruising around the Cottages. (1983, August 31). *Frontline*, pp. 13.

Gevisser, M. (1994). A different fight for freedom: A history of South African lesbian and gay organisation from the 1950s to the 1990s. In M. Gevisser & E. Cameron (Eds.), *Defiant desire: Gay and lesbian lives in South Africa* (pp. 14–89). Braamfontein, SA: Ravan Press.

Gevisser, M., & Matebeni, Z. (2012). *Forest Town – Michele. Joburg Tracks Sexuality in the City.* Museum Africa, Newtown Johannesburg.

Grunkel, H. (2010). *The cultural politics of female sexuality in South Africa.* New York, NY: Routledge.

Herman, P. (2016, November 22). Bid to start process to repeal apartheid-era laws fails in parliament. Retrieved from www.news24.com/SouthAfrica/ News/bid-to-start-process-to-repeal-apartheid-era-laws-fails-in-parliament -20161122

Hoad, N. (2010). Introduction. In N. Hoad, G. Reid, & K. Martin (Eds.), *Sex and politics in South Africa* (pp. 14–25). Johannesburg, SA: Double Storey Publishers.

Hoctor, S. (2013). The offence of being found in disguise in suspicious circumstances. *Obiter, 34*(2), 316–21.

Jespersen, R. R. (1976). The jurisprudential problem of apartheid. *Texas Southern University Law Review, 4,* 323–41.

Khangale, N. (2004a, September 12). Rival gay group out to get drag queens. Retrieved from www.iol.co.za/news/south-africa/rival-gay-group-out-to-get -drag-queens-1.222338

Khangale, N. (2004b, September 20). Drag queens are now safe from Joburg cops. Retrieved from www.iol.co.za/news/south-africa/drag-queens-are- now-safe-from-joburg-cops-222216

Knoesen, E., & Wolmarans, R. (2006). The people have spoken. In S. De Waal & A. Manion (Eds.), *Pride: Protest and celebration* (pp. 158–9). Johannesburg, SA: Jacana Media.

Lock Swarr, A. (2012). *Sex in transition.* New York, NY: SUNY Press.

Lyon, D. (2001). *Surveillance society: Monitoring everyday life.* Philadelphia, PA: Open University Press.

Marks, S. (1997). What is colonial about colonial medicine? And what has happened to imperialism and health? *The Society for the Social History of Medicine, 10*(2), 205–19.

Memeza, M. (2006). *A critical review of the implementation of the Regulation of Gatherings Act 205 of 1993: A local government and civil society perspective.* Johannesburg, SA: Freedom of Expression Institute.

Mothapo, M. (2017, February 2). 1850 pieces of legislation being investigated – Office of the Chief Whip. Retrieved from www.politicsweb.co.za/politics/1850-pieces-of-legislation-being-investigated–off

Puar, J., & West, L. (2014, February 4). Regimes of surveillance. Retrieved from http://cosmologicsmagazine.com/jasbir-puar-regimes-of-surveillance/

Sears, C. (2008). Electric brilliancy: Cross-dressing law and freak show displays in nineteenth-century San Francisco. *Women's Studies Quarterly, 36*(3/4), 170–87.

Sears, C. (2015). *Arresting dress: Cross-dressing, law, and fascination in nineteenth-century San Francisco.* Durham, NC: Duke University Press.

Somerville, S. B. (2000). *Queering the color line: Race and the invention of homosexuality in American culture.* Durham, NC: Duke University Press.

Stanley, E. A. (2011). Fugitive flesh: Gender self-determination, queer abolition and trans resistance – An introduction. In E. A. Stanley & N. Smith (Eds.), *Captive genders: Trans embodiment and the prison industrial complex* (pp. 1–15). Oakland, CA: AK Press.

Stern, A. M. (2010). Gender and sexuality: A global tour and compass. In A. Bashford & P. Levine (Eds.), *The Oxford handbook of the history of eugenics* (pp. 173–91). Oxford, UK: Oxford University Press.

Stohr, F. O. (1928, September 9). Homosexuality. *Journal of the Medical Association of South Africa, II*, 455–60.

Stryker, S. (2008). *Transgender history.* Berkley, CA: Seal Press.

Stuart, D. R. (1967). Presumed intention in criminal law. *South African Law Journal, LXXXIV*(3), 256–60.

Tizmaghz, J. (2014, November 7). Malaysia: Court victory for transgender rights. Retrieved from www.hrw.org/news/2014/11/07/malaysia-court-victory-transgender-rights

Tsampiras, C. (2008). Not so "gay" after all – Constructing (homo)sexuality in AIDS research in the *South African Medical Journal*, 1980–1990. *South African Historical Journal, 60*(3), 477–99.

van Heerden, D. (1981, June 12). "Moffie-jagters" bedreig noe al speurders ook. *Rapport*, p. 15.

Wolmarans, R. (2004a, September 15). Police to rain on drag queens' parade. Retrieved from http://m.mg.co.za/article/2004-09-15-police-to-rain-on-drag-queens-parade

Wolmarans, R. (2004b, September 17). Drag queens can march, after all. Retrieved from https://mg.co.za/article/2004-09-17-drag-queens-can-march-after-all/

Legislation

Births, Marriages and Deaths Registration Amendment Act, no. 51 of 1974.
Cape Colony Police Offences Act, no. 27 of 1882 SVIII(2).
Criminal Law Amendment Act of Natal, no. 10 of 1910 S6(2)E.
Orange River Colony Police Offences Ordinance, no. 21 of 1902 S25(2).
Immorality Amendment Act, no. 21 of 1950.
Immorality Amendment Act, no. 57 of 1969.
Prohibition of Disguises Act, no. 16 of 1969.
Regulation of Gatherings Act, no. 205 of 1993.
Transvaal Law on Masks, False Beards or Other Disguises, no. 2 of 1891 S1.

Cases

National Coalition for Gay and Lesbian Equality v. The Minister of Justice 1998
    (12) BCLR 1517 (CC).
R v. Lesson, 1906 20 EDL 183.
R v. Ntokile Zulu, 1947 (1) SA 241 (N) 1947 (1) SA.
S v. Kola, 1966 (4) SA 322 (A) 1966 (4) SA.

SECTION TWO

Identification, Regulation, and Colonial Rule

Surveillance systems rely on being able to identify who or what is being tracked, although identification can serve profoudly different and often ambivalent ends. In the case of civil registration programs, for example, in particular of births, deaths, and marriage, Simon Szreter and Keith Breckenridge (2012) argue convincingly that identification systems *enable* access to social goods, and that non-registration "is an even more disabling birthright lottery than the inequalities that go with registration" (p. 1). Still, as David Lyon (2009) stresses, while identification is central to practices of citizenship, "today's modes of citizenship, as represented by ID card regimes, are aimed at the *exclusion* of certain proscribed groups" (p. 17, emphasis in original). These positions are not mutually exclusive; both capture key dimensions of identification practices.

Historically, advances in areas like public health or social welfare programs that relied on systems of identification and registration led to falling mortality rates, greater universality in access to care, and the mitigation of harsher forms of capitalist exploitation. However, as we discussed in chapter 1 with the burning of archival records during the Paris Commune of 1870, state record-keeping was also often experienced as threatening and coercive, prompting some Communards to propose alternative forms of identification that could overcome the contradictions of those rooted in liberal citizenship models of the nation state. In 1871, for example, the Union of the Women's General Assembly granted the Russian radical Elisabeth Dmitrieff, who had been reporting back to Marx on the Commune, the title of "citizen of Paris, while waiting for the universal Republic to give her the letters of naturalization necessary to make her a citizen of humanity" (quoted in Ross, 2015, p. 29). This was a utopian form of identification that very deliberately bypassed the state, pointing instead towards a radically egalitarian and participatory vision of a community encompassing all of humanity. It was a refusal, in other words, of the social sorting performed by most forms of surveillance.

The universalist ideal of the Commune was contradicted most strongly in practices of colonial governance. The passport, for instance, regulated migration across borders, but access to colonizing states was largely denied to colonial subjects who were not recognized as citizens. They produced some people as "national," and others not. The recent Windrush scandal in the United Kingdom, prompted by an immigration crackdown that caught up many of the over 400,000 people who had migrated from the Carribean between 1948 and 1970, shows the persistence of these often racist inequities. As British colonial subjects they had not required official documentation during their initial immigration, but this lack of documentation subsequently left many exposed to state persecution, with some losing jobs and homes, and at least 80 deported despite having immigrated as citizens.

Ian Warren and Darren Palmer's examination in chapter 5 of how older British legal practices travelled to and were transformed in the nineteenth century in the penal colony of Australia shows us a much earlier instance of the use of differentiated systems of identification and registration to "manage" population movement and rights. The concepts of attainder and civil death, the latter intersecting in interesting ways with Orlando Patterson's (1982) conception of social death discussed in chapter 1, formed the basis of the ticket of leave system that provided the identificatory foundation for rewarding convicts who were deemed obdient, and for the differentiation of convicts from free settlers. This complex surveillant assemblage, Warren and Palmer argue, was central to a variety of criminological shifts that linked colony and metropole. Uma Dhupelia-Mesthrie and Margaret Allen also engage with Australia, but in chapter 6 their comparative frame is with South Africa. They trace the emergence in both contexts of similar systems of non-citizen identification that enabled the restriction and regulation of the mobility of Asian migrants and residents. The South African domicile certificates and permits/certificates of identity, as well as the Australian Certificate Exempting from Dictation Test, enabled forms of racializing surveillance built in particular on visual and biometric markers.

As discussed earlier in the book, biometric surveillance played an especially important role in the dynamics of the colony-metropole. Such practices also foreground the extent to which surveillance systems were hybrid in nature, sustaining colonial relations even while they both drew on existing practices amongst colonized peoples and exported colonial surveillance back to the metropole. As Jacob Steere-Williams suggests in chapter 2, British rule in India presents a striking example of such interactions. Indeed, British rulers drew heavily on existing domestic systems of surveillance, information gathering, fingerprinting, and knowledge production, with local practices co-opted and transformed in the service of the empire (Bayly, 1996). In part, as we see in Australia and South Africa, identification practices were about policing and limiting mobility in general, but central to Britain's self-proclaimed "civilizing" mission in India was the suppression of "banditry." Under a variety of provisions, most notably in the Thuggee campaign of the 1830s and the implementation of the Criminal Tribes Act of 1871, entire tribes were deemed criminal and forced to register with local officials. Biometric techniques, first tattooing and then fingerprinting, were central to the enactment of these forms of social sorting. In India as well we find the early use of criminal photography that, as in Europe in more circumscribed ways, cemented highly racialized notions of the "habitual criminal" (Anderson, 2004; Singha, 1993, 2015).

Chapters 7 and 8 further take up these colonial practices of identification and trace their development in twentieth century contexts. First, Midori Ogasawara considers the implementation of fingerprinted ID card systems in the Japanese colonial occupation of Northeast China in chapter 7. Building on an analysis of the Koseki system in Japan itself, Ogasawara foregrounds the role of biometrics as a colonial technology for watching over the movements of racialized Others. She argues that these identification systems served a dual purpose, on the one hand enabling the repression of those deemed risky, and on the other enabling the exploitation of colonial labour. Ogasawara uses the notion of a "state of exception" to analyse this colonial expansion and Japan's nation-building as simultaneous processes of modernization. Ahmad H. Sa'di also uses the idea of the state of exception in his examination of Israeli surveillance of Palestinians in chapter 8, contrasting it with the panoptic model. He contends that the Military Government established after the declaration of the state of Israel in 1948 deployed a host of forms of surveillance to differentiate and repress Palestinian citizens, who were thereby excluded from full citizenship participation. This was premised on an underlying desire to expel the Palestinian people that had only been partially implemented during the process of Israeli state formation. The chapters in this section thus powerfully illustrate the specific characteristics of the complex practices of identification deployed in surveillance regimes in colonial contexts.

REFERENCES

Anderson, C. (2004). *Legible bodies: Race, criminality and colonialism in South Asia.* Oxford, UK: Berg.
Bayly, C. A. (1996). *Empire and information: Intelligence gathering and social communication in India, 1780–1870.* Cambridge, UK: Cambridge University Press.

Lyon, D. (2009). *Identifying citizens: ID cards as surveillance.* Cambridge, UK: Polity Press.

Patterson, O. (1982). *Slavery and social death: A comparative study.* Cambridge, MA: Harvard University Press.

Ross, K. (2015). *Communal luxury: The political imaginary of the Paris Commune.* London, UK: Verso.

Singha, R. (1993). "Providential" circumstances: The Thuggee campaign of the 1830s and legal innovation. *Modern Asian Studies, 27*(1), 83–146.

Singha, R. (2015). Punished by surveillance: Policing "dangerousness" in colonial India, 1872–1918. *Modern Asian Studies, 49*(2), 241–69.

Szreter, S., & Breckenridge, K. (2012). Recognition and registration: The infrastructure of personhood in world history. In K. Breckenridge & S. Szreter (Eds.), *Registration and recognition: Documenting the person in world history* (pp. 1–36). Oxford, UK: Oxford University Press.

5 The Penal Surveillant Assemblage: Attainder and Tickets of Leave in Nineteenth-Century Colonial Australia

IAN WARREN AND DARREN PALMER

England's penal experiments from the seventeenth to the nineteenth century offer important insights into multiple aspects of early modern individual and collective surveillance. This chapter outlines the basic character of a surveillant assemblage (Haggerty & Ericson, 2000) that served distinct regulatory objectives in imperial Britain and colonial Australia during the nineteenth century. These processes stretch further back in time, to encapsulate the underlying philosophy of early modern English criminal justice, the use of transportation as a criminal punishment, and distinct patterns in the reception of English law throughout its colonies. The enduring impacts of these developments are ripe for critical examination using contemporary sociolegal research. While an extensive body of criminology origin stories highlights stark levels of brutality and control within Australia's emerging penal state (Hughes, 2003; Neal, 1991), the implications of this expansive imperial and colonial surveillance apparatus are multifaceted, nuanced, and contain many uniquely Southern criminological elements (Carrington & Hogg, 2017; Carrington, Hogg, & Sozzo, 2016), which shaped penal and surveillance practices in Australia in quite unique ways, while influencing various aspects of modern British and Irish criminal justice and penal reform.

Various historical forms of identity authentication, including convict musters and bodily inspections, and the use of technologies such as photography, fingerprinting, and documents recording anthropometric criteria, are important precursors to many contemporary surveillance practices (Bennett & Lyon, 2008; Caplan & Torpey, 2001; Ruggerio, 2001). Most forms of mass surveillance enable nation states and their agencies to record knowledge about individuals to advance preventative and pre-emptive governance and securitization objectives (Amoore, 2008). Equivalent processes are also evident in the roles of surveillance during the formative years of Australia's colonial development. However,

unlike the simultaneous protection of settler populations and more stringent governance of colonized people evident in South Africa (Breckenridge, 2014) and India (Sengoopta, 2003), the primary emphasis of
early forms of Australian surveillance was to distinguish free settlers from
transported convicts. These measures foreshadow in quite interesting
ways later surveillance practices directed at Asian migrants, as outlined
by Uma Dhupelia-Mesthrie and Margaret Allen in the following chapter.
This basic form of social sorting also served various purposes in Britain
and Ireland throughout the late nineteenth and early twentieth centuries, as well as a range of day-to-day administrative functions within Australia's "open prison" system from the early 1800s (Roberts, 2005, p. 97).
In fact, these layers of surveillance are central to understanding the relationship between domestic crime control and colonization practices in
the Old World, and quite distinct, yet interconnected, state formation
and governance objectives in early colonial Australia. These modes of
colonial self-governance were initially semi-autonomous, and subject
to various forms of oversight from London. However, their respective
legal, economic, social, policing, and surveillance implications varied
within Australia's penal settlements, and differed in both space and time
from other sites of British transportation in North America and the West
Indies (Morgan & Rushton, 2004; Smith, 1947).

Our origin story of surveillance and Australian colonial penality has two
main threads that are connected to early modern British and Irish criminal justice philosophy. The first involves the concepts of attainder and civil
death, which had their origins in medieval English criminal law (Dayan,
2011). Civil death stemmed from the law of attainder, which involved the
state's power to stigmatize a felon as a "consequence of the judgment
which condemned the criminal for a *capital offence*," and resulted in the
"forfeiture etc. of lands, goods and personal rights, and corruption of
blood" (Bills of Attainder, 1867, p. 74, emphasis in original). Attainder
required a certain degree of surveillance in Britain and Ireland to ensure
those sentenced to transportation for a fixed or indeterminant period and
who returned prematurely received the stipulated penalty of execution.
However, the convict "'taint' or 'stain'" (Brown, 2002, p. 322) extended
to various forms of surveillance and law in Australia's open prison, to
reinforce the reality that transportation remained a formal punishment
implemented under the authority of the British Crown. These interconnected systems also represent part of an important shift in Western penal
and surveillance history that contributed to an individualized, knowledge-
based system of convict reform.

Our specific concern is to document several modes of imperial oversight that applied to emerging forms of state formation in early colonial

Australia. Our analysis is based on the detailed examination of select legal and penal histories from the seventeenth century onwards, including archival research on transportation from England to both North America and Australia, biographies of noted legal and social reformers (Morris, 2002; White, 1976), and contemporary scholarship linking principles of attainder to a range of "secondary punishments" (Shaw, 1953, p. 16; see also Edgely, 2010), collateral criminal penalties (Colgate Love, Roberts, & Klingele, 2016), and other forms of shadow carceral surveillance (Beckett & Murakawa, 2012; Chin, 2012; Palmer & Warren, 2014; Warren & Palmer, 2018). First, we outline the relationship between sociolegal conceptions of space and time, and Foucauldian theories of penal transition and governance in the seventeenth and eighteenth centuries. Second, we explain the relationship between attainder and civil death in early modern English law. Third, we identify key problems stemming from the application of the doctrine of attainder, and its implications on surveillance, citizenship, property rights, mobility, and the ability to testify in colonial Australian criminal trials. Fourth, we describe the ticket of leave system as a localized element of this broader transnational legal and surveillant assemblage that focused on rewarding convict discipline through a form of conditional freedom, which was largely unrecognized in Britain and Ireland. We briefly conclude by reinforcing the importance of this and related origin stories examining the intersecting elements of early modern transnational penality, surveillance, and crime.

Transportation and Contemporary Surveillance Theories

Penal transportation between the seventeenth and nineteenth centuries involved a patchwork of strategies founded upon the exclusion of convicted felons from Britain and Ireland (Hirst, 2008; Hughes, 2003; Morgan & Rushton, 2004, 2013; Smith, 1947). As with any historical phenomenon, the interrelated surveillance and regulatory aspects of this penal strategy are highly diffuse. Morgan and Rushton (2004) identify considerable variations in the implementation of transportation as a criminal punishment in different English and Scottish local government regions throughout the eighteenth century. Similarly, Smith's (1947) landmark study of transportation to North America and the West Indies demonstrates how transportation coexisted with both voluntary indentured service contracts and involuntary slavery administered mainly by private owners. It appears many Britons were willing to contract out their freedom for a fixed term of up to four years in search of a better life, unless they were abducted by the agents of shipping companies or large North American landowners and forced into indentured transportation.

While similar exclusionary imperatives were also at play in the establishment of Australia's penal settlements, their links to commercial slavery
or indentured servitude are less evident.

Space and time are crucial dynamics in the governance and surveillance practices associated with imperial transportation. These facets of
"internal legal pluralism" (Valverde, 2015, p. 175) involve multiple forms
of governance that coexist at quite distinct jurisdictional scales to deal
with the same phenomenon. How these processes operate in discrete
spatial and temporal settings reveals complex jurisdictional synergies
and tensions that are given meaning through the differing objectives of
law and its variegated forms of enforcement. In the present context, the
discrete roles of law in imperial Britain and fledgling colonial settings
contain many interrelated facets, including similar languages, underlying motives, and relatively unified conceptions of sovereign authority. Yet the geographic and temporal separation from Britain, which
involved twelve to eighteen months of sea travel, is perhaps one of the
most salient factors influencing the early development of a uniquely Australian law. This temporal separation, combined with an undeveloped
and unfamiliar topography and climate, reveals why colonial Australian
law might have produced several adaptations to British common law that
were only reined back at key moments when English parliamentary and
executive oversight was possible.

Transportation also accompanied a philosophical shift in the emergence of modern penal science. Foucault (1977) describes the gradual
transition from the highly discretionary, arbitrary, and inconsistently
applied forms of sovereign authority that characterized medieval criminal justice, to emerging philosophies aimed at promoting individual
self-discipline. Transportation reflected both of these developments.
On the one hand, throughout the eighteenth and nineteenth centuries,
the use of discretionary Crown powers to pardon individuals convicted
of a growing range of capital felonies increased markedly, resulting in
a significant reduction in the number of executions for criminal behaviour (Foucault, 1977; Hay, Linebaugh, Rule, Thompson, & Winslow,
1975). On the other, judges and executive authorities in Britain began
to operate "like a 'pilot steering between the rocks'" (Foucault, 1977,
p. 97), setting the initial conditions of geographic exclusion, while recognizing the importance of using individualized criminal punishments
that reflected specific offender and offence characteristics. However,
a variety of motives behind these developments operated according
to the distinct scales of penal administration within England and its
colonies, which were often determined by spatial and temporal considerations. Thus, while Australian penal transportation is often viewed

as a brutal form of servitude administered by violent colonial authorities and private masters (Hughes, 2003), revisionist histories also reveal many success stories of individual reform, economic productivity, active citizenship, and nation-building (Carrington & Hogg, 2017; Hirst, 2008). Geographic isolation arguably promoted strong incentives for convict self-discipline, while various surveillance routines overseen by local authorities sought to achieve pastoral objectives aimed at shepherding most felons into self-sufficient, economically productive, and law-abiding lives (Lippert, 2016). Many of these positive legacies of penal transportation are underacknowledged (Carrington & Hogg, 2017), yet are aligned with various forms of convict surveillance that served interconnected and sometimes conflicting or incompatible objectives in England and Australia.

Throughout this period positive scientific methodologies and bureaucratic forms of governance were also becoming more refined and better equipped to obtain greater knowledge about people. In England, mass population surveillance helped to foster greater centralization of government in the areas of health, security, and policing (Hacking, 1990; Higgs, 2004, 2011; Rose-Redwood, 2006). The mentality of promoting the collective social good through targeted forms of governance was underpinned by various forms of surveillance considered necessary in the formative years of Australian state formation. Early Australian convict governance involved many surveillance routines that encapsulated convict labour productivity, crime and social order, marriage, social networks, and the policing of geographic spaces where convicts lived under conditional freedom, or in closed environments to promote their reform. The ticket of leave was an important component of a broader publicly administered surveillant assemblage that is commonly viewed as the historical precursor to modern parole (Chan, 1991; Morgan, 1992). It replaced extant forms of physical disfigurement, such as the branding of recidivists (Foucault, 1977), burning laypeople on the base of the left thumb to indicate they had received benefit of the clergy (Kercher, 2003), or forcing non-incarcerated criminals to wear "badged" clothing (Higgs, 2004; Winter, 2013). As we show, the ticket of leave system, and the surveillance apparatus it generated and contributed to, was sophisticated and closely aligned with both centralized and highly devolved police functions (Lippert, 2016; McMullan, 1998; Pasquino, 1991). However, its genesis was only possible through overarching transportation policies, which required their own forms of surveillance to ensure convicts would be aptly deterred by the prospect of execution if they prematurely returned to Britain and Ireland.

Attainder, Civil Death, and English Criminal Law

Colonial transportation was only possible because of profound shifts in English governance between the fifteenth and the early nineteenth centuries. The emergence of scientific principles of causation, reason, and knowledge-based intervention (Hacking, 1990) produced a series of governmental technologies that enhanced the localized collection and consolidation of knowledge about individuals. Centralized analysis and storage processes enabled greater efficiency in the collection of taxes, which funded various public services established for the collective good, such as military and policing agencies (Higgs, 2004; McMullan, 1998). Early forms of mass population surveillance enabled states "to constitute their subjects and territories as 'governable'" (Rose-Redwood, 2006, p. 471). This was achieved through "the disciplinary strategy of 'individualisation'" that enhanced the state's ability to record various aspects of citizen conduct, including the use, transfer, and sale of land and localized forms of income generation, which gradually evolved into more sophisticated "biopolitical techniques of 'totalization'" that sought to "regulate and manage entire populations ... as objects of governance" (Rose-Redwood, 2006, p. 472).

Developing incrementally over several centuries, these governmentalities (Lippert, 2016) involved manual techniques of knowledge production, including the quantification, classification, cross-matching, and expert interpretation of information about people and their productive activities. This enabled economic, demographic, geographic, and behavioural trends to be identified and governed through informed, evidence-based processes. Examples of these governance technologies include the mapping of lands to improve agricultural yields, "the spatial practice of house numbering" to assist tax collection (Rose-Redwood, 2006, p. 470), and bookkeeping (Urton, 2009). These routines of data recording helped to foster more efficient centralized governance and audit procedures to fund English military and imperial expansion (Connell, 2007; Higgs, 2004, 2011).

By the eighteenth and nineteenth centuries these technologies were entrenched aspects of English statehood and governance that logically permeated the emerging administrative structures of the colonized world (Kalpagam, 2000). Adaptations of these processes in the colonies also fed back into imperial governance, often through intricate networks of British officials undertaking administrative duties in multiple colonial settlements (Laidlaw, 2005). However, the centralization of different forms of mass population data in England was also highly staggered. For example, sophisticated data about crime, criminals, and prisoners that were collected in local English parishes from the mid-seventeenth century garnered little national attention until the mid-nineteenth century (Shoemaker & Ward, 2017). Scientific approaches to understanding crime and criminality also contradicted the highly arbitrary methods of administering justice throughout Britain, which continued to invoke torture to signify the sovereign's vengeance as the ultimate victim of crime, or the use of discretionary pardons that were immune from independent judicial scrutiny, yet demonstrated the sovereign's good grace and benevolence (Foucault, 1977; Hay et al., 1975).

The principle of attainder operated alongside these developments, and provided legal legitimacy for British and Irish transportation policies. It reflects the state of flux between arbitrary generalized criminal punishments that could be tempered by the highly selective use of Crown pardons, and a more scientific approach to individualized sentencing that emerged during the mid- to late nineteenth century. Attainder reflects the Crown's power to exact vengeance by excluding felons from the realm. In the sixteenth and seventeenth centuries, it was used throughout Britain and Ireland to compel military service and increase centralized state wealth by enabling the confiscation of all property owned by people convicted of treason and a growing range of capital felonies (Hay et al., 1975; Morgan & Rushton, 2013). The most extreme form of forfeiture was death authorized in a parliamentary bill of attainder, which served as "a political weapon and efficient mechanism of economic and land dispossession levelled against enemies and insurrectionists" (Follis, 2014, p. 176). Bills of attainder enabled the Crown to automatically pronounce guilt and punishment against specified individuals or groups without the prospect of judicial review. They were widely criticized, particularly in colonial North America, where attainders were also employed and eventually prohibited at both federal and state levels under the 1787 Constitution (Steilen, 2016), because they were viewed as "contrary to the course of common law" for retroactively criminalizing "the past acts of individuals" while making no attempt to govern "the future conduct of the community" (Bills of Attainder, 1867, p. 81). Their arbitrariness mirrored the processes governing the automatic forfeiture of tainted property to the state in cases involving wrongful death or other forms of offending involving contraband goods and chattels (see Finkelstein, 1973).

The status of *civiliter mortus*, or civil death, was the main consequence of common law and statutory attainder. Civil death automatically applied when a death sentence for treason or an increasing range of capital felonies enacted between the sixteenth and the seventeenth century was commuted by the British Crown. There were three major consequences of civil death: the "forfeiture of property to the king; corruption of blood ... and the extinction of civil rights, [including] the incapacity to perform any legal function" authorized by the Crown (Dayan, 2011, p. 44). Civil death had important spatial and temporal aspects (Valverde, 2015) that removed a convict's legal personality while he or she remained on Crown territory. It prevented an individual's involvement in civic activity and completely removed any prospect of formal legal protection by the state. Corruption of blood stripped the felon's rights as citizen, and had intergenerational effects by removing the ability to own, inherit, or transmit property to one's descendants (Rhodes, 2010). As property was central to participation in legal and political affairs (Hay et al., 1975), "state-sanctioned degradation" created a form of "negative personhood" that was at once endorsed and propelled by the law's "focus on *personal identity*, [and] the terms by which personality is recognised, threatened or removed" (Dayan, 2011, p. xii, emphasis added). Thus, "[t]he person convicted of felony is alive in fact but dead in law" (Dayan, 2011, p. 4), and was effectively outcast from English and Irish communal, political, and propertied life. The civilly dead were unable to engage in financial transactions or sign contracts, and ultimately barred from any form of protection under Crown law. Royal pardons and transportation offered the only form of grace that enabled convicted and attainted British felons a modicum of legal and physical personhood. This conception of "civil death" arguably prefigured later forms of necropolitics (Mbembe, 2003; also refer to the discussion of necropolitics in the book's introduction) that became central to the elaboration of systems of colonial rule not only in the British Empire, but elsewhere as well (see Midori Ogasawara's examination of Japan's colonial surveillance systems in Northeastern China in chapter 7), albeit without the explicitly racializing dimensions entailed in those systems of rule.

Attainder, Property, and the Reception of English Law

The stigma of attaint was enforced through geographic exile. Those convicted of treason or felony and declared civilly dead had no option for a social existence aside from voluntary departure from Britain or Ireland. This is the main legal justification for penal transportation, initially to the United States, then to other colonial outposts such as Australia. A similar geographic logic also provided the legal foundation for England's imperial penal and economic expropriation of Australian land, with the legal classification of *terra nullius* enabling the entire continent to be governed by England without the need for a treaty with its Indigenous inhabitants (Edgely, 2010). In fact, unlike many of the 168,000 transported convicts who retained some basic rights in an extremely fragmented process of Australian reception of English common law (Neal, 1991), Indigenous Australians technically held no legal, human, or land rights under the *terra nullius* doctrine until formal citizenship was granted by a general referendum of Australian citizens in 1967 (Warren & Palmer, 2018). Therefore, Indigenous peoples occupied a vague subterranean space somewhere between civil and actual death, while transported convicts at least had some legal capacity to survive as people in the face of the stigma of their attaint and geographic exile from England.

Attainder created an "ambiguous relationship between English common law" and the emerging laws and practices throughout the empire (Follis, 2014, p. 179). Transportation was ultimately considered an act of mercy on behalf of the British Crown. It replaced a sentence of death with transportation for life or fixed periods of between seven and fourteen years (Kercher, 2003). However, although attainder could also be enforced in Australia, it was often impractical to do so rigidly. The legal status of attainder was generally articulated on dubious and inconsistent grounds that reflected "politics and the organisation of power" in both England and the emerging process of Australian colonial governance (Brown, 2002, p. 315). As with England's highly discretionary use of Crown pardons to validate transportation and populate the colonies, the transposition of common law of attainder into Australia was discretionary, selective, ad hoc, and lacked recourse to a coherent body of English or locally developed legal precedents. Multiple English and Australian common law adjudicators adopted quite divergent interpretations of the same problem that produced inconsistent, yet broadly compatible, ends in both jurisdictions.

A combination of fragmented manual record-keeping at various points in the transportation process, and time delays that prevented prompt verification of original conviction and sentencing records held in England, made it difficult for Australian authorities to determine which convicts were attainted for capital offences that had been commuted to transportation for life or a period of years. Also, few Australian convicts were physically incarcerated, which meant they intermingled routinely with free settlers and colonial administrators. Attainder was, however, of immense political and legal significance, as propertied free settlers, called Exclusives, sought to strictly enforce its hereditary impacts by excluding "those with the convict taint from positions of status and power," even though many "convicts, emancipists [former convicts] and their children" gradually accumulated considerable financial wealth (Neal, 1991, pp. 18–19). The reception of the attainder doctrine under Australian common law revealed the social and spatial ambiguities associated with demands for its strict application as a central element of the English procedure for implementing criminal punishment that were ultimately reconciled by local courts when determining the practical consequences of civil death in light of specific demographic, economic, social, and geographic requirements in each colony (Kercher, 2003).

The first New South Wales court ruling allowed a husband and wife who were convicted and transported to successfully sue for the value of personal property that was either lost or stolen during transit from England (Brown, 2002; Kercher, 1995; Neal, 1991). However, Australian courts also upheld prohibitions that prevented attainted convicts from acquiring property or exercising other positive legal rights, mainly to rectify the "deficient supply of free labour" from Britain and Ireland (Therry, 1863, p. 507). This enabled colonial administrators to use unpaid convict labour on early public infrastructure projects, the establishment of a private assignment system, and the fulfilment of other localized economic and land acquisition objectives that were ultimately overseen at a distance by the London Colonial Office and British Parliament (Laidlaw, 2005). These developments also served a pragmatic role in fostering penal discipline and self-discipline amongst convicts in an open prison where the majority of prisoners were not physically incarcerated.

The property of Australian convict labour was initially vested in England's Crown representatives. Colonial governors retained the ultimate right to deploy convicts on publicly administered road gangs (Atkinson, 1979; Karskens, 1986), in manufacturing industries and lumber yards (Robbins, 2000), or in public service roles that could not be filled by free settlers or bureaucratic appointments made in London. Many were appointed as police constables (Neal, 1991) or assigned to free settlers and deployed as agricultural labourers, a task that was very loosely supervised. Women were commonly assigned to supervised government work in clothing mills or as the domestic servants of wealthy landowners (Daniels, 1998; McCabe, 1999). A small proportion of convicts who reoffended in the colonies were assigned to frontier townships for "rehabilitative banishment" (Roberts, 2005, pp. 103–4) to encourage self-discipline through geographic isolation, although this could provide opportunities for those with skills as tradesmen, government clerks, or butchers to fill labour shortages in these regions. Private masters acquired qualified proprietary rights over convicts and their labour for a specified period of years that were ultimately subject to completion of the convict's sentence and possible return to Britain or Ireland. The Crown reserved limited scope to revoke these private licences under common law if the convict was granted a ticket of leave or formal pardon, when there was evidence of physical brutality by the master, or if the prisoner was violent or otherwise unsuited to the assignment (Kercher, 1995; Neal, 1991). Although a certain degree of geographic liberty was necessary to develop early Australian infrastructure and commercial projects, local authorities were careful to distinguish forced government labour and private assignment from indentured servitude or slavery (Neal, 1991). However, English political reformers critical of transportation and the assignment system were more than willing to make this connection (Hirst, 2008).

Many early colonial inheritance disputes, such as *Septon v. Cobcroft* (1833), required Australian courts to examine the evidentiary weight of official documents that often failed to accurately record the convict's personal details or original English offence. In this case, the official shipping indent indicated the convict was sentenced to transportation for life, but there was no further statement of whether this was a commuted sentence of death that imposed attainder. The New South Wales Supreme Court allowed the jury to infer that transportation for life was a sentence for a capital felony (Kercher, 2003), given the obvious gaps in available manual records and prospect of an eighteen-month delay in obtaining the original conviction record held in London.

Attaint was most easily established if the felon was sentenced to death for a crime committed in the colonies (Kercher, 2003). However, such cases could also undermine the administration of colonial justice if all legacies of civil death that applied under English law were strictly incorporated into Australian common law. Thus, two suspects accused of murder at the Port Macquarie penal settlement in New South Wales unsuccessfully argued that the only witnesses were unable to testify because they were attainted felons and thus *civiliter mortus*. Nevertheless, the New South Wales Supreme Court endorsed their right to testify, as it was in the interests of "the due administration of justice ... [to] lay down a rule which shall be applicable to the state and condition of the community in which we live" (*R. v. Gardener and Yems*, 1829, p. 49; see also Kercher, 2003). This outcome would have been unlikely if the same scenario arose before an English court. However, subsequent interpretations of the same rule met different objectives depending on the nature of the case and political sentiment at the time. As Brown (2002) notes, "various ideologies and practices of exclusion are historically specific and variable ... rediscovered ... refashioned and recreated, clothed in a popular legitimacy ... [and] stitched up in contemporary sentiments of desert, forfeiture, impurity and exclusion" (p. 314). The discretion exercised by early Australian judges was a spatial and temporal necessity of colonial governance. This enabled the consequences of attainder and civil death imposed under English law to be recognized according to local needs and conceptions of justice (Kercher, 2003). Different scales of legal, bureaucratic, and judicial administration in both England and colonial Australia (Valverde, 2015) could thus alternately remove or restore a felon's civic and citizenship rights, either by adopting a uniform or generalized approach to the attainder doctrine, or through novel constructions that reflected an individual's personal circumstances, behaviour, geographic isolation, and various other practical requirements in an emerging colonial state.

Tickets of Leave and the Colonial Surveillant Assemblage

While attainder had potential intergenerational legal and political consequences within England and in the major penal settlements of New South Wales and Tasmania, the overall number of Australian convicts who were attainted is unknown. However, continuous surveillance of all classes of convict was necessary to reinforce the penal character of their geographic exile from Britain and Ireland. Personal information about each convict's background and conduct before and during the term of transportation was collected at regular intervals by private assignees and various public agencies. These data flowed in numerous directions, and varying forms, within and between the penal colonies, and ultimately back to England (Winter, 2013). Governors were obliged to prepare annual Returns, called Blue Books, which were lodged with the London Colonial Office and tabled in the British Parliament. These documents consolidated financial and social data from colonial census, police, court, educational, agricultural, and commercial records. As expanded reporting criteria were demanded from London and data analysis became more sophisticated over the course of the nineteenth century, these consolidated dossiers led to detailed comparisons of common economic, social, and governance metrics between the colonies that were collected locally and transferred through discrete layers or scales of governance for different forms of analysis (Higgs, 2004, 2011; Valverde, 2015). This important part of Australia's penal surveillant assemblage (Haggerty & Ericson, 2000) had the sophistication of a modern digitized "police state ... fastening on the dossiers kept on every convict" (Neal, 1991, p. 31).

Individualized convict dossiers were routinely compiled and distributed between networks of private and government officials. When consolidated, this information provided the basis for correcting errors in existing records, refining the content of rules, regulations, and other modes of formal governance, or developing more efficient and streamlined approaches to convict management. This process commenced once each vessel arrived in New South Wales or Van Diemen's Land. Government officials boarded the vessel to conduct an initial muster before any convicts or crew could disembark. This important disciplinary technique was also a self-legitimating form of "surveillance of surveillance" (Palmer & Warren, 2016, p. 180), aimed at identifying "any error that may have crept into the indents and assignments of the convicts that are transmitted from the Secretary of State's office to the governor of New South Wales" (House of Commons, 1822, p. 14). In full view of the captain, the ship's crew, and the accompanying surgeon, each convict was

> asked his [*sic*] name, the time and place of his trial, his sentence, native place, age, trade and occupation; and the answers are compared (and corrected if necessary) by the description in the indents and in the lists transmitted from the hulks [in England]. After ascertaining the height of each convict by actual admeasurement, and registering it in several columns, as well as the colour of the hair, eyes, the complexion or any particular mark that may tend to establish the identity of each convict, an inquiry is made respecting the treatment that each has received during the passage, whether he has received his full ratio of provisions; whether he has any complaint to make against the captain, his officers and crew; and lastly, whether he [*sic*] any bodily ailment or infirmity. A further inquiry is made of the surgeon, respecting the conduct of each convict during the passage, and whether he has any bodily infirmity that may prevent him from being actively employed. (House of Commons, 1822, pp. 13–14)

Musters could take several hours or even days, and represented the start of an ongoing series of weekly musters, headcounts, and checks during government or private assignments that continually monitored and matched the convict's identity, behaviour, and labour productivity with data from the first muster and the vessel's original indent. Weekly public musters were coordinated by a police constable in each district, and were followed by a compulsory evening church service for those on non-government assignments. Absentees and those deemed to have engaged in "disorderly conduct" or of "uncleanly appearance" would be reported to a magistrate the following day (House of Commons, 1822, p. 78). These surveillance routines continually re-verified the same information recorded in the initial muster, and when consolidated would ultimately determine eligibility for a ticket of leave. Data routinely compiled by local officials was brought to the office of the Chief Superintendent of Convicts, where it would be cross-matched and stored with previous records. Centralized storage also facilitated intercolonial surveillance, which was vital to monitoring people convicted of an offence in Australia who required secondary transportation to a more secure and geographically remote penal settlement.

In exceptional cases, an official was sent from London to comprehensively audit these processes of convict administration based on observations and detailed scrutiny of available police, court, and administrative records. This was the case with the 1822 Bigge Report, a British Royal Commission that undertook a systematic review of Australian colonial governance arrangements. However, most routine forms of surveillance sought to oversee convict discipline or apprehend escapees, and, as with most surveillance systems, the recording practices of government officials and private assignees. The so-called runaway lists provided a supplementary surveillance mechanism that was a sign of convict ill-discipline, and would invariably block a ticket of leave or conditional pardon. These lists were consolidated at the office of the Chief Superintendent of Convicts from regular audits prepared by local "commandants, magistrates, and the superintendents and overseers of state institutions and work gangs" (Roberts, 2005, p. 99). The lists were then published on a weekly basis in the *Sydney Gazette* and documented the names of missing convicts as well as various details that supplemented the original shipping and muster records, including "each individual's ship of arrival, number of weeks missing, age and native place, physical description and where/who absent from" (Roberts, 2005, p. 99).

Most absentees were convicts undertaking fleeting recreational activities who voluntarily returned to their government or private assignments. However, some were actively protesting brutal physical discipline, reduced rations, the failure to provide appropriate clothing or blankets, or the inadequate enforcement of protective regulations aimed at preventing labour exploitation by private masters and government officials (Neal, 1991). The Bigge Report considered a ticket of leave to be a privilege that equated to private assignment. However, in some regions multiple industries were operated by private entrepreneurs who would also hold official government or magisterial posts (Roberts, 2005). Thus, accountability for questionable forms of convict treatment was limited, although the Blue Books enabled "some central checks to be placed on colonial jobbery and corruption" from London (Laidlaw, 2005, p. 171). The runaway lists also promoted the non-governmental surveillance of escapees through a reward or bounty system for their successful apprehension and return. Anomalies in local recording practices, though, meant some absentee records were not forwarded for consolidation and publication, while "many cases did not warrant reporting" at all (Roberts, 2005, p. 112).

While tickets were issued at the discretion of the Governor, there was an expectation they would be granted for continuous good behaviour after two-thirds of a fixed sentence, or after ten to twelve years had been served on a life sentence. As with a conditional pardon, the Governor's decision was based on recommendations from government officials or private assignees, and written records demonstrating the convict had the means of independent support and no record of offending, dishonesty, or associations with people of bad character. Exemplary government service, including active contributions to exploring the nation's frontier, could be grounds for a favourable decision. The original English offence would have limited bearing on the outcome, but marriage was usually viewed favourably for men, and was virtually a prerequisite for women (Daniels, 1998). These discretionary executive decisions were not subject to judicial review, with negative decisions largely driven by labour demands in specific areas or industries rather than merit.

Local demands for physical labour in the construction of roads, early rail networks, and other forms of essential infrastructure shaped the availability of tickets of leave, as well as the gender dynamics of surveillance in both public and private employment assignments. Limited work opportunities had impacts on both skilled and unskilled male convicts, while women, who were generally first-time offenders convicted of theft or robbery in Britain and Ireland and not attainted, were generally confined to private domestic assignments. In some cases, women, married couples, or people from privileged social and educational backgrounds who demonstrated a capacity for self-support would be granted tickets of leave as soon as they arrived in Australia.

Many women experiencing violent or unpalatable private assignments would abscond or deliberately offend to seek more suitable reassignments, after mandatory confinement in a female factory. These closed environments served as arrival depots for unaccompanied women awaiting assignment, but also acted as important places of refuge from the violence, abuse, and discipline of outside life (Daniels, 1998). As with places of secondary punishment in Macquarie Harbour (Tasmania), Moreton Bay (New South Wales), and Norfolk Island, which housed convicts who reoffended and were sentenced to additional punishments by local courts, female factories promoted discipline through intensive observation, categorization, medical and behavioural record-keeping, and physical separation from family, friendship networks, and each other. This rigid institutional surveillance classified and labelled women according to observed standards of behaviour and deportment to determine suitability for release into respectable assignments. Inscribed clothing and head shaving were used to demarcate the incorrigible from those nearing release, while routine inspections of work patterns, general behaviour, and demeanour were recorded and formed part of the official dossier on each woman. For the emerging state, private assignment or marriage was the ultimate objective to reduce public expenditure on the female factory system (House of Commons, 1822; McCabe, 1999). Many women simply vanished from public records after marriage, indicating its importance as a disciplinary mechanism that could help to avert official labelling as an incorrigible, which was usually imposed for public drunkenness or suspected prostitution (Daniels, 1998).

A ticket of leave was an incentive for convict self-discipline that provided some respite from the weekly routines of surveillance experienced by unticketed convicts. Ticket of leave musters were conducted in each district on a monthly basis to authenticate the convict's work, behaviour, and associations that were specified on the ticket (Aislabie, 1857; Doulman & Lee, 2008). Figure 5.1 presents the standard front side of a ticket of leave, which resembles a contract between the convict and the colonial state, and contained basic identity information routinely collected during the convict musters. These documents generated new forms of surveillance that "endorsed internal movement between districts in the colony" (Doulman & Lee, 2008, p. 12) and required annual renewal to verify compliance until the full sentence was served, or a full pardon was issued by the Governor. The Bigge Report indicates that between 1810 and 1820, 2,319 tickets of leave were granted in New South Wales, compared with 1,367 conditional pardons, a ratio of almost 2:1 (House of Commons, 1822, p. 120). In some cases, new ticket holders could obtain small land grants and minimal salaries from their district's police fund to cover the costs of establishing qualified independence from government or private assignments (House of Commons, 1822). Tickets could also be revoked for the full duration of the original sentence at the Governor's pleasure for proof of offensive and improper conduct, fraudulent behaviour, any formally recorded criminal charges, or failure to comply with mandatory reporting requirements contained on the ticket, which included notifying police of a change of work or place of residence.

Tickets of leave were important evidence in colonial legal disputes where an attainted party or a presumed heir was attempting to assert a property right (Kercher, 2003), or in criminal trials to establish defences involving good character. For example, *R. v. Kelly* (1841) involved the alleged murder of an Indigenous youth by a ticket-of-leaver defending his home, which was provided by an agricultural company, during a raid by "a large tribe of blacks ... with bad intention" (Sydney Gazette, 1841, p. 2). Favourable testimony from the company's bailiff affirmed "the blacks have stolen maize, and are a nuisance to the settlement," but depicted Kelly as an honest worker who had never caused harm (*R. v. Kelly*, 1841, n.p.). The jury convicted Kelly of murder, but the sympathetic testimony and his unblemished record as a ticket-of-leaver led

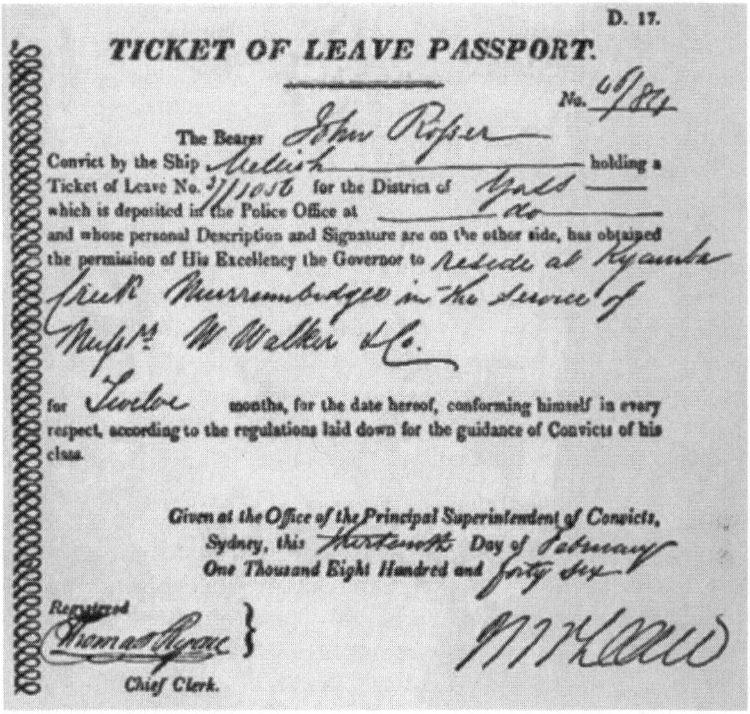

Figure 5.1 Australian ticket of leave passport, 1846. https://commons.wikimedia
.org/wiki/File:Ticket_of_Leave_Passport.jpg?uselang=en-gb.

to a recommendation from the judge that the death sentence be commuted to a term of imprisonment, although the precise outcome in this
case remains unknown. However, private and government masters would
often withhold a favourable recommendation for a ticket of leave, particularly for convicts with specialized work skills, to ensure productivity
in certain government and private industries was maintained (Robbins,
2000). The Bigge Report noted, and criticized, the inconsistency of this
practice, and recommended that lists of all tickets granted and withdrawn
should be communicated to and published in London to add transparency to this discretionary process (House of Commons, 1822).

As the nature of colonial governance and convict surveillance
matured, a "marks system" (Morris, 2002; White, 1976) helped temper
the discretionary nature of decisions to grant or deny tickets of leave.
This intricate surveillance system is largely attributed Alexander Maconochie during his service as commandant of the Norfolk Island penal

settlement, although there is clear evidence it was routinely adopted much earlier in the female factories during the 1820s (Daniels, 1998). Informed by a mutual obligation philosophy, the marks system determined eligibility for a ticket of leave based on the detailed recording of labour productivity and good behaviour, rather than the length of time served by the convict. This required intricate surveillance of a convict's work patterns whilst in relative confinement, and the itemization of rewards given to "the prisoner for his [*sic*] industry and behaviour and withdrawing the required equally precisely for misbehaviour and sloth" (Morris, 2002, p. 179). Thus, "by the accumulation of 'marks' measuring progress toward completion of the allotted work, the prisoner should gain improved living conditions including a 'ticket-of-leave' so that his work could continue in the community rather than in prison" (Morris, 2002, pp. 180–1). The marks system was a scientifically measured form of individualized surveillance that determined eligibility for a graded system of tickets of leave on Norfolk Island, then to specified locations in Sydney, then to ultimate freedom on the expiration of a fixed sentence. However, this method of expanding a convict's citizenship rights (Brown, 2002, 2008) for good conduct and productivity was considered antithetical to the stigma of the English law of attaint according Royal Commissioner Bigge, who reported to British Parliament that

> instances have occurred in which convicts holding tickets of leave have held and purchased land in New South Wales, and that they have been admitted to maintain actions in the courts of justice, although in strictness of law, they are disabled from so doing, by reason of their conviction ... I do not conceive it to be necessary in the present state of the colony, to encourage this class of convicts to acquire it [property] more expeditiously than they are disposed to do, or to assume any other relation than that of tenants, if they cultivate land, or in that of labourers working for their own profit. (House of Commons, 1822, p. 171)

Conclusion

Although brief, wide-reaching, and open to ongoing development and critique through further archival research, this chapter reveals an intent behind the intricate and multilayered colonial surveillant assemblage that is highly discretionary, "paternalistic, and intimate" (Lippert, 2016, p. 85). Our origin story identifies key aspects of a colonial surveillance apparatus that was a strategic and multilayered tool to instil discipline, labour productivity, and law-abiding behaviour amongst a flock of convicted British and Irish felons, many of whom were attainted and civilly dead. Various sites of discretionary executive decision making enabled qualified freedom from colonial surveillance routines, in a multilayered penal bureaucracy sustained by intricate domestic and transnational data flows. This penal governmentality conflated prisoner integration with productive labour (Robbins, 2000), and conditional freedom with relaxed surveillance. It also generated periodic adaptations of English law, as the consequences of attaint drifted in and out of Australia's political consciousness and early colonial legal precedent (Brown, 2002; Kercher, 2003).

As individual and collective recording criteria became more formalized and detailed, convict administration could be seen as a more rational and evidence-based biopolitical form of colonial governance. However, this also skewed the focus of surveillance to achieve new discretionary social and disciplinary objectives. The Bigge Report shows how independent auditing of colonial governance might have tightened the fragmented reception of England's attainder laws under Australian common law, though more evidence is needed to determine whether and how data obtained at various sites of the penal surveillant assemblage were ultimately used (Shoemaker & Ward, 2017). There are also important questions regarding whether these surveillance routines actually provided greater opportunities for convicts to integrate into Australian or British society. Further, origin stories of Indigenous non-personhood in Australian law and life are legion, as these communities were largely ignored by this surveillance apparatus, and consequently shunned by persistent regulatory abuse, forced displacement from traditional lands, or simply abandoned by overt governmental neglect. Indigenous experience reflected a necropolitical social death that, as noted earlier, shared some characteristics with civil death, but was also quite distinctive (Mbembe, 2003).

Convicts received a modicum of accountability from London that was largely viewed in terms of metrics, tables, grids, audits, and economic outputs. These data-driven governmentalities were often geographically and temporally distanced from day-to-day colonial administration. In an interesting twist, the very surveillance processes needed to administer social sorting in an open prison were lacking when tickets of leave were adopted in England as transportation was incrementally abolished in New South Wales in 1853, Tasmania in 1857, and Western Australia in 1868. This coincided with the erosion of the attainder doctrine in England, as the "presumption ... that every man you meet is a free man" began to emerge (Aislabie, 1857, p. 14). However, the intensive "police surveillance, tests of reformation, the preparation of licensees for release ... and the establishment of a central repository of criminal records"

(Bartrip, 1981, p. 165) that were entrenched and uniquely southern criminological elements of Australian convict governance (Carrington & Hogg, 2017; Carrington, Hogg, & Sozzo, 2016) took many decades to develop in England, although were comparatively more robust in Ireland. This encouraged English criminals who were released on licence to actively conceal their identities and "roguery" (Aislabie, 1857, p. 9) to readily evade the piecemeal forms of police surveillance (Murray, 1857). Although the prospect of relatively ubiquitous convict surveillance was normalized in colonial Australia, it remains debatable whether these processes tempered the "absolute control and repression of individuality" of the modern prison environment (Morris, 2002, p. 181), given that they sanctioned various forms of social sorting, geographic exclusion, and qualified rights associated with attainder and civil death.

ACKNOWLEDGMENTS

We thank Angela Daly, Raewyn Connell, Emily van der Meulen, and Rob Heynen for important suggestions that were central to the arguments we raise in this chapter.

REFERENCES

Aislabie, W. I. (1857). *The ticket-of-leave system, in Australia and in England.* London, UK: Effingham Wilson, Royal Exchange.

Amoore, L. (2008). Governing by identity. In C. Bennett & D. Lyon (Eds.), *Playing the identity card: Surveillance, security and identification in global perspective* (pp. 21–36). Milton Park, UK: Routledge.

Atkinson, A. (1979). Four patterns of convict protest. *Labour History, 37,* 28–51.

Bartrip, P. W. J. (1981). Public opinion and law enforcement: The ticket-of-leave scares in mid-Victorian Britain. In V. Bailey (Ed.), *Policing and punishment in nineteenth century Britain* (pp. 150–81). London, UK: Routledge.

Beckett, K., & Murakawa, N. (2012). Mapping the shadow carceral state: Towards an institutionally capacious approach to punishment. *Theoretical Criminology, 16*(2), 221–44.

Bennett, C., & Lyon, D. (Eds.). (2008). *Playing the identity card: Surveillance, security and identification in global perspective.* Milton Park, UK: Routledge.

Bills of Attainder. (1867). *The Western Jurist,* vol. 1. Des Moines, IA: Mills & Company, Steam Printers.

Breckenridge, K. (2014). *Biometric state: The global politics of identification and surveillance in South Africa, 1850 to the present.* Cambridge, UK: Cambridge University Press.

Brown, D. (2002). Prisoners as citizens. In D. Brown & M. Wilkie (Eds.), *Prisoners as citizens: Human rights in Australian prisons* (pp. 308–25). Annandale, NSW: Federation Press.

Brown, D. (2008). Giving voice: The prisoner and discursive citizenship. In T. Anthony & C. Cunneen (Eds.), *The critical criminology companion* (pp. 228–39). Annandale, NSW: Hawkins Press.

Caplan, J., & Torpey, J. (Eds.) (2001). *Documenting individual identity: The development of state practices in the modern world.* Princeton, NJ: Princeton University Press.

Carrington, K., & Hogg, R. (2017). Deconstructing criminology's origin stories. *Asian Criminology, 12*(3), 181–97.

Carrington, K., Hogg, R., & Sozzo, M. (2016). Southern criminology. *British Journal of Criminology, 56*(1), 1–20.

Chan, J. (1991). Decarceration and imprisonment in New South Wales: A historical analysis of early release. *University of New South Wales Law Journal, 13*(2), 393–416.

Chin, G. J. (2012). The new civil death: Rethinking punishment in the era of mass conviction. *University of Pennsylvania Law Review, 160*(6), 1789–833.

Colgate Love, M., Roberts, J., & Klingele, C. (2016). *Collateral consequences of criminal conviction: Law, policy and practice.* Eagan, MN: Thomson Reuters.

Connell, R. (2007). *Southern theory: The global dynamics of knowledge in social science.* St Leonards, NSW: Allen & Unwin.

Daniels, K. (1998). *Convict women.* St Leonards, NSW: Allen and Unwin.

Dayan, C. (2011). *The law is a white dog: How legal rituals make and unmake persons.* Princeton, NJ: Princeton University Press.

Doulman, J., & Lee, D. (2008). *Every assistance & protection: A history of the Australian passport.* Annandale, NSW: Federation Press and Australian Government, Department of Foreign Affairs and Trade.

Edgely, M. (2010). Criminals and (second-class) citizenship: Twenty-first century attainder? *Griffith Law Review, 19*(3), 403–37.

Finkelstein, J. J. (1973). The goring ox: Some historical perspectives on deodands, forfeitures, wrongful death and the Western notion of sovereignty. *Temple Law Quarterly, 46*(2), 169–290.

Follis, L. (2014). Of friendless and stained men: Grafting medieval sanctions onto modern democratic law. In S. Glanert (Ed.), *Comparative law: Engaging translation* (pp. 173–90). Milton Park, UK, and New York, NY: Routledge-Glasshouse.

Foucault, M. (1977). *Discipline and punish: The birth of the prison.* New York, NY: Vintage/Random House.

Hacking, I. (1990). *The taming of chance.* Cambridge, UK: Cambridge University Press.

Haggerty, K. D., & Ericson, R. V. (2000). The surveillant assemblage. *British Journal of Sociology, 51*(4), 605–22.

Hay, D., Linebaugh, P., Rule, J. G., Thompson, E. P., & Winslow, C. (1975). *Albion's fatal tree: Crime and society in eighteenth-century England*. London, UK: Verso.

Higgs, E. (2004). *The information state in England: The central collection of information on citizens since 1500*. Basingstoke, UK: Palgrave Macmillan.

Higgs, E. (2011). *Identifying the English: A history of personal identification 1500 to the present*. London, UK: Continuum International.

Hirst, J. (2008). *Freedom on the fatal shore: Australia's first colony*. Melbourne, VIC: Black Inc.

House of Commons. (1822). *Report of the Commissioner of Inquiry into the State of the Colony of New South Wales* (Bigge Report). House of Commons, UK: London. Retrieved from http://gutenberg.net.au/ebooks13/1300181h.html#ch-10

Hughes, R. (2003). *The fatal shore*. London, UK: Vintage.

Kalpagam, U. (2000). Colonial governmentality and the "economy." *Economy and Society, 29*(3), 418–38.

Karskens, G. (1986). Defiance, deference and diligence: Three views of convicts in New South Wales road gangs. *Australian Journal of Historical Archaeology, 4*, 17–28.

Kercher, B. (1995). *An unruly child: A history of law in Australia*. St Leonards, NSW: Allen & Unwin.

Kercher, B. (2003). Perish or prosper: The law and convict transportation in the British Empire, 1700–1850. *Law and History Review, 21*(3), 527–84.

Laidlaw, Z. (2005). *Colonial connections 1815–1845: Patronage, the information revolution and colonial government*. Manchester, UK: Manchester University Press.

Lippert, R. K. (2016). Governmentality analytics and human rights in criminology. In L. Weber, E. Fishwick, & M. Marmo (Eds.), *The Routledge international handbook of criminology and human rights* (pp. 80–90). Milton Park, UK: Routledge.

Mbembe, A. (2003). Necropolitics. *Public Culture, 15*(1), 11–40.

McCabe, K. (1999). Assignment of female convicts on the Hunter River 1831–1840. *Australian Historical Studies, 29*(113), 286–302.

McMullan, J. L. (1998). Social surveillance and the rise of the "police machine." *Theoretical Criminology, 2*(1), 93–117.

Morgan, G., & Rushton, P. (2004). *Eighteenth-century criminal transportation: The formation of the criminal Atlantic*. Basingstoke, UK: Palgrave Macmillan.

Morgan, G., & Rushton, P. (2013). *Banishment in the early Atlantic world: Convicts, rebels and slaves*. London, UK: Bloomsbury.

Morgan, N. (1992). Parole and sentencing in Western Australia. *Western Australian Law Review, 22*(1), 94–122.

Morris, N. (2002). *Maconochie's gentlemen: The story of Norfolk Island and the roots of modern prison reform*. Oxford, UK: Oxford University Press.

Murray, P. J. (1857). *Not so bad as they seem. The transportation, ticket-of-leave and penal servitude questions, plainly stated, and argued on facts and figures, with some observations on the principles of prevention: In a letter addressed to Matthew Davenport Hill, Esq, QC, Recorder of Birmingham.* London, UK: W. & F. G. Cash; Dublin, IE: W. B. Kelly.

Neal, D. (1991). *The rule of law in a penal colony: Law and power in early New South Wales.* Cambridge, MA: Cambridge University Press.

Palmer, D., & Warren, I. (2014). The pursuit of exclusion through zonal banning. *Australian and New Zealand Journal of Criminology, 47*(3), 429–46.

Palmer, D., & Warren, I. (2016). The "security of security": Making up the Australian intelligence community 1975–2015. In R. K. Lippert, K. Walby, I. Warren & D. Palmer (Eds.), *National security, surveillance and terror: Canada and Australia in comparative perspective* (pp. 177–98). Cham, CH: Palgrave Macmillan-Springer.

Pasquino, P. (1991). Theatrum politicum. The geneaology of capital – Police and the state of prosperity. In G. Burchell, C. Gordon & P. Miller (Eds.), *The Foucault effect: Studies in governmentality* (pp. 105–18). Chicago, IL: The University of Chicago Press.

Rhodes, A-M. (2010). Blood and behavior. *ACTEC Law Journal, 36*(1), 143–77.

Robbins, W. M. (2000). The lumber yards: A case study in the management of convict labor, 1788–1832. *Labor History, 79*(2), 141–61.

Roberts, D. A. (2005). A "change of place": Illegal movements on the Bathurst frontier, 1822–1825. *Journal of Australian Colonial History, 7,* 97–112.

Rose-Redwood, R. S. (2006). Governmentality, geography and the geo-coded world. *Progress in Human Geography, 30*(4), 469–86.

Ruggerio, K. (2001). Fingerprinting and the Argentine plan for universal identification in the late nineteenth and early twentieth centuries. In J. Caplan, & J. Torpey (Eds.), *Documenting individual identity: The development of state practices in the modern world* (pp. 184–96). Princeton, NJ: Princeton University Press.

Sengoopta, C. (2003). *Imprint of the Raj: How fingerprinting was born in colonial India.* London, UK: Pan MacMillan.

Shaw, A. G. L. (1953). The origins of the probation system in Van Diemen's land. *Historical Studies: Australia and New Zealand, 6*(21), 16–28.

Shoemaker, R., & Ward, R. (2017). Understanding the criminal: Record-keeping, statistics and the early history of criminology in England. *British Journal of Criminology, 57*(6), 1442–61.

Smith, A. E. (1947). *Colonists in bondage: White servitude and convict labor in America 1607–1776.* Chapel Hill, NC: The University of North Carolina Press.

Steilen, M. (2016). Bills of attainder. *Houston Law Review, 53*(3), 767–907.

Sydney Gazette. (1841, September 14). Untitled and unauthored case summary. p. 2.

Therry, R. (1863). *Reminiscences of thirty years' residence in New South Wales and Victoria. With a supplementary chapter on transportation and the ticket-of-leave system.* London, UK: Sampson Low, Son and Co.

Urton, G. (2009). Sin, confession, and the arts of book- and cord-keeping: An intercontinental and transcultural exploration of accounting and governmentality. *Comparative Studies in Society and History, 51*(4), 801–31.

Valverde, M. (2015). *Chronotopes of law: Jurisdiction, scale and governance.* Milton Park, UK: Routledge-Glasshouse.

Warren, I., & Palmer, D. (2018). Southern criminology, zonal banning and the language of urban crime prevention. In K. Carrington, R. Hogg, J. Scott & M. Sozzo (Eds.), *The Palgrave handbook of criminology and the global south* (pp. 183–201). Cham, CH: Palgrave MacMillan/Springer.

White, S. (1976). Alexander Machonochie and the development of parole. *The Journal of Criminal Law and Criminology, 67*(1), 72–88.

Winter, S. (2013). The global versus the local: Modeling the British system of convict transportation after 1830. In M. C. Beaudry & T. G. Parno (Eds.), *Archaeologies of mobility and movement: Contributions to global historical archaeology,* vol. 35 (pp. 133–49). New York, NY: Springer.

Cases Cited

R. v. Gardener and Yems, Supreme Court of New South Wales, 6 April 1829.

R. v. Kelly, Supreme Court of New South Wales, 8 September 1841.

Septon v. Cobcroft, Supreme Court of New South Wales, 30 October 1833.

6 Controlling Transnational Asian Mobilities: A Comparison of Documentary Systems in Australia and South Africa, 1890s to 1940s

UMA DHUPELIA-MESTHRIE AND MARGARET ALLEN

The apartheid state of South Africa (1948–1993) placed its residents and especially its opponents under total surveillance. As Dlamini (2014) has written, even the most "mundane" and "trivial details" were gathered (pp. 121–2). Such a state required the use of many technologies, bureaucratic dedication, and the expansion of security systems. We are concerned here with an earlier South Africa, from the 1890s to the 1940s, where there was *selective* surveillance of Asians. South Africa was not unique in this ambition and thus this chapter draws on comparisons with Australia. In both these contexts, surveillance was aimed at controlling Asian mobility into and outside these countries, and authorities were only interested in particular aspects of their lives. The chapter highlights the role that the transnational mobility patterns of Asians had in producing this surveillance as well as the technologies that were utilized and their effectiveness.

The period with which we are concerned (the later nineteenth through to the early/mid-twentieth century) was marked by considerable movement within the British Empire but, as Hyslop (2015) has noted, mobility was "not everywhere, and not for everyone" (p. 264). Motivated by a desire to close off entry to Indian immigrants, the British colony of Natal passed the Immigration Restriction Act (1897), which imposed a writing test in a European language on new immigrants. This had a ripple effect across South Africa – the Cape Colony followed in 1902 as did the newly conquered Transvaal Colony in 1908. When the South African colonies united to form the Union of South Africa in 1910, the writing test was retained in the Immigration Regulation Act (1913). In addition, regulations under this Act constituted all "non-Europeans" (except for contracted African mine-workers) as prohibited immigrants (Perbedy, 2009).

McKeown (2008) has urged us to see Asian exclusions in this period as a global pattern. The Natal system influenced the distant colonies of

Australasia in the 1890s, where dictation tests (as the writing tests were known there) were incorporated into immigration legislation by New South Wales, Western Australia, Tasmania, and New Zealand. When the Commonwealth of Australia was constituted in 1901, among its first legislative acts was the Immigration Restriction Act (1901). The targets of the dictation test were Asians, Africans, and Pacific Islanders. These linked exclusionary immigration histories of Australia and South Africa have been well told (Huttenback, 1973; Martens, 2006). Lake (2005) added a further transnational dimension by pointing to the prior influence on Natal of writing tests in the American South for American Blacks seeking voter registration. Influences wove their way in multiple directions as borne out by the close emulation in the title of the Cape Colony's Immigration Act (1902) of its Australian predecessor. There were local particularities too. The Cape Colony went a step further than Australia and the other South African colonies by passing the Chinese Exclusion Act (1904), which specifically excluded the further immigration of all Chinese as a race, while also requiring the registration of every Chinese person already in the colony.

This chapter compares the paper systems developed by Australia and South Africa to control the mobility of the Asian population already resident in both countries at the time of the passage of the exclusionary legislation, focusing also on the visual technologies used to document bodily characteristics and markers. Comparisons are made between systems in Natal and the Cape Colony prior to and after South African Union in 1910 on the one hand, and immigration restriction in the Australian colonies both pre-federation and after the establishment of the Federated Commonwealth of Australia in 1901 on the other. Both countries conceded the right of Asian immigrants to remain in the country but placed restrictions on how they could move between these territories and their home countries. What Robertson (2010) refers to as "documentary regime[s] of verification" evolved in response (pp. 10–11). The issue of a domicile certificate or the later South African permit/certificate of identity and Australian Certificate Exempting from Dictation Test were preceded by application forms through which the newly established immigration bureaucracies accumulated information. Central to this surveillance was the ability of the state to identify the individual on the move and establish his or her right both to travel and to return. This chapter thus sits at the intersection of identification and surveillance studies.

Our chapter responds to the call made by Caplan and Higgs (2013) to examine the paperwork materializing out of identification projects, and that of Hull (2012), who emphasizes that we "look at documents rather than through them" (p. 12). Contemporary surveillance theory developed by Haggerty and Ericson (2000) offers us insights into surveillance by both state and non-state actors and the development of a surveillant assemblage, where information originates from "scattered centres of calculation" (p. 613). In contemporary surveillance, the many components of the physical body of the surveillant subject are rendered into "a data double" (Haggerty & Ericson, 2000, p. 613), however, our concerns here are with another century where information gathered by the immigration bureaucracy was linked with information from authorities such as the police as well as from character referees. The physical presence of the body was mandatory in this surveillance – the body had to be seen. The centrality of the body to regimes of surveillance is a theme that runs through this book, linking eugenics strategies and conceptions of "risky bodies" (Midori Ogasawara, chapter 7) to long histories of biometric police practices (Matthew Ferguson, Justin Piché, & Kevin Walby, chapter 12). We thus adapt the idea of a data double by referring to the creation of a paper double through the different practices by which migration was regulated: first the domicile certificates adopted in different ways in both South Africa and Australia, and then the South African permit/certificate of identity and the Australian Certificate Exempting from Dictation Test.

In each case these documents relied on visual technologies in recording bodily markers and characteristics deemed significant for identification. Haggerty and Ericson (2000) have argued that in contemporary society there is a "levelling effect on hierarchies" as all people are drawn into surveillance systems (p. 614). In the case of immigration surveillance, however, this levelling effect was not evident, and we find most applicable to our work Lyon's (2003) idea of surveillance as involving a "social sorting" in which "deep discrimination occurs" (p. 1). Browne's (2012) elaboration of "racializing surveillance" (p. 72) is also particularly instructive for us. She refers to "those moments when enactments of surveillance reify boundaries and borders along racial lines, and where the outcome is often discriminatory treatment" (p. 72). We seek, then, to provide a historical case study that examines how discrimination, sorting, and racializing manifested themselves. Lyon's (2007) argument that surveillance systems involve a "web of relations" (p. 20), including power relations as well as "strategies and subterfuges of those under surveillance" (p. 27), is taken up in the final section below.

Resident Asians and the Domicile Certificate

Australia and South Africa implemented similar surveillance practices, but in quite different demographic contexts. In 1904 the total population of South Africa was 5,174,827, with whites constituting 21.6 per cent, "coloureds" (those considered of mixed descent) 8.6 per cent, Asians 2.4 per cent, and Africans 67.4 per cent (Beinart, 1994, p. 261). These percentages remained relatively unchanged for decades. It was only in Natal where, due to the importation of indentured labour from India between 1860 and 1911, Indians at 9.1 per cent of the population in 1904 outnumbered whites, who were at 8.8 per cent (Meer, Gains, Marie, Motala, Motala et al., 1980, p. 16). In Australia, by contrast, whites were the dominant group. In 1901 the total population was 3,773,801 with whites constituting around 96 per cent, Chinese 0.8 per cent, Indians 0.1 per cent, Syrians 0.04 per cent, Japanese 0.09 per cent (Yarwood, 1964, p. 163), and Pacific Islanders 0.26 per cent (South Sea Islanders Fact Sheet[National Archives of Australia]). Indigenous people were possibly around 1.5 per cent of the population, but this is only an estimate as they were not counted until the 1971 census.

In both Australia and South Africa, the Asian population was predominantly male, as Chinese and Indian men themselves preferred circular migration, leaving wives and children in China and India. In 1904, of the 2,457 Chinese in the whole of South Africa, only 23 were female (Yoon Yung Park, 2008, p. 19), and of the 8,489 Indians in the Cape Colony in 1904, 841 were female (Dhupelia-Mesthrie, 2014, p. 638). These skewed ratios remained until the number of wives began increasing in the 1930s. In Australia in 1901, the Chinese population was 98.5 per cent male, Indians 99.2 per cent male, Japanese 93.4 per cent male, and Syrians 62 per cent male (Yarwood, 1964, p. 163).

In both countries, white settlers deployed language that painted Asians as representing a civilization irreconcilable with that of whites, contaminating the body of the nation and threatening white livelihoods (trade and work). The hostility of whites was out of proportion to the actual size of the Asian population. Once immigration restriction laws were passed, the domicile certificate emerged in both countries as an initial attempt to manage the process of circular migration. In our comparison of the domicile certificates that emerged in both countries we argue that the colonial state used this process to sort out who was entitled to be a holder and who was not. It then developed identification systems so that the holder could be properly identified on return from India or China. The South African colonies learned from each other, while the federated government of Australia drew on earlier systems in operation in the colonies, and even drew on American practices in regulating Chinese entries. Most importantly, there were marked similarities between the systems developed in Australia and the Cape Colony, which indicated the spread of ideas across the British Empire.

Studies of the use of the writing/dictation test in the immigration leg-islation of both South Africa and Australia have revealed the role of the imperial offices in London in communicating knowledge about practices (Martens, 2006). Further, as McKenzie (2004) emphasizes, "the imperial network was above all a network of information. It was held together by the movement of people and paper: publications from throughout the empire, private letters and endless official documentation" (p. 7). Lester (2002) has argued that the British colonies were "nodes within an imperial network" where settler papers circulated and a "trans-global British settler identity" was created. While both McKenzie and Lester focused on the nineteenth century, their conclusions are relevant. We know that Australian authorities consulted Imperial Conference records and Sir Arthur Lawley, the former administrator of Matabeleland, about the strength of the Crown's 1858 proclamation of equality to its Indian subjects (NAA A1 1904/10519). However, at this stage of our work, and unlike the clear evidence of links between India and South Africa found by Jacob Steere-Williams in chapter 2, firm evidence of knowledge shar-ing relating to the documentary systems developed between the two countries or of interventions made from London has not come to light.

For all the shared characteristics in the paper systems of the two coun-tries, differences also emerged as each experimented with technologies of identification. That various versions of the domicile certificate that evolved indicate the learning process of the state, an adaptability and variation evident in the wide range of documentary systems discussed elsewhere in this book as well. We argue that identification systems devel-oped for populations deemed "criminal" came to be regarded as suitable for the Asian population as they were considered, to use Cole's (2002) phrase, "'suspect' and alien" (p. 3). This resulted in specific discrimina-tory treatment, and was built on a growing focus on the surveillance of Asian bodies. In this respect we can find strong similarities with the racializing surveillance apparatuses deployed in other contexts, with domicile certificates echoing especially strongly Japanese (see again Midori Ogasawara, chapter 7) and Israeli (see Ahmad H. Sa'di, chapter 8) colonial practices. The dynamics were quite different in other ways, however, primarily because with Australian and South African domicile certificates, migrants rather than Indigenous populations were the focus of state surveillance, and these made up a small minority of the over-all population. In this section we examine the different yet overlapping ways in which states learned to implement the domicile certificate in South Africa and Australia.

MacDonald's (2007) account of the paperwork developed in Natal, the first of the British colonies to exclude Asians with the writing test, points to significant unpreparedness and no clear plan for how Indians resident in the South African colony could travel and return without being declared prohibited on account of the writing test. The Immigration Restriction Officer then "hit on the idea of a Domicile Certificate" (MacDonald, 2007, p. 128). This was motivated by the officer's consideration that "the Government could not imprison a man here without his being allowed to [sic] an opportunity of visiting his native land" (2007, p. 128). The domicile certificate containing the applicant's name, place of birth, occupation, residential address in Natal, as well as whether he was accompanied by a wife and children was remarkable for its lack of attention to how the holder would be recognized by the immigration officer on his return. To secure the certificate, the applicant could present himself at any magistrate's office and provide an affidavit from a person (later two persons) that would attest to the applicant having resided in South Africa for two continuous years before the passage of the Immigration Restriction Act (1897). Individuals who had left the colony before the passage of the Act and qualified could also apply. The Immigration Restriction Department did not retain a copy of the document issued. The consequence, MacDonald argues, was extensive fraud, a market in domicile certificates, and migration by people not entitled to enter.

No doubt learning from the Natal experience, the Cape Colony in 1903 paid more attention to identifying the domicile certificate holder, specifically by introducing the recording of bodily markers. The certificate issued and signed by a magistrate (later by immigration officials once the process was centralized in 1905) provided for the holder's name, age, sex, place of birth, and occupation with additional requirements for a signature and left and right thumbprints. The certificate had a section for "distinguishing marks," which was to be filled out "in the case of Asiatics and Persons otherwise difficult to identify" (IRC 1/1/243 5500a, certificate 1904). Thus began the surveillance of Asian bodies by South African immigration authorities and their marking out for special attention. Certificates issued in 1905 and 1906 include a photograph pasted in the left-hand corner and information about a wife and children who may accompany the holder. For example, Shaga [Chaganlal] Bhuvan's certificate of 1906 recorded: "Scar top left wrist. Scar inside right forearm. Deformed finger in right hand. Height 5 ft 7 inches. Scar corner right eye" (IRC 1/1/46 1076a). In these descriptions, the name of the person seemed of least importance. Thus, Mutra Singh was named Martha Singh on his certificate, and Chaganlal was rendered as Shaga (IRC 1/1/145 3501a, domicile certificate, 1906; 1/1/46 1076a, domicile certificate, 1906).

The recording of bodily markings on the certificate reflected the verbal portrait, based on the work of Alphonse Bertillon of Paris (Cole, 2002), that had long been used in Europe to identify people deemed criminals, although the South African domicile certificate only recorded height and bodily marks. It also drew on the triple technologies of photography, anthropometry, and dactyloscopy, which had become the standard practice to identify prison inmates in Bengal by the mid-1890s (Pinney, 1997). In Europe since the second half of the nineteenth century, the photograph became what Allan Sekula described as the "truth apparatus" of the state (quoted in Hayes, Silvester, & Hartmann, 2002, p. 114). The "optical empiricism" of the camera spread through colonial territories to acquire information about colonial subjects (Hayes, Silvester, & Hartmann, 2002, p. 114). As Tagg (2009) has argued, states "seized" on photography to make it serve the desire for "compulsive knowledge" while retaining an inherent distrust of its "productivity" (p. 16), although Pinney (1997) notes that "the face ... proved an insufficiently stable and quantifiable object" (p. 70). Despite this instability, Cape authorities retained the photograph to identify Asians for decades. Its imposition as evidence of identity and as a surveillance method of the free Asian population was particularly unique in the South African colonies, with photographs used to identify Indian and Chinese indentured workers in Natal and the Transvaal, respectively (Breckenridge, 2014; Singha, 2000).

In India, photographs were supplemented with, to use the words of William Herschel, the governor of Bengal, "the penetrating certainty of fingerprints" (quoted in Pinney, 1997, p. 20). Fingerprinting is one of the best examples of "transnational knowledge exchanges" (About, Brown, & Lonergan, 2013, p. 9), and the spread of this technology via India and London to South Africa has been well documented (Breckenridge, 2014; Singha, 2000). The South African thread begins with the arrival of Edward Henry to head the police department in Johannesburg between 1900 and 1901. An inspector general in Bengal, he had worked on classifying fingerprints, work initially begun by two Indian police officers, Azizul Haq and Hem Chandra Bose. Breckenridge (2014) observes that "wherever Henry went in South Africa fingerprint repositories sprouted like mushrooms" (p. 22). Criminals, migrant African labourers, unfree Chinese and Indian indentured labourers, and the free Asian population became the target of this technology. In the Transvaal, the free Asian population were required after 1902 to provide thumbprints on registration certificates (indicating rights to residence), and this escalated to the demand of ten fingerprints in 1906. While it has been argued

that the more liberal Cape Colony was a reluctant embracer of the new
technology (Breckenridge, 2014), this is true only in so far as there was
reluctance to fingerprint Indigenous African workers. Conversely, there
was enthusiastic use of this practice against the much smaller Asian population. That administrators were new to this technology is indicated in
the instructions on the Cape domicile certificate: "the thumbprint can
be taken by lightly inking the ball of the thumb on an Ink Pad … and
then pressing it lightly on the paper" (IRC 1/1/145 3501a, domicile certificate, 1906).

The use of visual markers of identity was central to the complex
process by which domicile certificates were obtained, which was much
more difficult for Asians in the Cape compared with Natal. There the
applicant had to signal "The intention to permanently settle in South
Africa … and an intention to abandon domicile elsewhere" with the latter determined by ownership of property "or the presence of a wife and
family" (G 63–1904, *Report on the Working of the Immigration Act of 1902*,
p. 49). The length of residence in South Africa was unspecified. Officials
cautioned that

> an Asiatic, who had originally come to South Africa for a time with the
> intention of accumulating money and after remaining a year or so, had
> returned to his native country with his accumulated savings, would not, on
> his subsequently attempting to return to South Africa, be admissible on the
> ground of previous domicile. (G 63–1904, *Report on the Working of the Immi-
> gration Act of 1902*, p. 49)

Legal guidelines in 1905 made clear that a man with business interests
in the colony but whose wife and children were in India could not be eligible; nor was a single unmarried male who lacked any vested interests.
Men with vested interests, whose families were in the colony and who
had lived there for seven years, were deemed the ideal candidates for
success (IRC 1/1/117, 2887a, opinion of C.P Crewe, 2 February 1905).

Applications for domicile marked the moment of entry of Indians into
the files of the bureaucracy, which captured information about arrival in
the colony, years of residence, occupation, ownership of property, business interests, and the presence or absence of wives and children. This
was their initiation into the archive of surveillance. Singha (2000) makes
a distinction between "indirect surveillance" involving the collection of
information and "direct surveillance" involving the close examination of
the body at the port (pp. 81–2). In the Cape there was direct surveillance
in the information-gathering stage by police tasked with seeing the applicant and verifying details, for it was beyond the capacity of magistrates

and other officials to do this. Anwar Khan, a shop assistant in the colony, sought a certificate in December 1905, intending to visit his parents in India. The investigating police inspector confirming Khan's place of employment, that he owned no property, was single, and did not intend marrying reported: "I have seen him," adding that he "appears to be of a respectable caste. I do not know anything against him" (IRC 1/1/147 3549a). The application, however, was refused. The consequence for such unsuccessful applicants was their departure to India without securing rights for return – thus there was sorting of those considered undesirable by the state.

For the few successful applicants there was more seeing to be done as the body became a site of evidence to be captured on paper. Another round of seeing occurred on return to the colony. Port officers who retained the paper double of the departing passenger were instructed to permit re-entry into South Africa only on assuring that "his thumbprints and mark of identity accord with the Certificate of Domicile in your possession" (IRC 1/1/242 5493a, Note of Medical Officer of Health, 24 December 1904). Natal followed the Cape's example in 1906 with a new certificate of domicile including the signature of holder, distinctive marks, and left and right thumb impressions but no photograph (Mac-Donald, 2007).

We shift now to Australia, where the newly federated government grappled with similar issues. Although the Australian public and their government wished for a "white Australia," the state conceded that some long-resident "Asiatics" might have an entitlement to domicile status (Yarwood, 1964). As in South Africa, where the Cape Colony drew on the experience of Natal to improvise its system, the newly established federal government of Australia drew upon the earlier practices of the Australian colonies, with the former Victorian colonial administrative system forming the basis for the emerging system. Under the Victorian colony's Chinese Immigration Restriction Act (1888) and Chinese Act (1890), Chinese residents "of many years" could be exempted from the restrictions on their entry and re-entry. Applications for exemption, like those for naturalization, required a residence of some years, certification of good character with references from European Australians, a police report, and "a certificate from a Justice of the Peace, Police Magistrate or warden identifying them as the applicant" (Couchman, 2006, p. 133).

As in South Africa, Australian systems of identification emerged out of transnational borrowings, and involved visual technologies. Couchman (2006) argues that the colonial systems of immigration restriction "contributed to a wider transnational 'conversation' with other white-settler colonies" about regulation of Chinese immigration (p. 121). American discussions about using photography to identify Chinese people (see Pegler-Gordon, 2006) also inspired Australian administrators, with the federal government additionally drawing on the hand-printing and written physical descriptions formerly used in New South Wales and the Northern Territory of South Australia. The technology of fingerprinting, seen as "particularly well-suited for the identification of members of 'other' races" (Cole, 2002, p. 139), spread to Australian police forces along similar circuits sketched above relating to South Africa (see also Allen, 2013). However, we emphasize that Australian immigration authorities stood apart from practices elsewhere in the empire by using handprints rather than fingerprints until around 1919. The source of the federal practice appears to be the New South Wales Chinese Restriction Act (1888) and Immigraton Restriction Act (1898), where the customs service, possibly drawing from Herschel's (1916) early work, employed handprints on the back of exemption certificates (see Bagnall, 2015; NAA A1, 1919/12390).

While it was widely accepted throughout the empire that handprints were of little use for identification purposes, Atlee Hunt, the Secretary of the Department of External Affairs, resisted pressures from police to move to thumbprints. He argued that the goal for immigration authorities, unlike the police, was not to trace and identify "a particular individual." Rather, "We use the print of the hand as an adjunct to other means of identification for the purpose of readily testing whether an individual who presents a certain paper is the same person to whom that paper was issued" (quoted in Allen, 2013, p. 82). Hunt's confusing distinctions mark out Australia from other places in the empire. The Australians required the inking of the hand and the taking of up to four impressions, which was more invasive than fingerprinting (NAA AP214/9, p. 272). Its effect on the surveillant subject must have been considerable.

Like the Cape, eligibility for a domicile certificate was limited and sorted out those Asians with residency of five years, deemed of good character, and with capital, property, and/or stock of at least £200 from labourers or itinerant hawkers with little capital. There was no concession for Asians who were briefly out of the country when the new legislation was introduced (Allen, 2011a). There were also remarkable similarities with Cape officials in how Hunt defined "domicile" in 1902:

> Most of the Hindoos, Chinese etc. ... are not domiciled in the sense in which the word must be understood in future. These men have, most of them, come here leaving their wives and families in their native countries, to whose support they contribute from time to time, and to whom they intend to return as soon as they have accumulated sufficient money. (Yarwood, 1964, p. 68)

Until 1903, this notion of domicile was applied strictly with most applications refused. Initially, those granted domicile certificates could bring in wives and minor children, but, except in rare circumstances, this concession was discontinued in March 1903, making Australia's policy far harsher than that of South Africa (Allen, 2011b; Yarwood, 1964).

Bucksis Singh's application for a domicile certificate provides insight into the process of information gathering and surveillance in the context of Australian migration practices. A hawker from Brisbane, he provided details about his nationality, his date and place of birth, date of arrival in Australia, and places of residence and occupations in his Statutory Declaration. While his wife and ten-year-old son were living in the Punjab, he declared a wish to live in Australia permanently, explaining, "I am a British subject" (NAA: J3115, 76). Singh furnished details about his half share in a wagon and four horses worth £50 and foodstuffs of approximately £200. His referee, John Stewart, who had business dealings with him over the previous five years, testified to his good character.

As in the Cape Colony, what followed was a police report about Singh's character and the truthfulness of his statements. Investigations revealed that local residents found him to be of good character. The police were unable to sight his horses and wagon, which were in distant Toowoomba, but reported he did have £419, 5 shillings, and 6 pence in the Government Savings Bank. Singh's documentation was then sent by the Brisbane Collector of Customs 1,300 kilometres south, to Melbourne, for approval by Atlee Hunt. On approval, Singh paid the £2 fee. His domicile certificate, numbering 04/24, was given to the Tide Surveyor, who handed the certificate to Singh on board as he departed Brisbane for Bombay. Handing the certificate to the applicant on board ship offered a further opportunity for identification and prevention of fraud with the substitution of a different traveller. While Singh travelled, a duplicate certificate was held at the Brisbane customs office, thus requiring Singh to match his paper double upon return.

The Australian domicile certificate went further than its South African counterpart, revealing a state more bent on capturing as much information as possible to prevent fraud, including through extensive visual markers. Information from Singh's application form about his family, his date of arrival in Australia, his occupation, his property and its value, and his referees was recorded along with the date of departure and ship taken. There was a similar descriptive list detailing his nationality, birthplace, age, height, eye colour, complexion, and build. He had no particular marks recorded, although typically, as was the case in South Africa, scars on hand or face were recorded, but not on the torso. Unlike South African documents, however, on the reverse of Singh's certificate was a print of his left hand. It also included side and frontal photographs, which were similar to the criminal mugshots that had become common by this time.

Bucksis Singh's certificate was one of a series of iterations of the domicile certificate as the Australian state determined the best means of identification and surveillance, with the focus on visual technologies (see Bagnall, 2017). The earliest versions included photographs and verbal bodily descriptions but no handprint. Then, in 1902, photographs were dispensed with and handprints added. A year later, both photographs and handprints were required, and duplicates of photographs were retained in the customs office. Photographs were initialed by the customs officers and a customs stamp was placed "partly on the photograph and partly on the certificate" in order to foil attempts to substitute another photograph (NAA AP214/9, p. 25, 44). The process of identification was tightened further when Atlee Hunt instructed customs officers to provide police with photographs of applicants when they were asked to produce reports (NAA AP214/9, p. 55).

When Singh returned from India, his papers were sent to his port of arrival from Brisbane for his identification. On board ship, he would have endured humiliating sorting, mustered on deck with other Asians away from the rest of the passengers while his documents and identity were checked (Day, 1996). His handprint was taken once again. His papers were then sent back to the Brisbane office and his rather worn domicile certificate was stored with its pristine duplicate (now cancelled) in customs files (NAA: J2482, 1904/24) where it could be referred back to when he made subsequent applications. Such filing systems and the successful retrieval of documentation were thus crucial to identification and surveillance (Caplan & Torpey, 2001; Tagg, 2009). In what could not have been a coincidence, unlike Natal, both Australia and the Cape Colony moved away from the domicile certificate in 1905 and 1906, respectively, to newer documentation.

The Certificate Exempting from Dictation Test, and the Permit and Certificate of Identity

In 1905 Australia amended the Immigration Restriction Act (1901) to replace the domicile certificate with a Certificate Exempting from Dictation Test as the main identification and travel document (NAA AP 214/9, pp. 85–6). This remained virtually unchanged into the 1950s. The Cape Colony also abolished the domicile certificate and inaugurated a new permit system with the passage of the Immigration Act in 1906. The significance of the Cape permit is that when the Union of South Africa was formed and the new Immigration Regulation Act (1913) replaced earlier colonial systems, the Cape permit formed the basis of the new certificate of identity issued to Asians countrywide who wished to travel abroad. This system remained in operation well into the 1940s.

The Australian certificate was valid for three years, although it could be renewed and often was. The Cape permit was valid for just one year, with renewal possible only through special pleading. The later South African certificate of identity was valid for three years. If the Cape permit or South African certificate of identity expired it meant that the holder could no longer return, whereas the Australians were not perturbed by long absences. In fact, in 1905, Alfred Deakin, Minister of External Affairs stated:

> The longer the persons ... desire to remain away the better we are pleased. If a man wished to go away for twenty years, we should be glad to allow him to do so ... We merely desire to keep control, to take care that the man is not leaving, and proposing to dispose of his permit [for fraudulent use by another person]. (Commonwealth of Australia, Parliamentary Debates, House of Representatives, December 6, 1905)

Indeed, in Australia the new Certificate Exempting from Dictation Test allied with the 1903 restriction on the entry of wives and children of Asians saw the state developing a new form of social sorting. Hitherto, the domicile certificate, available to a limited number of more prosperous prohibited immigrants, allowed them to bring their families into Australia. This policy had the potential for the development of a continuous population of Chinese, Indians, and other prohibited immigrants. With the abolition of the right to bring in wives and children, the spectre of "this rising generation" was banished. The eligibility for the Certificate Exempting from Dictation Test was much wider than for the domicile certificate and allowed for numerous extensions, thus amounting to an acknowledgment of domiciled status. The numbers of prohibited immigrants, however, declined as the men aged, died in Australia, or returned to pass their declining years among family in their homelands. The Indian population dropped from 4,681 in 1901 to 3,653 in 1911 with the Chinese falling from 30,452 to 22,753 during the same period (Yarwood, 1964, p. 163).

In South Africa the shift to the new documentation was focused on the *temporary* right of travel that it granted rather than on the permanent rights that a domicile certificate signalled. The new system in both countries could be interpreted as beneficial in allowing for the departure and return of individuals who under the domicile system would not have been allowed to travel. Possession of capital and property was no longer a requirement, and in Australia there was no impediment if the applicant's family resided abroad. In the early years of the permit system in the Cape, permit applications were generally refused on the grounds that applicants had families in India, but a more liberal policy emerged in 1908–1909. Yet, the permit system, like the domicile certificate before it, was used to reduce the size of the Indian population as unsuccessful applicants went back to India with no chance of a return and successful applicants had to meet the limited time for re-entry stipulated on the permit. As a result of both the domicile certificate policy and the new permit system, the Indian population in the Cape dropped from 8,489 in 1904 to 6,606 in 1911 (Dhupelia-Mesthrie, 2014, p. 638). Surveillance therefore had a more sinister effect than simply being a system monitoring mobility; it enabled practices of social sorting.

We examine this new system especially to illustrate the singling out of Asians for discriminatory treatment. In the case of Australia, the procedures developed required the repeated identification of the person in question: when an applicant was granted the certificate; when he or she left the port of embarkation or passed through other Australian ports; and upon return to Australia. These repeated demands, which included the refusal to hand the certificate to applicants until they were on board ship, are perfect examples of Browne's (2012) racialized surveillance encompassing discriminatory treatment. The file of the Sym Choon family is demonstrative of the repetitive and detailed procedures of surveillance and identification, and indicative of "a systemic mistrust" of racialized people, including the Australian born, as well as "a presumption that non-whites were prepared to become implicated in acts of illegality" (Jones, 1998, p. 79).

In March 1921, Sym Choon, Adelaide greengrocer and importer of Chinese goods, applied for a Certificate Exempting from Dictation Test (NAA: D400, SA1956/9039). Born in Hong Kong, Choon had been in Australia for about twenty-eight years. For the first time since he had arrived, he planned a visit to Hong Kong. He was required to provide information similar to that demanded for the domicile certificate, with the exception of details about property and capital. Choon's European referees were a general merchant and his bank manager. He submitted six photographs, three full face and three profile. He also applied for certificates for his Australian-born children, Gladys and Gordon, to accompany him. They too provided photographs and individual references including from a high school principal and a local merchant.

What followed was a careful checking of Choon's identity and to a lesser extent that of his children. He was interviewed by the customs watchman to ensure that he was indeed the same person whose photograph was supplied. The interview was mandatory even though the watchman had "personally known the applicant for the past twelve years" (NAA: D400, SA1956/9039). The referees too had to look at the photographs and attest to these being of the man they provided references for. The Acting Collector of Customs for South Australia approved the issue of the certificates after receiving reports from the Detective Inspector. Each time the applicants travelled they faced the same humiliating and detailed process of application and identification.

Sym Choon's certificates and those of his family for this trip are not extant, but figure 6.1 shows Sym Choon's subsequent Certificate Exempting from Dictation Test, stamped to prevent the substitution of a different photograph. Here we can see that the certificate bore details of the holder's age, height, complexion, build, and distinguishing bodily marks. The profile and front photographs were similar to those required of people deemed criminals, except they could be taken at studios. Sym Choon's thumbprints are on the rear. The Australians had persisted in the use of the handprint until around 1919, after which thumbprints became the norm. The handprint, indicating a significant capture of the body, took up large space on the back of earlier forms (see figure 6.2). The details of the vessel the holder took were recorded as were the dates of departure and return. The Sym Choon family were only given the certificate on board ship.

It was not only the physical bodies of the Sym Choon family that were under surveillance but also their paper doubles. Officers had to account for the location of the certificates at all times, tracking them carefully though the postal system. Since the family was leaving from Sydney, the Adelaide office was duly advised by the New South Wales office when Gladys and Gordon had been handed their certificates before their ship departed. Sym Choon's certificate was returned to Adelaide as it had arrived in Sydney after his departure. A letter with his photograph affixed authorizing his return was then sent to him in Hong Kong. On their return to Australia in August 1921, the family's duplicate certificates were forwarded from Adelaide to Melbourne, their port of arrival, to verify their identities. Subsequently, the originals and duplicates were returned to Adelaide for cancellation and filing.

The surveillance of the paper double might continue after the death of the human it represented. Dorothy, Sym Choon's daughter, had to account for his certificate when he died during another trip to Hong Kong in 1922. The suspicion was that it had been sold or passed on to someone else who might use it fraudulently to enter Australia. The Inspector reported that Dorothy "thought she had brought it back to Australia ... but could not find it" (NAA: D400, SA1956/9039). In fact

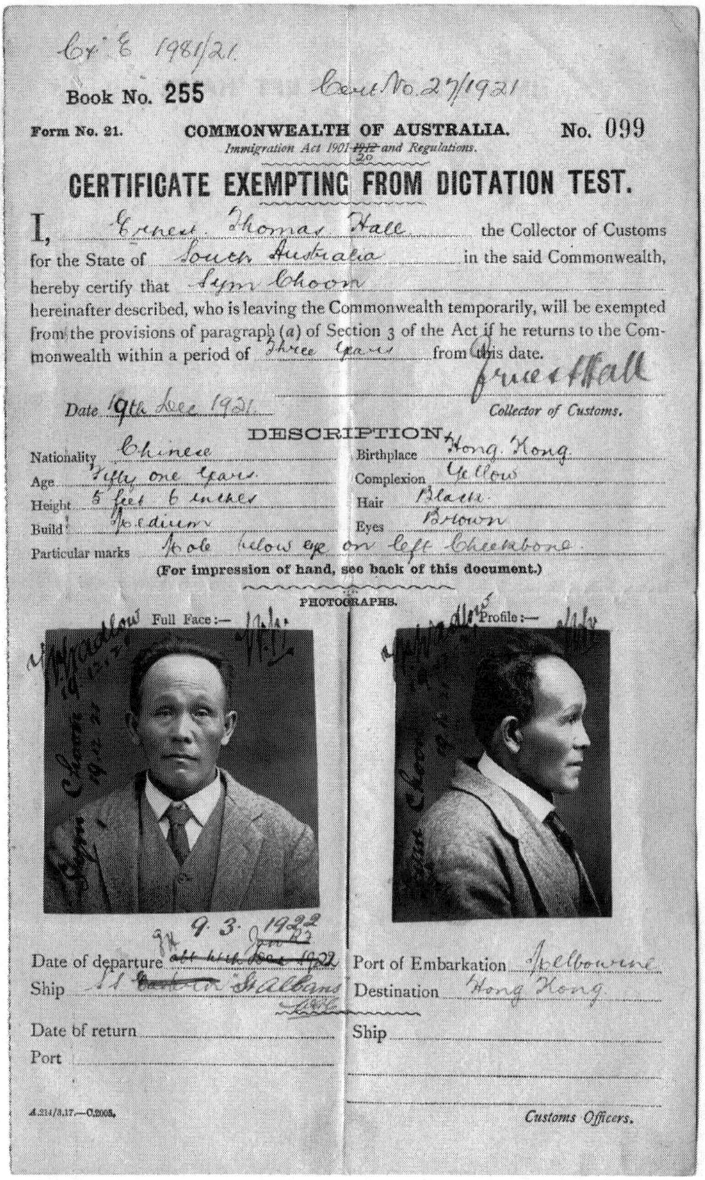

Cr. E 1981/21.

Book No. 255 Cert No. 27/1921

Form No. 21. COMMONWEALTH OF AUSTRALIA. No. 099

Immigration Act 1901-~~1912~~ 20 and Regulations.

CERTIFICATE EXEMPTING FROM DICTATION TEST.

I, Ernest Thomas Hall the Collector of Customs for the State of South Australia in the said Commonwealth, hereby certify that Sym Choon hereinafter described, who is leaving the Commonwealth temporarily, will be exempted from the provisions of paragraph (*a*) of Section 3 of the Act if he returns to the Commonwealth within a period of Three Years from this date.

Ernest Hall

Date 19th Dec 1921. *Collector of Customs.*

DESCRIPTION.

Nationality Chinese Birthplace Hong Kong

Age Fifty one Years Complexion Yellow

Height 5 feet 6 inches Hair Black

Build Medium Eyes Brown

Particular marks Mole below eye on left Cheekbone

(For impression of hand, see back of this document.)

PHOTOGRAPHS.

Full Face:— Profile:—

Date of departure 9. 3. 1922 Port of Embarkation Melbourne

Ship SS St Albans Destination Hong Kong

Date of return Ship

Port

A.214/3.17.—C.2005. *Customs Officers.*

Figure 6.1 Certificate Exempting from Dictation Test, Sym Choon, Australia, 1921. Migration Museum Collection HT 2009.233, History Trust of South Australia. Donated by Shaan Sym Choon. Note: This certificate was held by the Sym Choon family until they donated it to the Migration Museum in 2009.

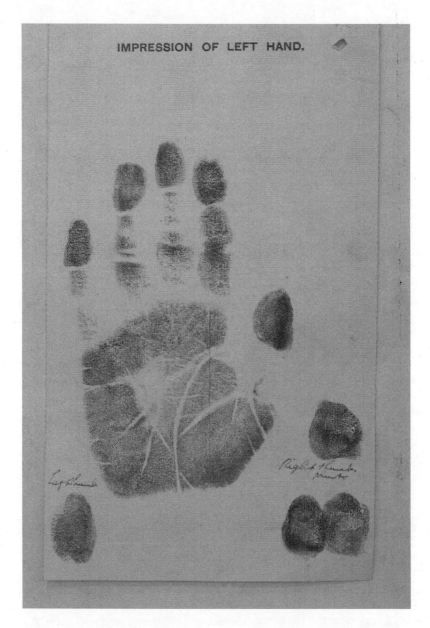

Figure 6.2 Handprint and thumbprints of Jewan Singh, Australia, 1921. Note: This image dates from a year or more after handprints were abandoned but curiously includes both a handprint and the thumbprints of Jewan Singh.

the Sym Choon family retained the certificate, in defiance of the requirement that they return it (see figure 6.1). The details of Sym Choon's certificate were now included on a list of certificate holders who had died
in China, a list circulated to all offices to guard against fraudulent use
by other Chinese. The Australians issued similar documentation to all
Asians, whether Chinese, Indian, Japanese, Syrians, or others. This form
of racial sorting was also a point of contestation, however, in which the
racialized boundaries of the "Asian" were strategically challenged. This
was evident in the successful bid by Syrians to persuade the bureaucracy
that they were not Asian. After 1920, Syrians thus shared the freedoms of
mobility enjoyed by white Australians (Jones, 1998).

South Africa deployed similar forms of racial sorting as Australia, but
in their case we find the development of a much more detailed set of
racial categorizations. The Cape Colony's Immigration Department used
the permit system it inaugurated in 1906 to create distinct racial hierarchies with differential treatment, tied in every case to the differential use
of visual technologies and bodily markers. While illiterate whites would
be unable to pass the writing test, they were allowed to apply for a permit to safeguard their entry; a permit which did not have their photograph, thumbprints, or descriptions of their body. The darker-skinned
Portuguese-speaking working-class Madeirans, however, who often lived
among racialized peoples in poor districts, were considered not quite
white. They were thus fingerprinted and their body marks recorded,
but they were not required to supply photographs when applying to visit
Madeira (PIO 20, 2451e, permit, 1912). The permit for Indians was different again. They were labelled "Asiatics" and required to provide photographs and thumbprints, and had their physical descriptions noted, as
shown in figure 6.3. Syrians and Turks were treated as borderline categories between Asians and whites, and, while requiring permits, these were
minus photographs or thumbprints (IRC 1/1/248 5621a, permit, 1910;
IRC 1/1/238 5348a, permit 1910).

While the Chinese were seen as self-evidently Asian, they were issued
with a different permit from Indians under the Chinese Exclusion Act,
demonstrating how systems of surveillance enabled the construction of
an intricate racialized order. The Chinese were strictly monitored both
in their internal as well as external movements (see Harris, 2014), an
extreme surveillance made possible by their very small numbers in the
Cape. Every Chinese person in the Cape had to be registered annually
under the Chinese Exclusion Act, and this remained in force till the
1930s. They were required to have an exemption certificate entitling
them to reside in the Cape, which bore their thumbprints and record
of distinguishing marks. They had to remove clothing so that marks on
their abdomen and back could be recorded.

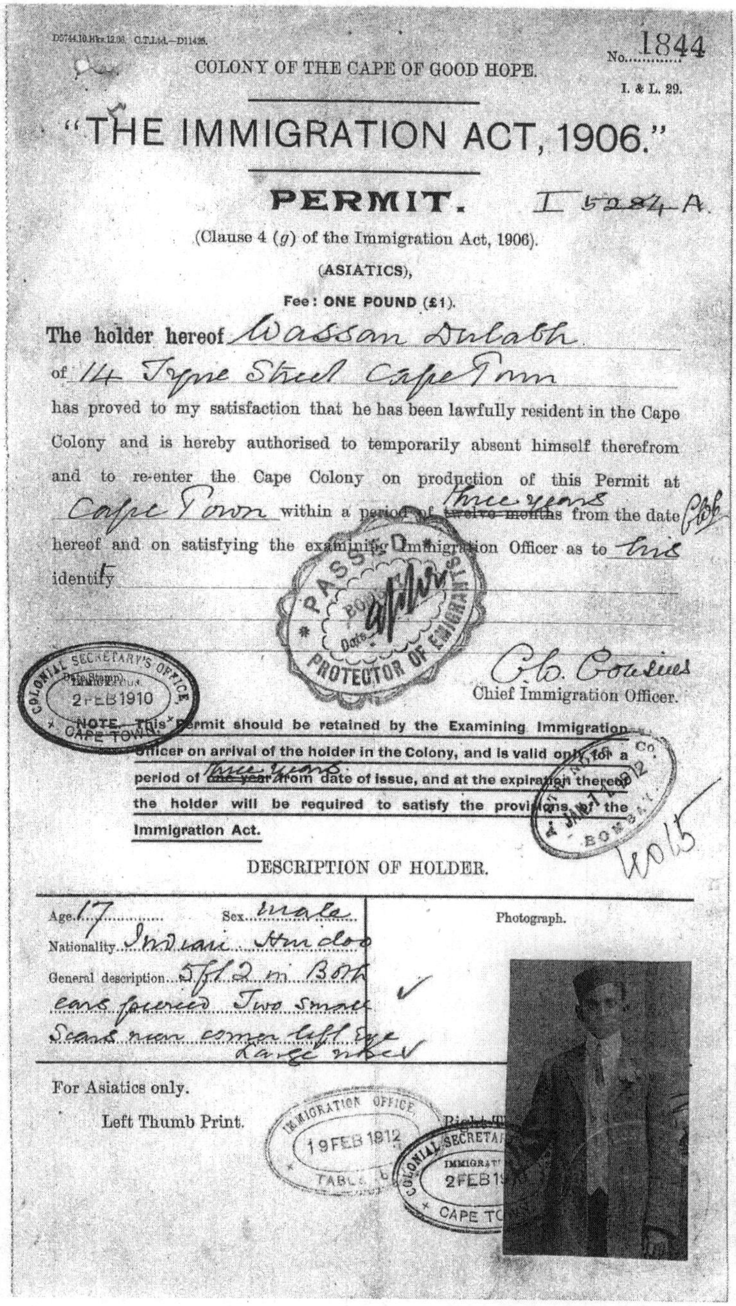

D5744.10.Hk 12.08. C.T.Ltd.—D11426.

No. 1844

COLONY OF THE CAPE OF GOOD HOPE.

I. & L. 29.

"THE IMMIGRATION ACT, 1906."

PERMIT. I 5284 A.

(Clause 4 (*g*) of the Immigration Act, 1906).

(ASIATICS),

Fee: ONE POUND (£1).

The holder hereof Wassan Dulabh

of 14 Tyne Street Cape Town

has proved to my satisfaction that he has been lawfully resident in the Cape Colony and is hereby authorised to temporarily absent himself therefrom and to re-enter the Cape Colony on production of this Permit at Cape Town within a period of ~~twelve months~~ three years from the date hereof and on satisfying the examining Immigration Officer as to his identity

Clo. Cousins
Chief Immigration Officer.

COLONIAL SECRETARY'S OFFICE 2 FEB 1910 CAPE TOWN

PASSED PROTECTOR OF EMIGRANTS BOMBAY

NOTE. This Permit should be retained by the Examining Immigration Officer on arrival of the holder in the Colony, and is valid only for a period of ~~one year~~ three years from date of issue, and at the expiration thereof the holder will be required to satisfy the provisions of the Immigration Act.

BOMBAY JAN 1912

DESCRIPTION OF HOLDER.

Age 17 Sex male

Nationality Indian Hindoo

General description 5ft 2 in Both ears pierced Two small Scars near corner left Eye Large nose

Photograph.

For Asiatics only.

Left Thumb Print. Right Thumb Print.

IMMIGRATION OFFICE 19 FEB 1912 TABLE BAY

COLONIAL SECRETARY IMMIGRATION 2 FEB 19 CAPE TOWN

Figure 6.3 Permit of Wassan Dullabh, Cape Colony, 1910.

Applications for such certificates captured their biographical details, and if they moved residence they had to inform the magistrate immediately. To travel overseas, a permit was issued provided there was proof of a certificate of exemption. There was thus a double paper requirement. In the case of Chan Tson, for example, the exemption number 38 was attached to him for his entire life. His internal movements from one district to the next produced a flurry of correspondence headed "Notification of Chinaman Leaving the District" (IRC 1/2/3 38c), requiring his appearance before the magistrate of the new district. When Tson sought to visit China in 1916 his application was recommended by the Port Elizabeth Chinese Association. As in the case with Indians, Chinese permits bore the photograph of the holder, thumbprints, and descriptions.

Applications for permits, as with the domicile certificate process, required the capture of select biographical information with references from at least three individuals who could attest to the veracity of the details pertaining to residence in the Cape. Unlike Australia, the Cape decided to conserve police resources and dropped the police reports on applicants. This would have repercussions, as we shall see. The permit issued to Asians captured details suggesting little regard for the holder. Makan Gosai's permit bore the following: "large lips. Scowling eyes. Upper parts of ears pierced" (IRC 1/1/233 5306a). His photograph revealed a mark placed on his forehead as was customary for Hindus. There was some surveillance of the body as on his return the immigration officer observed: "mark on forehead has been removed (only ink stain)" (IRC 1/1/233 5306a). Bhana Meta's permit noted his "large protruding ears" (IRC 1/1/247 5595a). The immigration officer ticked off details on observing the similarities between the body before him and the paper double. The official stamp was placed partially over the photograph so that it could not be substituted, and was additionally stamped on issue, in India on departure, and on re-entry into South Africa. The files give one the impression of a system of effective surveillance with permits retained by the department on return.

After Union in 1910, when the new certificate of identity was adopted, the description of bodily marks ultimately gave way as fingerprinting and fingerprint matching had become more effective, as shown in figure 6.4. Union meant better resources and better coordination of systems than the smaller colonial governments could manage. Applicants were allocated an individual file number, which made for quick retrieval. Over the decades, files thickened with the detailed records of circular migration and as the state improved its knowledge about applicants, wives, and children who might be eligible for entry. Australia saw a similar thickening of the networks of surveillance and information gathering. In both

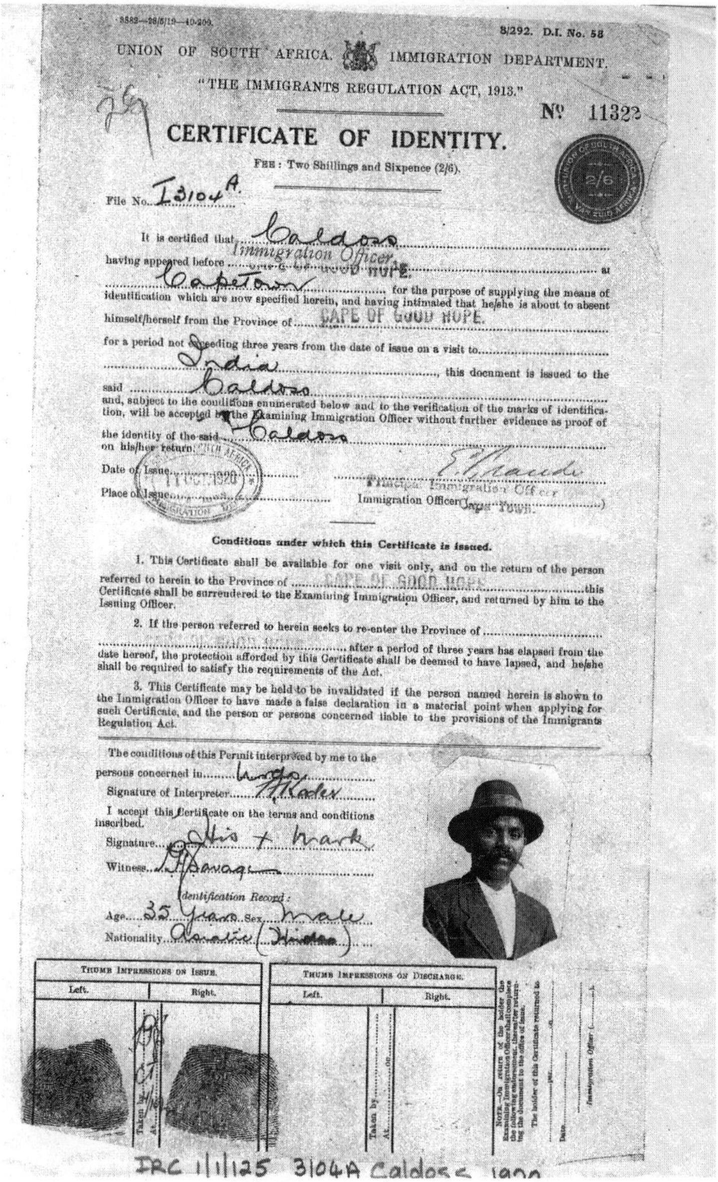

3382—28/5/19—40-200. 8/292. D.I. No. 58

UNION OF SOUTH AFRICA. IMMIGRATION DEPARTMENT.

"THE IMMIGRANTS REGULATION ACT, 1913."

No 11322

CERTIFICATE OF IDENTITY.

FEE: Two Shillings and Sixpence (2/6).

File No. I3104A.

It is certified that Caldoss having appeared before Immigration Officer Cape of Good Hope at Capetown for the purpose of supplying the means of identification which are now specified herein, and having intimated that he/she is about to absent himself/herself from the Province of CAPE OF GOOD HOPE for a period not exceeding three years from the date of issue on a visit to India, this document is issued to the said Caldoss and, subject to the conditions enumerated below and to the verification of the marks of identification, will be accepted by the Examining Immigration Officer without further evidence as proof of the identity of the said Caldoss on his/her return.

Date of Issue 11 OCT 1920

Place of Issue

Principal Immigration Officer Immigration Officer (Cape Town.)

Conditions under which this Certificate is issued.

1. This Certificate shall be available for one visit only, and on the return of the person referred to herein to the Province of CAPE OF GOOD HOPE this Certificate shall be surrendered to the Examining Immigration Officer, and returned by him to the Issuing Officer.

2. If the person referred to herein seeks to re-enter the Province of CAPE OF GOOD HOPE after a period of three years has elapsed from the date hereof, the protection afforded by this Certificate shall be deemed to have lapsed, and he/she shall be required to satisfy the requirements of the Act.

3. This Certificate may be held to be invalidated if the person named herein is shown to the Immigration Officer to have made a false declaration in a material point when applying for such Certificate, and the person or persons concerned liable to the provisions of the Immigrants Regulation Act.

The conditions of this Permit interpreted by me to the persons concerned in Hindu

Signature of Interpreter

I accept this Certificate on the terms and conditions inscribed.

Signature His + mark

Witness

Identification Record:

Age 35 years Sex Male

Nationality Asiatic (Hindoo)

Thumb Impressions on Issue.		Thumb Impressions on Discharge.	
Left.	Right.	Left.	Right.
Taken by ... At ...		Taken by ... At ...	

Note.—On return of the holder the Examining Immigration Officer shall complete the following endorsement, thereafter returning the document to the office of issue.

The holder of this Certificate returned to ... per ... on ...

Date ... Immigration Officer (...)

IRC 1/1/125 3104A Caldoss 1920

Figure 6.4 Certificate of identity of Caldoss, South Africa, 1920.

cases the result was the establishment of a profoundly discriminatory documentary system based on racialized social sorting. The considerable power of the state is evident here, but despite the impression of a seamless system of total surveillance, acts of agency and resistance emerged that sought to undermine the system.

Surveillant Subjects and the Documentary System

Asians complied with the system of certificates and permits because they had little choice. The desire to secure travel documents so that they could visit their families overrode the difficulties they endured. While various forms of resistance developed to challenge these practices, including everything from organized political protests to individual acts of disruption, the scope for such resistance varied considerably between South Africa and Australia. In South Africa, a range of Chinese political associations arose across the Cape Colony to protest fingerprinting and body examinations, but these were not successful (Yap & Man, 1996). Indian political organizations in the Cape accepted the permit system, for it allowed Asians to travel (Dhupelia-Mesthrie, 2017). They did not take an aggressive stand against the use of photographs such as that taken by Mohandas Gandhi against the registration of Asians in the Transvaal between 1902 and 1906 (Dhupelia-Mesthrie, 2013). There are, however, examples of individual Indians who tried, in the initial period, to avoid providing photographs on the grounds that this was against Islam. The Chief Immigration Officer found, after consultation, that there were no religious grounds for objection and that the particular protestor was "up to devilment" (IRC 1/1/6 125a, application 1907). Yet individuals shaped how they wished to look; they dressed in their best and offered studio photographs. Once wives and children joined husbands on their travels, they provided beautiful family group pictures. Some applicants developed quirks like always being photographed in a trademark hat (IRC 1/1/275 6059a). While in Australia certificate photographs were more akin to photographs required of the criminal justice system with the insistence on front and profile views, there were likewise some examples of individual agency, such as with Sher Mohamed, photographed sporting a bow tie (Allen, 2013), indicating his modern and respectable status.

Gandhi had a stronger impact on the Cape's fingerprinting project. He led a major passive resistance campaign against the Transvaal's ten fingerprint project for Asians from 1906 (Breckenridge, 2014), which prompted the Cape authorities to suspend the taking of fingerprints for permits between 1906 and 1910. Gandhi, however, lost the battle against fingerprinting by the time of Union, which had repercussions for Asians throughout the South African colonies. From 1911 onwards application forms, permits, and passenger arrival forms of the Cape routinely bore thumbprints. Ten fingerprints, even those of little infants accompanying their parents to India, were placed on record.

The situation in Australia was somewhat different, with the federal government proving much stronger than the Cape Colony in controlling fraud. The Australians involved the police and also handed the certificate on board ship to the departing passenger. In the Cape Colony, private agents collected permits for passengers and there was widespread fraud by these agents and colluding immigration officers with permits issued to those not entitled (Dhupelia-Mesthrie, 2017). If an individual never returned to South Africa from India, his permit and biographical details were available for appropriation, with the help of agents, by other individuals. The Australians, in contrast, sought more actively to stem such appropriation, on occasion circulating lists of those who had died abroad.

Rather than a system of surveillance whereby the state captured a complete set of accurate data on Asian travellers, individuals at the Cape supplied manufactured narratives of the self that they perceived the state wished to receive. Most claimed to arrive in 1900 or 1901 just before the Immigration Act (1902) came into effect. When the state denied permits because wives were in India, individuals claimed they were single. Some created biographies (such as the death of a wife), which they were able to sustain for several decades but which fell apart when they tried to bring in minor sons (Dhupelia-Mesthrie, 2013). Biographies indicated that applicants had long periods of residence in South Africa, but places of work and residences could be invented. These claims were made possible because the immigration department came to depend on references from a select group of Indians who also functioned as agents and interpreters and who gained the trust of officials and corroborated the fictional accounts (Dhupelia-Mesthrie, 2017). Photographs were sent to agents from India by hopeful Indians, with fingerprints (in collusion with corrupt officials) inserted on forms to create false documents. The market in permits was extensive, covering a wide network spanning the Indian Ocean (MacDonald, 2012).

Individuals who created false biographies avoided detection as long as they sustained their narratives. Bhoodia Birjee's false biography and fraudulently obtained permit served him well for a decade, but a desire to visit India led to his deportation (IRC 1/1/290 6257a). His permit application of 1911 did not correspond with his 1921 application in its details about his age and height. Furthermore, the fingerprints on the 1911 permit were not his. Laloo Dajee's file provides an example of a man who successfully kept details of his personal life from the authorities

(IRC 1/1/253 5720a). His file creates the impression of a lonely man in South Africa with a wife and children in India from whom he was separated for long periods of time. He voluntarily confessed at the age of seventy to having had six children with a local woman. Laloo possibly sought to have official recognition for his local children, but the historian would like to believe that, as his life was coming to an end, he had a laugh at officials at their lack of knowledge about his local life.

The Australian system was stronger from the start. Authorities there insisted on only white referees who were also visited by the police. When it became possible from 1911 for Indians who had left without documentation to apply for readmission, many used the Punjabi firm W. M. Shah as agents (Allen, 2011a). Applicants needed to construct a narrative about their former residence, a narrative "legible" to the authorities, often supported by invoices, receipts, voters' rights, references, and other documentation as evidence of prior residence. While W. M. Shah was warned "not to support cases where people wrongly claim to have resided in Australia" (Jones, 1998, p. 137 n.71), it is likely that some slipped through the net. There were examples of stowaways jumping ship before documentation was checked and deserting crew gaining illegal entry. Some syndicates of smugglers, who worked with corrupt officials forging or buying certificates, were uncovered and presumably others prevailed (Allen, 2011a; NAA, A1 1913/4976; Yarwood, 1964). Indeed, in Australia there was nothing like the South African evidence of extensive fraud.

There are, however, examples of agency in the Australian files where applicants questioned the need for the constant supplying of referees each time they applied (Allen, 2013). Dorothy, the Australian-born daughter of Sym Choon, successfully "objected to having her thumb prints taken and did not anticipate any trouble whatever as she was well known in Sydney and Melbourne" (NAA: D400, SA1956/9039). Furthermore, as noted above, she retained her late father's certificate. While the Department of External Affairs conceded that "persons born or naturalized in this country are not to be regarded as immigrants ... the practice of requiring all non-white residents to apply for CEDTs [Certificate of Exemption from the Dictation Test] was retained" (quoted in Jones, 1998, pp. 59–60). The agency and potential resistance of applicants was thus strongly contained by the racialized structures of surveillance and documentation within which they were acting.

Conclusion

Though separated by thousands of miles, Australia and South Africa were linked by the network of empire. Both British territories adopted

similar restrictive legislation directed against Asians. Through immigration legislation, they sorted out those who could write in a European
language and those who could not, and excluded the latter. In fact,
Australia, unlike Natal and the Cape Colony, manipulated the dictation
test to ensure failure by Asians. Both countries also sorted out resident
Asians in terms of who could qualify for travel documents. Their systems
racialized Asians in a variety of ways, from the specific requirements for
identity verification to their treatment by immigration officials when
they arrived and departed.

The transnational mobility patterns of Asians led to the development
of these early elaborate forms of identification that involved lengthy processes of information gathering and the use of technologies to visually
document the subject on the move. This chapter has shown how the photograph had a compelling place on documents of mobility and how the
fingerprint – and in the case of the Australians, the handprint – created
the impression of a fully captured individual. The effect of such documentation on the traveller must have been considerable. The advantage
of the photograph is that it could be circulated amongst officials, as
indeed it was between distant Australian ports, but this circulatory capacity of the photograph also was used to advantage by fraudsters, as the
South African examples show. The fingerprint was a much more effective
technology of identification. Seeing the subject in person and recording
and keeping images, filed so they were easily retrievable, remained a
central part of the state arsenal of both countries. The photograph could
not be the main technology of surveillance nor could it be eliminated –
it had to form part of a range of other strategies of surveillance.

This chapter also points to differences among states in pre-Union
South Africa, as well as between Australia and South Africa more broadly.
Cape authorities created much more detailed and differentiated racial
hierarchies. Within pre-Union South Africa, the colony of Natal was
weaker than the Cape Colony in its attention to identity verification
while the Cape system failed at the human level of implementation. The
extent of fraud meant that paper doubles were anything but that. The
struggle over fingerprinting led by Gandhi also stalled the implementation of what eventually became an effective tool of identification. Despite
this, the reality is that Asians who wished to travel outside of South Africa
fell under a system of selective surveillance.

The Australians, alert from the start to fraud, created elaborate verification systems and control over the movement of documents – even tracking
down the documents of the dead. The details on the permissory travel
documents were far more extensive than the South African documents
and there was sustained use of the police. There was no mass campaign

such as that launched by Gandhi to impede implementation. In contrast to Australia, it was only two decades after Natal first introduced restrictive legislation that the South African verification of identity became its most effective. The South African colonies were weak on their own – unification ensured better resources and a more efficient surveillance system.

REFERENCES

Archival Sources

NATIONAL ARCHIVES OF AUSTRALIA (NAA)
A1 1904/10519 Question by Mr Higgins MP.
A1 1913/4976 Inspector F. W. Gabriel – Visit to Hong Kong.
A1 1919/12390 Bunt Singh Readmission.
AP214/9 Register containing "Immigration Restriction Act 1901" and related Correspondence.
D400 SA1956/9039 Sym Choon family – CEDT.
D596 1921/3865 Jewan SINGH – Indian – application for Certificate of Exemption.
J2482 1904/24 Bucksis Sineh [*sic*] of Brisbane.
J3115 76 Application for Certificate of Domicile for Bucksis Singh from Brisbane.

HISTORY TRUST OF SOUTH AUSTRALIA
Migration Museum Collection Sym Choon, Certificate Exempting from Dictation Test HT2009.233, Donated by Shaan Sym Choon.

WESTERN CAPE ARCHIVES AND RECORDS SERVICES, CAPE TOWN, SOUTH AFRICA. INTER REGIONAL DIRECTOR SERIES (IRC)
IRC 1/1/6 125a, file of Abdool Aggee.
IRC 1/1/20 476a, file of Esmail Bikha.
IRC 1/1/46 1076a, file of Chagan Lal.
IRC 1/1/117 2887, file of Hassan Suleiman.
IRC 1/1/145 3501a, file of Mutra Singh.
IRC 1/1/147 3549a, File of Anwar Khan.
IRC 1/1/233 5306a, file of Makan Gosai.
IRC 1/1/238 5348a, file of Ameen Milag.
IRC 1/1/242 5493a, file of Bhima Lala.
IRC 1/1/243 5500a, file of Ibrhaim Ismail.
IRC 1/1/243 5503a, file of Ahmad Moffee.
IRC 1/1/253 5720a file of Laloo Dajee.

IRC 1/1/275 6509a, file of Makan Kara.
IRC 1/1/248 5621a file of Nora Sawoma.
IRC 1/1/290 6257a file of Boodia Birjee.
IRC 1/2/3 38c, file of Chan Tson.

WESTERN CAPE ARCHIVES AND RECORDS SERVICES (WCA), PRINCIPAL
IMMIGRATION OFFICER SERIES (PIO)
PIO 20, 2451e, file of Manuel Freitas.

GOVERNMENT PUBLICATIONS
G 63–1904, *Report on the Working of the Immigration Act of 1902 (Government Printers, Cape Town, 1904).*

Secondary Sources

About, I., Brown, J. & G. Lonergan (Eds.). (2013), *Identification and registration practices in transnational perspective: People, papers and practices.* New York, NY: Palgrave Macmillan.
Allen, M. (2011a). Shadow letters and the Karnana letter: Indians negotiate the White Australia Policy, 1901–1921. *Life Writing, 8*(2), 187–202.
Allen, M. (2011b). Family stories and "race" in Australian history. *Critical Race and Whiteness Studies e-journal, 7*(2), 1–13. Retrieved from https://acrawsa .org.au/wp-content/uploads/2017/12/CRAWS-No-7-Vol-2-2011.pdf
Allen, M. (2013). "Most painful and humiliating": The surveillance of Sher Mohamad. *Studies in Western Australian History, 28,* 63–83.
Bagnall, K. (2015). Anglo-Chinese and the politics of overseas travel from New South Wales, 1898 to 1925. In S. Couchman & K. Bagnall (Eds.), *Chinese Australians: Politics, engagement and resistance* (pp. 202–39). Leiden, Netherlands: Brill.
Bagnall, K. (2017). Form 21 certificates, 1902–1908, blog post 31 July, *The tiger's mouth, thoughts on the history and heritage of Chinese Australia.* Retrieved from http://chineseaustralia.org
Beinart, W. (1994). *Twentieth-century South Africa.* Cape Town, SA, and Oxford, UK: Oxford University Press.
Breckenridge, K. (2014). *Biometric state: The global politics of identification and surveillance in South Africa, 1850 to the present.* Cambridge, UK: Cambridge University Press.
Browne, S. (2012). Race and surveillance. In K. Ball, K. D. Haggerty, & D. Lyon (Eds.), *Routledge handbook of surveillance studies* (pp. 72–80). Oxford, UK: Routledge.
Caplan, J., & Higgs, J. (2013). Afterword: The future of identification's past. In I. About, J. Brown, & G. Lonergan (Eds.), *Identification and registration practices*

in transnational perspective: People, papers and practices (pp. 302–8). New York, NY: Palgrave Macmillan.

Caplan, J., & Torpey, J. (2001). Introduction. In J. Caplan, & J. Torpey (Eds.), *Documenting individual identity: The development of state practices in the modern world* (pp. 1–12). Princeton, NJ, and Oxford, UK: Oxford University Press.

Cole, S. (2002). *Suspect identities: A history of fingerprinting and criminal identification.* Harvard, MA: Harvard University Press.

Commonwealth of Australia, Parliamentary Debates, House of Representatives, December 6, 1905. Retrieved from https://historichansard.net/hofreps/1905/19051206_reps_2_30/#subdebate-6-0-s51

Couchman, S. (2006). *In and out of focus: Chinese and photography in Australia, 1870s–1940s* (Unpublished doctoral dissertation). Melbourne, VIC: La Trobe University.

Day, D. (1996). *Contraband and controversy: The customs history of Australia from 1901.* Canberra, AUS: Australian Government Publishing Service.

Dhupelia-Mesthrie, U. (2013). Cat and mouse games: The state, Indians in the Cape and the permit system, 1900s–1920s. In I. About, J. Brown, & G. Lonergan (Eds.), *Identification and registration practices in transnational perspective: People, papers and practices.* (pp. 185–202). New York, NY: Palgrave Macmillan.

Dhupelia-Mesthrie, U. (2014). Split households: Indian wives, Cape Town husbands and immigration laws, 1900s to 1940s. *South African Historical Journal, 66*(4), 635–55.

Dhupelia-Mesthrie, U. (2017). Engaging with the bureaucracy: Indian immigration agents and interpreters in Cape Town, South Africa (1902–1916). *South Asian Studies, 33*(2), 180–98.

Dlamini, J. (2014). *Askari: A story of collaboration and betrayal in the anti-apartheid struggle.* Auckland Park, Johannesburg: Jacana.

Haggerty, K. D., & Ericson, R. V. (2000). The surveillant assemblage. *British Journal of Sociology, 51*(4), 605–22.

Harris, K. (2014). Paper trail: Chasing the Chinese in the Cape, 1904–1933. *Kronos, 40,* 133–53.

Hayes, P., Silvester, J., & Hartmann, W. (2002). "Picturing the past" in Nambia: The visual archive and its energies. In C. Hamilton, V. Harris, J. Taylor, M. Pickover, G. Reid, & R. Saleh (Eds.), *Refiguring the archive.* Claremont, Cape Town: David Philip Publishers.

Herschel, W. J. (1916). *The origin of finger-printing.* London, UK: Oxford University Press.

Hull, M. S. (2012). *Government of paper: The materiality of bureaucracy in urban Pakistan.* Berkeley, CA: University of California Press.

Huttenback, R. (1973). The British empire as a "white man's country": Racial attitudes and immigration legislation in the colonies of white settlement. *Journal of British Studies, 13*(1), 108–37.

Hyslop, J. (2015). Oceanic mobility and settler-colonial power: Policing the global maritime labour force in Durban harbour c. 1890–1910. *The Journal of Transport History, 36*(2), 248–67.

Jones, P. (1998). *Alien acts: The white Australia policy, 1901–1939* (Unpublished doctoral dissertation). Melbourne, VIC: University of Melbourne.

Lake, M. (2005). From Mississipi to Melbourne via Natal: The invention of the literacy test as a technology of racial exclusion. In A. Curthoys & M. Lake (Eds.), *Connected worlds: History in transnational perspective* (pp. 209–30). Canberra, AUS: ANU E Press.

Lester, A. (2002). Colonial settlers and the metropole: Racial discourse in the early 19th-century Cape Colony, Australia and New Zealand. *Landscape Research, 27*(1), 39–49.

Lyon, D. (Ed.). (2003). *Surveillance as social sorting: Privacy, risk, and digital discrimination.* London, UK: Routledge.

Lyon, D. (2007). *Surveillance studies: An overview.* Cambridge, MA: Polity Press.

MacDonald, A. (2007). Strangers in a strange land: Undesirables and border-controls in colonial Durban, 1897–c. 1910 (Unpublished Master's dissertation). University of Kwa-Zulu Natal, Kwa-Zulu Natal.

MacDonald, A. (2012). The identity thieves of the Indian Ocean: Forgery, fraud and the origins of South African immigration control, 1890s–1920s. In K. Breckenridge & S. Szreter (Eds.), *Registration and recognition: Documenting the person in world history* (pp. 253–76). Oxford, UK: Oxford University Press.

Martens, J. (2006). A transnational history of immigration restriction: Natal and New South Wales, 1896–97. *The Journal of Imperial and Commonwealth History, 34*(3), 323–44.

McKenzie, K. (2004). *Scandal in the colonies: Sydney and Cape Town 1820–1850.* Carlton, VIC: Melbourne University Press.

McKeown, A. (2008). *Melancholy order: Asian migration and the globalization of borders, 1900 to c. 1950.* New York, NY: Palgrave.

Meer, Y. S., Gains, P., Marie, S., Motala, S., Motala, R., et al. (Eds.). (1980). *Documents of Indentured Labour: Natal 1851–1917.* Durban, Natal: Institute for Black Research.

National Archives of Australia, South Seas Islanders Fact Sheet. Retrieved from www.naa.gov.au/collection/fact-sheets/fs269.aspx

Park, Y. (2008). *A Matter of honour: Being Chinese in South Africa.* Auckland Park, Johannesburg, SA: Jacana.

Pegler-Gordon, A. (2006). Chinese exclusion, photography, and the development of US immigration policy. *American Quarterly, 58*(1), 51–76.

Perbedy, S. (2009). *Selecting immigrants: National identity and South Africa's immigration policies, 1910–2008.* Johannesburg, SA: Wits University Press.

Pinney, C. (1997). *Camera Indica: The social life of Indian photographs.* Chicago, IL: The University of Chicago Press.

Robertson, C. (2010). *The passport in America: The history of a document.* Oxford, UK: Oxford University Press.

Singha, R. (2000). Settle, mobilize, verify: Identification practices in colonial India. *Studies in History, 16*(2), 151–98.

Tagg, J. (2009). *The disciplinary frame: Photographic truths and the capture of meaning.* Minneapolis, MN: University of Minnesota Press.

Yap, M., & Man, D. L. (1996). *Colour, confusion & concessions: The history of the Chinese in South Africa.* Hong Kong: Hong Kong University Press.

Yarwood, A. T. (1964). *Asian migration to Australia: The background to exclusion 1896–1923.* Melbourne, VIC: Melbourne University Press.

Legislation

COLONY OF NEW SOUTH WALES
Chinese Restriction Act (4 of 1888).
Immigration Restriction Act (3 of 1898).

COLONY OF SOUTH AUSTRALIA
Chinese Immigration Restriction Act (439 of 1888).

COLONY OF VICTORIA
Chinese Immigration Restriction Act (1005 of 1888).
Chinese Act (1073 of 1890).

AUSTRALIA
Immigration Restriction Act (17 of 1901).

NATAL
Immigration Restriction Act (1 of 1897).

CAPE COLONY
Immigration Act (47 of 1902).
Immigration Act (30 of 1906).
Chinese Exclusion Act (37 of 1904).

UNION OF SOUTH AFRICA
Immigration Regulation Act (22 of 1913).

7 Bodies as Risky Resources: Japan's Colonial Identification Systems in Northeast China

MIDORI OGASAWARA

It is an error to imagine that civilization and savage cruelty are antitheses. On the contrary, in every organic process, the antitheses always reflect a unified totality, and civilization is an organic process ... In our times the cruelties, like most other aspects of our world, have become far more effectively administered than ever before. They have not and will not cease to exist. Both creation and destruction are inseparable aspects of what we call civilization.

Richard L. Rubenstein (1975, p. 92)

This chapter examines the characteristics and consequences of national registration and identification (ID) systems in modern Japan, how these systems constructed individuals as subjects of the modern nation state, and how ID techniques were transformed in the colonial zones of imperial Japan. In the metropolis, the population was required to register collectively in patriarchal family units under the Koseki system established in 1871. In Japanese colonies, the government implemented different techniques to identify the population on a more individual basis. In this chapter I look specifically at the case of Manchuria in Northeast China, which was occupied by Japan from 1931 to 1945. While a compulsory national ID card has never been introduced to the general population in Japan, the Japanese colonial government developed new cards to identify and classify workers and residents in occupied Manchuria, in particular using fingerprinting. Although Japan lost Manchuria and other colonies in the defeat of the Second World War, the same techniques were introduced to post-war Japan as the Alien Registration System. This chapter will examine why and how Japan implemented these novel ID and biometric techniques, which are now used commonly around the world, on the margins of the expanding empire.

The ID cards in colonized Manchuria were used for two primary purposes: to repress resistance and to mobilize labour power for production. As David Lyon (2009) states, "identification is the starting point of surveillance" (p. 4). In the contemporary context, national ID systems have become naturalized, with identification generally thought to be proof of a pre-existing identity. But identification needs to specify both who a person is and who they are not, with categorization and classification necessarily occurring in the process of identification (Caplan & Torpey, 2001; Jenkins, 2000). Many scholars have argued that these processes were key to the development of institutions and systems such as the nation state, capitalism, bureaucracy, war, and colonialism (Foucault, 1977; Giddens, 1981; Lyon, 2007; Marx, 1976; Weber, 1946; Zureik, 2011). Being included in or excluded from a category is crucial, because ID systems accord different treatments to different kinds of people through these categories, and give or deny entry to basic rights or privileges. In the specific context of colonial rule, ID cards helped Japanese authorities classify the population by sorting them into "desirable" and "undesirable" categories that could vary depending on if it was in the service of military occupation or colonial development. I thus argue that the Manchurian ID systems served the dual purposes of colonization, for both repression and production. Indeed, Chinese bodies were seen as fundamental *risks* to Japanese security, but at the same time, as profitable *resources*, namely, cheap labour power.

In order to reduce risks and increase profits, the colonizers used biometric markers, fingerprints specifically, as an ideal means to classify individual bodies and regulate their movements, whether in workplaces or residences. When compared to the metropolis, then, colonial ID systems paid particular attention to tracking the movements of individuals. For the colonized, it was often fatal to be classified as undesirable under military occupation. Perhaps the worst example, one that is emblematic of the broader colonial system, is that of the thousands of Chinese who were categorized as bandits and communists and sent to a secret military project called Unit 731, a facility for biochemical and poison gas warfare where prisoners were used as live subjects to test the impact of plague, dysentery bacillus, frostbite, and other diseases, or were dissected by Japanese medical experts for anatomical experiments (Jin, 2010; Matsumura, 2017; Yang, 2016). The classification of colonial populations enabled the identification of Others for scientific massacre in Manchuria. This chapter sheds light not only on the administrative use of surveillance for the management of risk and the control of labour, but also on the violent consequences of surveillance under Japanese colonial rule.

In the following, I will first introduce my theoretical framework for understanding the repressive and productive characters of Manchurian ID systems, as well as their violent effects, drawing primarily on the concepts of biopower (Foucault, 1978), inclusive exclusion (Agamben, 1998), and necropolitics (Mbembe, 2003). I then describe the historical background of Japan's modern nation-building that generated the national registration and identification system of Koseki in the metropolis. Next, the Manchurian ID systems will be explained and discussed in contrast to Koseki, highlighting the individualistic characteristics of biometric ID cards that paid special attention to racialized bodies. Finally, I will conclude with some implications of the Manchurian case for today's more sophisticated ID schemes. The dual purposes of the Manchurian ID systems continue to operate in many respects in postcolonial political and economic structures, and should be recognized in contemporary discussion of ID technologies. Though repression and production seem discrete, or even contradictory, the two were woven together in the colonial surveillance systems in much the same way that, as Rubenstein suggests in the epigraph to this chapter, savage cruelty is woven into civilization.

Theoretical Perspectives on Colonialism

Michel Foucault's theorization of power is useful in explaining how surveillance developed as a form of population control. Within the multidisciplinary field of surveillance studies, Foucault's (1978) notion of biopower has been one of the most common theories to illustrate sovereign intervention into individual lives for productive purposes. For Foucault, biopower consists of two poles of power, one working as discipline, an *anatomo-politics of the human body*, and the other intervening in the biological functions of a species's body, a *biopolitics of the population*. In biopolitics, sovereignty transformed its political rationale from destroying life to growing life: "Hence there was an explosion of numerous and diverse techniques for achieving the subjugation of bodies and the control of populations, marking the beginning of an era of 'biopower'" (Foucault, 1978, p. 140).

Biopower helps explain the development of capitalist economic systems and the growth of surveillance within nation states. It further accounts for a regulatory mechanism of population control: how the population is mobilized to produce more for markets and states, and, in turn, how this regulatory mechanism has brought wealth to populations and raised living standards in the West, a perspective also applicable, as we will see, to Japanese modernization and the development of national ID systems. Biopower is an indispensable element in the development of capitalism, Foucault asserts. In his later studies of governmentality, Foucault (1991) elaborates that this productive aspect of sovereignty supervises and intervenes in the population to arrange things for "an end which is 'convenient' for each of the things that are to be governed" (p. 95), and in order to maximize wealth. It gives rise to the knowledge of political economy, and is done by apparatuses of security, according to Foucault.

In this popular model of modern surveillance, there is a clear transition from the old power of the right to kill to the new power of the right to foster life within the frameworks of the nation state and capitalism. Thus, this power is not only repressive, but also productive. It is the dual character of modern power, where the productive side accounts for new public policies, most notably public hygiene, health care, and social insurance comprising the welfare state. Through this, it constructs the population as a new political object subjected to the interventions of power. Biopower, Foucault argues, entails a political process that guarantees relations of domination and the effects of hegemony.

The transition of the sovereign principle from destroying life to growing life is observable in historical trajectories of Western nation states. However, that transition is different, or at least differently experienced, among colonial populations. Unlike in the metropolis, colonized people(s) were all too often placed in a state of exception. Giorgio Agamben (1998) suggests that sovereignty has treated subaltern groups differently from the dominant groups through the legal and extralegal generation of a state of exception by which sovereignty can suspend the normal application of law to certain groups. Claiming the right to apply the state of exception, a sovereign power can act free from the juridical constraints of citizenry protections and habeas corpus, and even nullify laws.

Based on Foucault's biopower, Agamben analyses how productive power has divided and repressed particular populations, and legitimated repression through juridical rules. The state of exception allows the sovereign to treat the individual only as a biological being, stripped of the political, legal, or social status of a full subject. Agamben calls it "bare life," being exposed to the direct exertion of power, wherein the law is indistinguishable from state violence. While the application of law to the population is exclusive inclusion, the exception is inclusive exclusion, because the population in the state of exception is incorporated into direct sovereign control, with no protection. Agamben argues the state of exception to be the original activity of sovereign power throughout Western politics. From this perspective, colonies are located on the threshold between outside and inside imperial prosperity, since they are incorporated into the political economy of the Western power, but deprived of citizenship protections. This mechanism of inclusive exclusion is seen in the case of the Manchurian population at the colonial threshold; included by the Japanese empire, but excluded from legal protections, and exposed to state violence. Those "bare lives" were classified by sovereign power into the desirable and undesirable, with the latter group often determined as "invalid" and not permitted to live among the population.

Previous studies of colonial surveillance support Agamben's critical understanding of biopower (Breckenridge, 2014; Longman, 2001; McCoy, 2009; Young, 1998; Zureik, 2011). In different ways, scholars have argued that the colonizing power is more interested in controlling resources (for example, land, water, or minerals) and tends to neglect the population, which is where we find the clearest divergence from European concepts like "civil death," as discussed by Ian Warren and Darren Palmer in chapter 5. In colonial South Africa, for example, Breckenridge (2014) finds more of an absence of information on the population than a presence, finding in this administrative indifference "no will to know" (p. 115). The indifference towards and neglect of the lives of colonized people implies that population control does not necessarily mean *care* of the population, in the Foucauldian sense. Rather, as Zureik (2011) reports in his studies of Israeli settler colonialism, Palestinian individuals have often been exposed to extrajudicial brutality, torture, executions, and ubiquitous surveillance under the occupation. Arendt (1968) stresses that in the case of South Africa, colonizers did anything required to maintain their hegemony on the racial pyramid, even when their practices contradicted economic rationality and profits. The concept of biopower on its own, then, does not reflect the exclusionary practices of colonial management, where colonized bodies were recognized as less than individuals in the metropolis.

Facing the overwhelmingly violent experiences of colonization, Achille Mbembe (2003) suggests that the concept of necropolitics captures the ultimate expression of sovereignty, namely, dictating who *may* live and who *must* die in the colonies. Mbembe criticizes biopower as insufficient to account for contemporary forms of subjugation of life to the power of death. He exemplifies the formation of modern terror in slavery, apartheid, and the occupation of Palestine, all of which have attacked racialized populations:

> The most original feature of this terror formation is its concatenation of biopower, the state of exception, and the state of siege. Crucial to this

concatenation is, once again, race. In fact, in most instances, the selection of races, the prohibition of mixed marriages, forced sterilization, even the extermination of vanquished peoples are to find their first testing ground in the colonial world. Here we see the first syntheses between massacre and bureaucracy, that incarnation of Western rationality. (Mbembe, 2003, pp. 22–3)

Mbembe argues that colonies are zones where the guarantees of judicial order can be suspended any time in the service of "civilization," and the sovereign's necropower, or right to kill, is ubiquitous, an argument that echoes in part Arendt's (1968) earlier work. As Mbembe (2003) suggests, selective and systematic massacres of racialized groups were all first tested on colonial soil: "The colony represents the site where sovereignty consists fundamentally in the exercise of a power outside the law and where 'peace' is more likely to take on the face of a 'war without end'" (p. 23). Under such a normalized state of exception, surveillance as a regulatory technique of biopower works closely as violence. Bare life refers to the *living dead*, argues Mbembe.

These theoretical approaches offer a valuable starting point for understanding the surveillance systems implemented as part of Japan's colonial projects in Manchuria. Indeed, Japan's colonial ID systems contributed to the creation of racialized bodies, the classification of individuals into desirable and undesirable, and the rationalization of state violence against the latter. However, this necropolitical project – the exercise of sovereignty in the production of bare life – did not simply displace the workings of a Foucauldian biopower, the integration of people into a capitalist system of production. Colonial attempts to mobilize the population for production do not contradict the repressive principles applied against racialized Others, but in fact work together in elaborating systems of colonial rule. Therefore, this chapter examines nation-building and colonial practices as simultaneous processes of modernization. In what follows, the three theoretical perspectives outlined above guide my research on how repressive and productive forms of sovereignty gave birth to Manchurian ID card systems in the service of imperial prosperity.

Defining the Japanese and Others through Koseki

Japan's first centralized registration and identification system, Koseki, was generated in the period of westernization and modern nation-building. In the early nineteenth century, Western nations often sent ships to demand international trade with Japan, but the Tokugawa (Edo) regime, based on the isolationist policy it had followed since it came to power in 1603, refused it. Amongst the Western nations was the United States, which sent a battleship to Japan in 1853, forcing it to end its 250-year isolationist policy. Facing the global market and international relations, Japan's ruling-class warriors were divided over whether or not to seek westernization, and the country plunged into civil war. Victory was won by the side that restored the Emperor to sovereign power, the Meiji Restoration, in 1867. The Meiji regime launched a constitutional monarchy around which political, economic, and cultural institutions were hurriedly constructed in the Western style. The goal was to enable the recognition of Japan as a civilized nation state by its Western counterparts. The global market, nation-building, and modernization literally came together in Japan's path.

The Koseki Act of 1871 counted the people on the territory of Japan, and registered them as the subjects of the nation. Koseki became the fundamental basis for all kinds of national administration, including taxation, conscription, and compulsory education under the national principle of enriching the nation and strengthening the soldiers (Sato, 1991). Although some people resisted the implementation of the system, Koseki ultimately constituted the foundation of the new nation state, and simultaneously served in many ways to bring "the Japanese" as a national body into being (Endo, 2010; Ogasawara, 2008a). Instead of providing a legal definition of Japanese subjects or citizenship, the government substituted Koseki for the legal proof of nationality.

At the heart of the Koseki system was the establishment of a patriarchal family framework for the registration of people, which contrasted with Western registration systems that focused on the individual. Koseki gave each family member a hierarchical position, such as master of the family and his dependents (Sato, 1988). Bummei Sato (1991) argues that patriarchy was incorporated into the constitutional monarchy for "educational" purposes, by embodying an imagined biological connection among the Emperor and his subjects in which they are linked through bloodline. Because Koseki represents the biological connection among all Japanese recorded in the files, it symbolizes one ethnicity or race through bloodlines that eventually belong to the sacred Emperor. Using the rhetoric of the Japanese people as the Emperor's offspring, the state apparatuses spread this myth and taught that all Japanese owed the Emperor for his mercy, and should pay their debt to him as his loyal subjects. Koseki not only constructed the population of the new nation state, but also solidified national unity and fostered loyalty in individuals. The national registration indeed "made up people" (Hacking, 1986, 1990) through the category of the Japanese as loyal subjects, and encouraged them to monitor each other in everyday life.

Koseki's educational role clearly shows a disciplinary characteristic, one of the two poles of biopower, which works on individual bodies (Foucault, 1978). It enlists and sorts out members of the national imagined community (Anderson, 1991), but ultimately constitutes the Emperor's property list. A Japanese, who is expected to be a loyal child within the family order, a loyal student for the school, or a loyal worker for the company, should most importantly be a loyal subject of the Emperor. Koseki legitimated supreme loyalty to the state by ordering different loyalties, from interpersonal to public, within the outer frame of the Japanese people. Self-devotion and sacrifice were inscribed on docile bodies, increasingly through wartime nationalism, especially after the aggressive war in China began in 1931.

While Koseki defined the Japanese, it simultaneously defined Others as those who do not belong to the Emperor. Put differently, the Japanese were defined as the elite and supreme race in contrast with Other neighbouring Asians. Colonial ideology was in fact embedded in Japanese nation-building from the very beginning. Japan participated in the global market beginning in 1853, and soon after began to engage in imperial expansion. In the south, the Meiji regime invaded the Kingdom of Ryukyu and annexed it as the Okinawa Prefecture in 1879, which it remains today. In the northern island of Hokkaido, the government sent settlers and deployed strict assimilationist policies against the Ainu people. By the victory of Japan's first official war, the Sino-Japanese War of 1894–1895, the Qing dynasty of China ceded Taiwan to Japan, and Japan granted Korea independence from Chinese suzerainty. The Japanese-Russo War of 1904–1905 expanded Japan's efforts to block Russian competition for influence over Korea and Northeast China, and enabled Japan to establish political and economic hegemony in the region. Japan then annexed Korea in 1910. Along with the expansion of imperial territories, new populations were added to the Emperor's property lists: the Taiwanese and Koreans used to be Others, but now became "the Japanese," without any choice in the matter.

An important question is whether neighbouring Asians were included on the lists as equals, and whether the boundaries of Otherness disappeared in an imperial harmony. The short answer is no. Japan deployed family registration systems similar to Koseki in the colonies of Taiwan and Korea, but the Koseki system in mainland Japan and its counterparts in the colonies were strictly separate (Sato, 1991). Colonial subjects were not allowed to register on the mainland, even if they lived on the mainland or were married to people who were registered there and considered "Original Japanese" (Endo, 2010; Sato, 1991) Through this distinction in the national ID systems, Japan established a racial order among the populations. Koseki noted their Otherness and underpinned ethnic rankings devised by the empire: so-called Original Japanese on the top, Ainu and Okinawans second, then Taiwanese and Koreans, and at the bottom of the ranking, the smaller ethnic groups like Gilyak residing in Sakhalin, who came under Japan's rule later than other colonies (Sato, 1988).

It is obvious in the expansion of Koseki and the formation of a Japanese identity that the building of modern state systems of identification is deeply involved in imperial competition. Looking at the British context, Edward Higgs (2001) suggests that the rise of the internal Information State enabled external battles over colonial expansion, and, in turn, colonial battles consolidated nationality and citizenship. Japan similarly used Koseki to build new forms of knowledge about the population that enabled it to fight imperial wars and construct a national identity in opposition to Others. In other words, Koseki represents a simultaneous process of nation-building and imperial competition in modern Japan, which defines who the Japanese are and how they should be. The national ID system binds the two sides of internal classification and external expansion to establish a new social order. Internal and external politics together drew the boundaries of Japanese identity. Thus, the ID techniques in suzerainties and colonies should be examined together in an integrated fashion, rather than separated. The effects on the colonial side are more likely overlooked, unrecorded, and hidden in the history of civilized nation states.

The Colonial Contradiction

The Koseki system provided the basis on which a hierarchical colonial classification system was established, in the process knitting together the Japanese empire. Unlike in its other colonies, however, Japan did not deploy the Koseki system as it expanded its colonial reach to Northeast China. In this case, the establishment of Japanese rule was built on a fundamental contradiction, with Japan claiming that Northeast China remained an independent nation while simultaneously treating it as a Japanese colony. The contradiction shaped the occupation until its end in 1945, and the nature of this contradiction gave rise to novel ID systems that enabled both practices of repression and the exploitation of Chinese labour.

In the aftermath of the Japanese-Russo War (1904–1905), Japan took over Russian leases in Northeast China, including the management of the railways, and founded the South Manchuria Railway Company in 1906. The South Manchuria Railway Company, a joint agreement linking corporations and the government, owned the coal mines, the seaports, and the steelworks, in addition to building railways. The company's administrative services along the railways were especially valuable. Japan sought to control them to sustain its political economic interests, and stationed the military to secure its privileged position in the region. Japan entered the First World War (1914–1918) in part because it saw an opportunity to take over German leases in China as well, and kept sending troops in the name of protecting the Japanese residents in Shandong province, located in the south of Manchuria.

The Chinese Revolution of 1912, however, overthrew the Qing dynasty, and growing nationalism began to counter Japanese imperial power. The Chinese Nationalist Army and the Imperial Japanese Army fought in Jinan, the capital city of Shandong province in 1928, which alarmed the Chinese, who now saw Japan as a major invader competing with the British and other Europeans to divide the country (Ishikawa, 2010). The Manchurian Incident of 1931 was planned by some Japanese Army officials to break through this tense situation. On September 18, the Japanese Kwantung Army bombed a railway line near the city of Shenyang, and attacked the regional warlord Zhang Xueliang, blaming him for the bombing. The plot enabled the Kwantung Army to occupy three northeastern provinces within a few months (Ishikawa, 2010) and place 30 million people under its control (Mitter, 2013), although Japan never formally declared war against China. Kwantung Army officials gave up their original plot to annex this region to the territory of the Great Empire of Japan due to increasing Chinese resistance and international pressure against Japan's aggression (Yamamuro, 2004). As a compromise, in 1932 Japan declared the state of "Manchukuo,"[1] a place that would practise a "kingly way of governance" and "racial harmony" among five ethnic groups (Manchurian, Han, Mongolian, Korean, and Japanese) (Yamamuro, 2004, p. 10; see also Young, 1998). It appointed as head of state the Qing dynasty's last emperor, Puyi.

Despite the claim of independence, Manchukuo was clearly a puppet regime, as it allowed the Japanese army to occupy the area, and all administrative decisions were made through the army. The League of Nations did not admit Manchukuo, based on an investigation by Victor Alexander Lytton. Dissatisfied, Japan left the League. Even after the declaration, Manchukuo remained in a state of internal war, with the Kwantung Army combatting the armed resistance by China's Nationalists and Communists, as well as local groups. Thus, while presenting the cosmetic appearance of a nation state, Manchuria remained occupied, and was treated as a battlefield by the army. The internal resistance was also generally concealed and unreported to the public.

Louise Young (1998) calls Manchukuo "Japan's Total Empire," with the "total" in this context referring to "multi-causal and multi-dimensional, all-encompassing and, by the end, all-consuming" imperial power (p. 14). Japanese imperialism took every opportunity to harness the colonial population for its own supremacy while repressing it as a potential risk for the empire. According to Young, Manchukuo was unique as a colony in its total mobilization to benefit a sovereign power. Both capitalist and state systems were rapidly constructed by various agents of the imperial imagination on continental soil, seized from Others (Yamamuro, 2004; Young, 1998). This reflects Prasenjit Duara's (2003) discussion of how imperialism more broadly became a significant means of integrating and subordinating different classes into the nation state in the context of global competition. Colonial wars over territory and resources formed imperialist nationalism as a powerful ideology of the nation state, which made possible the mobilization of the population to enrich the state.

In addition, to justify military invasion, it was important for Japan to modernize Manchukuo as a quasi-independent Pan-Asian state. In fact, productive aspects of colonization, rather than repressive ones, attracted more Japanese to the land of opportunity. Building Manchukuo was a collaborative project of many different agencies (Mitter, 2000; Yamamuro, 2004; Young, 1998). Not only right-wing expansionists but also left-wing social reformers, leading scientists, and city planners enthusiastically participated in the project with their own visions and hopes. The novel technology of fingerprint ID card systems also developed as a collaboration between bureaucrats, engineers, businessmen, and soldiers, with the goal of promoting the Manchurian economy. These groups, espousing varying ideas and approaches, attempted to serve democratization, eliminate poverty, or even support the Chinese Revolution. However, those attempts were eventually demolished and incorporated into the accelerating militarism underlying Japan's conquest of China.

Japan's economic interests were in using the resources of Manchukuo to feed the metropolitan market, and as a market for goods imported from Japan. In addition, Japanese economists proposed to solve domestic agrarian issues through emigration to Manchukuo. Japanese peasants had suffered extreme poverty during the country's rapid industrialization, and to create a self-sufficient production sphere, or Japanese-Manchurian "bloc economy," more than a million Japanese agricultural emigrants, entrepreneurs, and soldiers crossed the ocean to the promised land (see Young, 1998). At the same time, the local Chinese population and migrants were mobilized to work in coal mines and other heavy industries to support Japan's ever-expanding territorial battles.

Manchurian economic development became Japan's top priority as it exploited opportunities in the Chinese market.

Under these circumstances of constructing modern capitalism and undeclared war, Japan implemented very different identification techniques from its other colonies, most notably Taiwan and Korea. While the colonial Koseki system attempted to construct "Japanese" by incorporating the Taiwanese and Koreans into the Emperor's property list, the ostensibly independent status of Manchukuo meant that Manchurian subjects could not be Japanese in Koseki. This produced a profound contradiction at the heart of the colonial project. As a state encompassing five ethnic groups, all groups should seemingly have received equal legal status as subjects of the nation. However, the Kwantung Army and Japanese settlers wanted to keep Japan's privileged position in the ethnic hierarchy as a ruling group. Because of this self-created contradiction between the purportedly independent and egalitarian state on the one hand, and the Japanese interests represented by the puppet regime on the other, who could be a subject in Manchukuo was not defined and inscribed in statutory law, remaining as unsolved problem in long debates among the Japanese bureaucrats and army officers (Endo, 2010). As a result, Manchukuo lacked a Nationality Act and a Constitution that defined who the nationals were, and what were their rights.

The large number of Chinese migrant workers coming from outside of the three northeastern provinces posed an additional challenge to systems of identification. The need for labour was driven by the expansion of heavy industries such as coal mining that were providing resources to Japan, and also by the demands of the expansion of aggressive wars in China, Southeast Asia, and the Pacific after 1941. Japan thus recruited and sent many Chinese workers to Manchuria, mainly from Shandong and Hebei provinces. These were most often single male workers, who changed workplaces from time to time in order to look for better opportunities, escape from slave-like work conditions, or return home during the harsh winter (Takano, 2016). The Koseki system was incapable of capturing the constant movements of numerous migrants, as it was based on static records of a settled population. Also, the patriarchal frame of Koseki did not easily encompass lone workers. If the Koseki system was designed to be educational, exerting a disciplinary power that sought to construct a population of loyal subjects, in the context of internal war this proved impossible. The Chinese population was instead more often seen by the Japanese as the enemy within. Thus, Japan wanted to watch over the movements of cheap labour power, but the social and political contexts meant that they were simultaneously seen as a risky source of resistance. A new "solution" was pursued for Manchurian population control, to restrain resistance on the one hand, and to mobilize labour power on the other.

Tracking the Movements of Others

In contrast to the collective Koseki system, based on patriarchal family units and for the educational purpose of identity politics, Japan employed individualistic techniques in occupied Northeast China. This involved two kinds of ID cards, both of which included fingerprints and a photograph of the individual (GSNF, 1987; Tanaka, 1987). The first ID card was issued to workers and migrants, and the second to residents. An important source of information on these ID systems comes from the Group Saying "No" to Fingerprinting (GSNF), which was organized in Japan to support non-citizens who refused fingerprinting for the post-war Japanese Alien Registration System. This group conducted field research in Northeast China in 1987 in order to find the origins of fingerprint identification in the Alien Registration System. The scholar Hiroshi Tanaka (1987) took part in this trip, with the group visiting major cities in the former Manchukuo: Shenyang, Fushun, Changchun, and Yanbian Korean Ethnic Autonomous Prefecture.

According to their research, the South Manchuria Railway had begun using fingerprints to identify its workers at the Fushun Mine in 1924, before the state of Manchukuo was declared in 1932 (GSNF, 1987; Tanaka, 1987). The Fushun Mine was one of the largest coal mines run by the South Manchuria Railway. The GSNF interviewed five men who worked for Fushun Mine and had undergone forced fingerprinting for their employee ID system. Those men explained that the Chinese employees each had ten fingers rolled with ink on paper upon their arrival at the mine, and the company used those fingerprints to determine whether they had escaped from other workplaces or had previously organized labour strikes. If the fingerprints were matched with "blacklists," the workers were denied employment or were sent back to their former workplace. Others escaped from the harsh, slave-like labour conditions found in the mine. If captured, they too were matched with the collected fingerprints and sent back to the mine and tortured.

Asako Takano's (2016) recent research found that Japanese bureaucrats had gathered information about the use of fingerprinting in British colonial management, with the government economic journal translating and publishing an English article about the fingerprinting of Chinese miners in South Africa in 1915. Fingerprinting technologies had been invented in India under the British Empire (Sengoopta, 2003; see also Dhupelia-Mesthrie & Allen, this volume chapter 6), tested again in South Africa (Breckenridge, 2014; Cole, 2001), and then travelled to Northeast China through the transnational colonial networks highlighted in a number of chapters in this book. However, fingerprinting was not welcomed by the locals. Tanaka (1995) found a record of a strike against fingerprinting at a woollen mill in 1926 in which Chinese employees demanded the Japanese company stop fingerprinting, and raise their wages. Despite opposition, however, the system developed to the extent that in 1937 the South Manchuria Railway did not hire 25 per cent of applicants because their fingerprints matched the blacklists (Tanaka, 1987).

In 1933, the Japanese Army established the Labour Control Committee, and began monitoring the labour force with a fingerprinted ID card system, issuing "work permits" to the regional population and "entry permits" to migrants (GSNF, 1987; Tanaka, 1987). This method spread widely when, in 1938 and working under government guidelines, the Manchurian Labour Industry Association began to issue a "labour card" to workers in major industries. This card applied to workers who were employed by a company with more than 13 employees, or in 27 kinds of industries, in 15 cities and 27 districts (GSNF, 1987). In accordance with the General National Mobilization Law, by the end of 1939, all workers in military or related industries aged 14 to 55 were required to register and carry the labour ID card. Most industries at this time were involved in military production, but in subsequent years the regions and industries subjected to the labour ID card system continuously expanded, and by January 1941 all workers in any industry or region employed by companies with more than 10 employees were covered. Beginning in April 1939, migrants were also required to register and provide their fingerprints. In order to implement this advanced new technology, in January 1939, the army proudly established the Fingerprint Management Bureau within the Policing Department. The army also founded the Centre for Training Fingerprint Technicians, which began with 58 students. The centre provided technicians for the bureau, which grew continuously, reaching 155 employees in the bureau headquarters and 227 in the local branches by 1943. Between 1934 and 1940, more than 5.2 million people were fingerprinted for the different Manchurian ID systems (GSNF, 1987).

While conducting research in 2016, I interviewed Lü Guiwen, who, when he was 20 years old, worked at the construction site of Fengman Dam in Jilin Province in 1941–1942. He confirmed the labour card system, saying "the Japanese took my ten fingerprints for their records." According to Lü, the Japanese guards watched over the Chinese workers every day, using a long club, whip, or gun. They hit Lü with the back of the gun when they thought he was slow. He escaped twice. The second time, he hid for one year and survived, though his chest is still bent from the punishment. In Fengman Dam, because of slave-like labour conditions, starvation, and cold, thousands of workers died and their bodies were discarded near the construction site. Their skeletons, discovered after the liberation, are preserved at the Jilin City Workers Memorial Hall. According to the exhibition panels, the skeletons are identified as those of men aged 20 to 40 years old, eight of whom have clear signs of torture and physical punishment. Those sites containing the unknown, abandoned victims of this fatal labour are called "ten-thousand-human hollows," and are located near former Japanese coal mines throughout Northeast China.

As we have seen, fingerprints were used as means of selecting workers deemed desirable, and who had no record of escapes or strikes in the past, but they were also used in military operations to identify members of guerrilla and anti-Japanese movements. Fingerprint identification became a powerful tool for the military tactics of searching, isolating, and killing the enemies of the Empire. The main tool in this fight was a second ID card specifically for residents, which was issued to people who lived in concentration villages that were built by the Japanese Army in strategic locations. The army developed the tactic of "separating bandits from innocents" (GSNF, 1987, p. 24) in the conflict zones in order to cut off the flow of materials and information between the people and the guerrillas (see also Kobayashi, 2008). Existing villages were often destroyed or burned by the army, and the villagers were forced to move to into what were called "concentration villages" of about one hundred houses, which were surrounded with mud walls that enabled the army to check the IDs of people entering. Construction of these strategic villages started around the Korean border by order of the Governor-General of Korea in 1933 (Tanaka, 1987). Accordingly, the first resident ID cards were issued to Koreans living in China near the border who were categorized as good and innocent. GSNF interviewed three men who lived in the collective village of this area. Each said that they were required to carry ID cards, with attached photograph and fingerprints, once they turned eleven years old. They had to show the card whenever they left the village to cultivate their fields or to serve the Japanese Army, and the cards were checked upon their return.

The collective villages eventually numbered 13,000, containing 5 million people (Tanaka, 1987), out of a total population of approximately 30 million (Yamamuro, 2004). Not only were ID cards checked at the gates to prevent guerilla members from visiting the villages, the army also used the data to identify targets on the battlefields. For example, under the "Special Sweeping Operation in the Southeastern Area," 590,000 cards were issued in the subjected area in 1939, and 740,000 in 1940 (GSNF, 1987). All fingerprints taken as part of this registration were to be matched with the watch list of the resistance by the Fingerprint Management Bureau. The army suggested in the "Guideline of Subjugation and Sweep" in 1941 that "the separation of bandits from innocents should be implemented through searching households, taking fingerprints, setting up checkpoints, and arrests. These measures should be undertaken suddenly and by surprise, regardless of time, place, and means" (GSNF, 1987, p. 24).

The two streams of labour and residence ID card systems in Manchuria finally merged into the National Passbook System in 1944 (Takano, 2016). The system began with targeting men 15 years of age or older who had not yet been conscripted, or in voluntary service for the nation, and who were employed in the industries appointed by the prime minister (Takano, 2016). The plan was that the whole population would be required to carry the passbook, but it remains unknown how widely the system was implemented. While conducting research in the Jixi City Museum in Heilongjiang Province, one passbook I saw was 26 pages long. More data were aggregated through the National Passbook System than the earlier labour card, building a more extensive profile, tracing past moves, checking duties as a national, and identifying skills and abilities as a human resource for the nation. The different streams of ID systems expanded toward covering the entire population, though there was a lack of clear definition of who nationals even were.

Bodies as Risky Resources

As described above, the biometric ID cards enabled not only the identification of individuals, but also, because they allowed for the matching of individual bodily characteristics to a larger database, made it possible to trace and intervene in the movement of individuals. Officers, employers, or soldiers could stop people at any given location, ask them to show their ID cards, and ask where they were going and why. The authorities could also record who went where, with whom, and why. Based on these individual records, the authorities could assume what kinds of people they were, classify the population into desirable and undesirable, and make decisions about whether to mobilize them or stop them. Biometric surveillance was both mass based and individualizing, and allowed for a much more active process of identification than that of the static Koseki system. In occupied Manchuria, which lay both outside and inside the empire, Japan adopted powerful techniques of surveillance that directly targeted individual, racialized bodies.

Under the circumstances of Japan's undeclared war on China, the primary purpose of the Manchurian ID systems was to distinguish between people labelled bandits or innocents and therefore pre-empt potential rebellion. As the chapters on policing and dissent in the final section of this book attest, the suppression of political radicalism, real or perceived, was central to most systems of state surveillance, although the dynamics were quite different in the United States and even Romania, where all citizens were perceived as potential threats, but without the colonial and racial dynamics structuring the situation in Manchuria. Thus, while for Japanese colonizers almost everyone in the entire population was a potential bandit or was supporting a bandit, they also needed innocents for labour power. Japan wanted Manchukuo to feed the metropolitan market, by exporting resources to and importing goods from Japan (Young, 1998). The authorities even forced villages and towns to provide a certain number of men to serve the nation during the severe labour shortages that resulted from the expansion of war (Takano, 2016). Lü Guiwen was one of those who were mobilized for forced labour.

Thus, the biometric ID card systems handled dual tasks: sorting the population and exploiting it as a resource. Chinese were distrusted in terms of their loyalty, but necessary as labour; they were a risk, but also a resource. Biometric ID cards had dual capabilities of watching over the movements of Chinese bodies as sources of resistance, while mobilizing the same bodies as profitable labour power, cheaper than Japanese workers, and a source of wealth. In the former process, the body is reduced to a fragment of data that tells a "truth" (Magnet, 2012; van der Ploeg, 1999) to the colonizers, and, in the latter, the body is reduced to a resource for the colonizers to appropriate (Longman, 2001). The bodies never fully become innocent, free, or able subjects in this ID system, because at the point of including the bodies, the ID systems categorize them as risky resources.

To understand such essentially despotic characteristics of colonial ID techniques, Foucault's notion of biopower is no longer adequate to stand on its own. The entire Manchurian population was placed into a state of exception, without any legal protection or rights, to be used merely as resources. While Agamben (1998) found his key example of bare life in the Nazi concentration camps, for the Japanese colonization of China, the ultimate case of inclusive exclusion can be found in Japan's Army Unit 731. The Japanese Kwantung Gendarmerie, the military police, regularly captured Chinese dissidents and then interrogated, tortured, and identified them. When the Gendarmerie classified the dissidents as bandits and found no possibilities to turn them into informants for Japan, it secretly transferred them to Unit 731 (Jin, 2010; Yang, 2016). This process was called the "special transfer" (Yang, 2016, p. 30). The

Gendarmerie developed the institutional criteria for the special transfer in 1938, as a chart of different categories including the "nature of crime," "personal history," "character," and "prospects" (Yang, 2016, pp. 31–2). Selected and transferred were the people classified as "unuseful" or "no possibilities of recovery (to become harmless)" for the Gendarmerie (Jin, 2010, p. 50). Unit 731 systematically used those people for bacteriological experiments and human dissection in a high-tech facility in the suburb of Harbin from 1936 to 1945. Their bodies were called "maruta," meaning "logs" in Japanese, the term capturing their status as living ingredients for the experiments. At least 3,000 people are estimated to have been killed (Matsumura, 2017; Yang, 2016), considered useful as resources, but, as "risky" bodies, not valid to live. If, as Jacob Steere-Williams argues in chapter 2, disease surveillance became central to systems of colonial rule, with colonized subjects seen as vectors of infection, here those subjects are themselves infected and thus rendered as living dead.

Because the Japanese Army destroyed the Unit 731 facilities, killed all prisoners, and burned and buried the documents when they left, it is hard to find which ID card specifically contributed to individual cases of special transfer. As the editors of this book addressed in chapter 1, the deliberate erasure of the negative consequences of colonialism has blocked our view of colonial surveillance and posed challenges for academic inquiry to this day. However, biometric ID systems were clearly part of a broader system of colonial surveillance. And the documents of special transfer discovered by the Chinese scholars evidence the systematic classification and use of Chinese bodies. As a result, colonial surveillance was most often experienced as threat, force, and violence among the Chinese. Although the ID systems were used for productive and repressive purposes for the Japanese, the same systems were experienced mostly only as repression among the Chinese, whether in workplaces, residential areas, or military facilities. The special transfer to Unit 731 is the ultimate case of colonial surveillance converted into unilateral violence. The systems of identification and classification helped to rationalize the naked brutality of colonial rule, and to maintain the systematic production of what Agamben (1998) calls bare life, and Mbembe (2003) described as the living dead. Sheer violence is, again, woven into the civilized techniques of surveillance. Risky resources can be economically useful, but are politically disposable and dispensable.

Conclusion

The colonial ID card systems collapsed in 1945, along with the Japanese Empire. The legal frame of modern patriarchy established by Koseki was also dismantled after the war as a juridical reform, due to its contribution to making loyal subjects for imperial militarism. However, both systems remained in post-war democracy in different forms. Koreans and Chinese who stayed in Japan after the war were unilaterally detached from Japanese nationality and officially labelled as Others in the Alien Registration System. The Alien Registration Law of 1952 required all "foreigners" over fourteen years old and staying in Japan for more than sixty days to be fingerprinted and registered, and to renew this registration every two years (Tanaka, 1995). In the aftermath of the Second World War, the majority of "foreigners" who resided in Japan were ex-colonials. The government issued the Alien Registration Card, making it compulsory to carry the card and show it to police if asked. Fingerprint identification was imported from the ex-colony of Manchuria, and officially legitimated as means of watching over the ex-colonials. Koseki was redesigned for the nuclear family, but preserved family registration, rather than replacing it by individual registration. It remains as the basis of all administrative services today. There is thus a continuation, rather than a cut-off, of both ID schemes between imperial and democratic Japan.

The Alien Registration System faced serious resistance after the 1980s. Many Koreans, Chinese, and other citizens refused to put their fingers in ink, seeing it as sign of criminality and stigma. This collective refusal became a diplomatic issue between Japan and South Korea, as well as the focus of lawsuits in Japan. Although those protesting the system never won in any court, the government announced the abolition of fingerprinting of all foreigners in 1998. However, since 2007, the Ministry of Law has fingerprinted all visitors to Japan at ports of entry, except for ex-colonials, following similar American anti-terror measures. Indeed, fingerprinting made a recent comeback during the "war on terror," after its long-awaited abolition. The Alien Registration System was merged into the new, centralized Resident Control System in 2012 (for the details, see Ogasawara, 2008b, 2012).

Japan implemented biometric ID cards in Manchukuo as a way to achieve the "kingly way of governance" (Yamamuro, 2004, p. 10), but the ID cards served only to assault and exclude the Chinese from governance of their own land. These events remain as a warning to identification systems today. In the context of the all-encompassing ID techniques implemented by multiple agencies of mass surveillance, their utopian dreams of technological possibilities need to always take note of the Others who are treated as risks to the system. The cruel experiments by Unit 731 in particular can be seen as a result of biopower to "*foster* life or *disallow* it to die to the point of death" (Foucault, 1978, p. 138, emphasis in original). The undesirable were eventually killed, but they were not allowed to die until they had produced useful knowledge (useful from the perspective of the colonizer) for imperial "science." However, on the side of the colonized, it was experienced as a direct result of necropower (Mbembe, 2003). A fundamentally asymmetric power between the colonizers and colonized should not be overlooked. For the colonizers, the process was rationalized with the institutional criteria to select and transfer the captured Chinese to Unit 731. But, for the colonized, the experiences can never be rational. It was the Japanese who unilaterally classified which Chinese were unworthy to live, selected some as bare life, and sent them to the bio-warfare facility. The biometric ID cards and colonial surveillance systems assisted such classifications of the population, ranging from using them as labour power to killing them in useful ways. Unit 731 is the ultimate result of colonial classification, where surveillance and violence closely work together. Such marginalized experiences of being surveilled are not only underrepresented in studies, but also often silenced and erased in the immediate site of experience by a deliberate policy of informing or concealing the facts.

Contemporary arenas of ID practices as surveillance, from anti-terrorism measures to refugee policies, should be also looked at against this horizon. Seventy years after the Japanese colonial period, the advancement of all-encompassing and all-consuming surveillance is prevalent in both state and corporate systems. Thanks to digital amena-bility, humans are contentiously reduced to bodies, numbers, or codes, not whole subjects, and are seen as suspect sources of dissent in global conflicts or pursued as sources of profit in neoliberal economies. They are, at best, treated as risky resources. It is no coincidence that formerly colonized countries are more likely to employ biometric systems for tracking their general populations, as this is where these systems were first implemented (Breckenridge, 2014). In systems such as "ID4Africa" (NIDA, 2016), they continue to provide business opportunities for mul-tinational corporations, while in contexts ranging from South African apartheid to the Rwandan genocide, they also legitimate classification and facilitate social inequalities and structural violence (Longman, 2001). In other words, we live with similar technologies, albeit elec-tronically updated, which classify our bodies as risks and resources for ever-expanding capital in a postcolonial era. Understanding historical experiences with colonial ID systems is more important than ever, to decipher current political economic relations of surveillance, and to unmask violent consequences of identification technologies. Civiliza-tion and cruelty are still not antitheses.

NOTE

1 "Manchukuo" is an English translation of 満州国 in Japanese, meaning the state of Manchuria. However, it is usually written as 偽満洲国 in Chinese, meaning the pseudo-state of Manchuria, since it was a puppet regime under total control of the Japanese Army. I thus use this term only in relation to specific Japanese imperial practices.

REFERENCES

Agamben, G. (1998). *Homo sacer.* Stanford, CA: Stanford University Press.
Anderson, B. (1991). *Imagined communities: Reflections on the origin and spread of nationalism* (Rev. and expanded ed.). London, UK: Verso.
Arendt, H. (1968). *The origins of totalitarianism.* New York, NY: Harcourt.
Breckenridge, K. (2014). *Biometric state: The global politics of identification and surveillance in South Africa, 1850 to the present.* Cambridge, UK: Cambridge University Press.
Caplan, J., & Torpey, J. (2001). Introduction. In J. Caplan & J. Torpey (Eds.), *Documenting individual identity: The development of state practices in the modern world* (pp. 1–12). Princeton, NJ: Princeton University Press.
Cole, S. A. (2001). *Suspect identities.* Cambridge, MA: Harvard University Press.
Duara, P. (2003). *Sovereignty and authenticity: Manchukuo and the East Asian modern.* Lanham, MD: Rowman and Littlefield.
Endo, M. (2010). *Nationality and Koseki in modern Japanese colonial governance: Manchuria, Korea, and Taiwan* (近代日本の植民地統治における国籍と戸籍満洲・朝鮮・台湾). Japan: Akashi Shoten.
Foucault, M. (1977). *Discipline and punish.* New York, NY: Vintage Books.
Foucault, M. (1978). *The history of sexuality volume one: An introduction.* New York, NY: Vintage Books.
Foucault, M. (1991). Governmentality. In G. Burchell, C. Gordon, & P. Miller (Eds.), *The Foucault effect* (pp. 87–104). London, UK: Harvester Wheatsheaf.
Giddens, A. (1981). *A contemporary critique of historical materialism.* London, UK: Macmillan Press.
Group Saying No to Fingerprinting (GSNF). (1987). Fingerprints in fake state of Manchukuo. *Institutes of Chinese Affairs, 41*(6), 16–27.
Hacking, I. (1986). Making up people. In T. C. Heller, M. Sosna, & D. E. Wellbery (Eds.), *Reconstructing individualism* (pp. 222–36). Stanford, CA: Stanford University Press.
Hacking, I. (1990). *The taming of chance.* Cambridge, UK: Cambridge University Press.

Higgs, E. (2001). The rise of the information state: The development of central state surveillance of the citizen in England, 1500–2000. *Journal of Historical Sociology, 14*(2), 176–97.

Ishikawa, Y. (2010). *Revolution and nationalism 1925–1945: Chinese modern and contemporary history III* (革命とナショナリズム). Japan: Iwanami shoten.

Jenkins, R. (2000). Categorization: Identity, social process and epistemology. *Current Sociology, 48*(3), 7–25.

Jin, C. (2010). *Collection of pictures on Japanese military germ warfare* (日本軍細菌戦争写真集). China: Inner Mongolia Culture Press.

Kobayashi, H. (2008). *The history of "Manchukuo"* (<満州>の歴史). Japan: Kodansha.

Koseki Act of 1871. National Archive of Japan. Retrieved from www.digital. archives.go.jp/das/image-j/M0000000000000832305

Longman, T. (2001). Identity card, ethnic self-perception, and genocide in Rwanda. In J. Caplan & J. Torpey (Eds.), *Documenting individual identity: The development of state practices in the modern world* (pp. 345–57). Princeton, NJ: Princeton University Press.

Lyon, D. (2007). *Surveillance studies: An overview.* Cambridge, UK: Polity.

Lyon, D. (2009). *Identifying citizens: ID cards as surveillance.* Cambridge, UK: Polity.

Magnet, S. A. (2012). *When biometrics fail: Gender, race and technology of identity.* Durham, NC: Duke University Press.

Marx, K. (1867/1976). *Capital,* vol. I. London, UK: Penguin Books.

Matsumura, T. (2017). The special transfer by the Kwantung Gendarmerie: To reveal the whole history of biowarfare Unit 731. In F. Ogino, T. Kojima, K. Eda, & T. Matsumura (Eds.), *Resistance and repression in "Manchukuo"* (満洲国における抵抗と弾圧) (pp. 287–348). Japan: Nihon Keizai Hyoronsha.

Mbembe, A. (2003). Necropolitics. *Public Culture, 15*(1), 11–40.

McCoy, A. W. (2009). *Policing America's empire: The United States, the Philippines, and the rise of the surveillance state.* Madison, WI: University of Wisconsin Press.

Mitter, R. (2000). *The Manchurian myth: Nationalism, resistance, and collaboration in modern China.* Berkeley, CA: University of California Press.

Mitter, R. (2013). *Forgotten ally: China's World War II, 1937–1945.* New York, NY: Houghton Mifflin Harcourt.

National Identification Agency of the Republic of Rwanda (NIDA). (2016). ID4Africa: The annual forum and expo on electronic identity in Africa. Retrieved from www.id4africaforum.com/2016/index.html

Ogasawara, M. (2008a). A tale of colonial age, or the banner of new tyranny?: National identification systems in Japan. In C. J. Bennett & D. Lyon (Eds.), *Playing the identity card* (pp. 93–111). Abingdon, UK: Routledge.

Ogasawara, M. (2008b). *ID troubles: The national identification systems in Japan and the (mis)construction of the subject.* Master's thesis. Queen's University.

Ogasawara, M. (2012). Identify and classify all population: The politics of surveillance in the common number system, resident card, and secrecy law (全人口を識別し、振り分けよ：共通番号、在留カード、秘密保全法にみる監視の政治). *Braku Kaiho, 666,* 64–73.

Rubenstein, R. (1975). *The cunning of history: Mass death and the American future.* New York, NY: Harper & Row.

Sato, B. (1988). *The reversed history of Koseki* (戸籍裏返し考). Japan: Akashi shoten.

Sato, B. (1991). *The life that Koseki watches over* (戸籍が見張る暮らし). Japan: Gendai shokan.

Sengoopta, C. (2003). *Imprint of the Raj: How fingerprinting was born in colonial India.* London, UK: Palgrave Macmillan.

Takano, A. (2016). *Fingerprint and modernity* (指紋と近代). Japan: Misuzu shobo.

Tanaka, H. (1987). The origin of fingerprinting. *Asahi Journal, 9,* 21–3.

Tanaka, H. (1995). *The foreigners in Japan* (在日外国人). Japan: Iwanami shinsho.

van der Ploeg, I. (1999). The illegal body: "Eurodac" and the politics of biometric identification. *Ethics and Information Technology, 1*(4), 295–302.

Weber, M. (1946). Bureaucracy. In H. H. Gerth & C. W. Mills (Eds.), *From Max Weber* (pp. 196–244). New York, NY: Oxford University Press.

Yamamuro, S. (2004). *Chimera: A portrait of Manchukuo* (revised ed.) (キメラ満洲国の肖像増補版). Japan: Chuoukoron shinsha.

Yang, Y. (2016). *Japan's biological warfare in China.* China: Foreign Language Press.

Young, L. (1998). *Japan's total empire.* Berkeley, CA: University of California Press.

Zureik, E. (2011). Colonialism, surveillance, and population control: Israel/Palestine. In E. Zureik, D. Lyon, & Y. Abu-Laban (Eds.), *Surveillance and control in Israel/Palestine: population, territory, and power* (pp. 3–46). Abingdon, UK: Routledge.

8 A State of Exception: Frameworks and Institutions of Israeli Surveillance of Palestinians, 1948–1967[1]

AHMAD H. SA'DI

From 1923 to 1948, what is currently the State of Israel and the Occupied Territories was governed by Great Britain under the League of Nations (later the United Nations) mandate system, which legitimated a form of quasi-colonial rule over territories, such as Mandatory Palestine, deemed not ready for independence. On November 30, 1947, the United Nations General Assembly adopted a partition plan that would see Mandatory Palestine divided into Palestinian and Jewish states, triggering a civil war that broadened into the Arab-Israeli War of 1948, during which the Jewish community (the Yishuv) declared the State of Israel in Mandatory Palestine. During the last stage of the 1948 War, and following the almost total cleansing of the Palestinians from the territories that were assigned to the Jewish state according to the United Nations partition resolution, Israeli leaders and military commanders decided to go ahead with the occupation of the parts of the Galilee that were allocated to the Palestinian state. This was to be done promptly in order to forestall any intervention by the international community. Consequently, a double movement military campaign was launched in order to reach the international borders of Palestine with Syria and Lebanon, thus leaving large Palestinian communities unremoved. Moreover, in the peace talks that followed the war, Israel laid claim to and subsequently received the Jordanian-controlled Triangle area. The Palestinian residents of these regions along with the small, scattered communities in some cities and Bedouin tribes in the Negev constituted a sizeable minority. Israeli leaders and military commanders viewed the existence of these Palestinians as momentary, expecting their transfer in an imminent second round of violence. And indeed various transfer plans were prepared.

With the passage of time, Israeli leaders continued to cling to the transfer option along with institutionalizing the ad hoc measures of surveillance, population management, and political control of Palestinians.

Given this, Palestinians were to be governed not by state bureaucracy and normative law, but rather by an Israeli Military Government (*Mimshal Zvai* in Hebrew). This military rule was based on ostensibly temporary emergency regulations that had originally been enacted in 1945 mainly to fight Zionist terrorism by the British Mandatory government. These regulations formed the basis for Israeli policy towards Palestinian citizens of Israel throughout the 1948–1967 period that I am examining here, and continued to shape Israeli practice after 1967. It was at this point that Israel occupied Gaza, the West Bank, and the Golan Heights, thus establishing control as well over a large non-citizen Palestinian population. While much remained the same, these occupations also changed the dynamics of surveillance, not least as a result of the establishment of Jewish Israeli settlements in those territories.

Military rule was implemented earlier, but the Military Government itself was officially established on September 3, 1948, and on May 14, 1950, Prime Minster David Ben-Gurion decreed that the various state ministries would deal with Arab issues only through the military governor (Kafkafi, 1998). The Military Government was headed by a military general who was responsible for all aspects of Palestinians' lives, and who was part of two hierarchies: the military and the civilian. On issues under military authority, he reported to the Chief of Staff; on civilian issues, he worked under the Minister of Defence. The area under military rule was initially divided into five regions, but was divided into three from 1950 onwards: the Northern District (the Galilee), the Central District (the Triangle), and the Southern District (the Negev) (Peretz, 1991), each of which was headed by a military commander. The Arab population that remained in the cities of Haifa, Jaffa, Lydda, Ramle, and Al-Majdal (before their transfer during the early 1950s) was concentrated in poor neighbourhoods and was put under military rule until July 1, 1949 (Ozacky-Lazar, 2002; Segev, 1998), when Jewish immigrants were settled in deserted Arab houses, thus converting some of these cities – which had all along been Arab cities – into mixed ones.

While some have suggested that the Israeli Military Government reflects a kind of panoptic model wherein the projection of power induces compliance among a large and segmented population through the establishment of extensive surveillance networks, this chapter argues that in many respects the Israeli government's surveillance of Palestinians does not in fact reflect the dynamics of Bentham's (1995) and Foucault's (1991) panopticon. In extending and entrenching the Mandatory state of emergency, power and surveillance were not routinized, but instead constituted a state of exception. Drawing on historical documents and archival sources, I make this argument first through an

examination of Israeli state policies and institutions, including the Histadrut, the umbrella labour organization that worked closely with the
state, and then I detail how these diverged from the panoptic model.
I follow this discussion with a consideration of the ways in which the
panoptic model remains useful in understanding some aspects of Israeli
surveillance, including its "intimate" forms, and in particular in terms
of how it enabled the social sorting (see Lyon, 2003) of Palestinians.
Building on these discussions of the nature of Israeli state surveillance,
I next turn to some of its implications. The state of exception entailed
forms of spectacular surveillance and punishment that were bound up
with what I call the pleasures of control. The state of exception, I argue
in conclusion, also produced an aestheticization of power that sought to
obscure both the realities and contradictions of Israeli surveillance and
the repression of Palestinians.

Structural Bases of Surveillance and Control

Michel Foucault (1991) argued that surveillance and population management in liberal societies usually take place in the grey area between
legality and illegality. The Israeli discourse of control, however, could
not be reconciled with a reasonable interpretation of a liberal legal
framework. As Midori Ogasawara demonstrated in the previous chapter
in the case of Japanese colonial rule in Manchuria, these kinds of contradictions emerged in many colonial contexts. Emergency laws therefore
were the only accommodating alternative. Indeed, Israel's first piece of
legislation passed soon after the state's declaration of independence was
the Law and Administration Ordinance (1948), which affirmed the continuity of the legal system that had existed hitherto, including the Mandatory Emergency (Defence) Regulations (1945), with the exception of
those laws that restricted Jewish immigration. The Law and Administration Ordinance further enabled the provisional government to declare a
state of emergency, to suspend or alter any law, and to extend the emergency regulations.
    The war preceding the establishment of the State of Israel saw the dec-
laration of a three-month state of emergency, although, as Walter Ben-
jamin has rightly suggested, such a regime of exception often becomes
the rule (see the discussion in Agamben, 2005). In fact, the state of
emergency in Israel has never been revoked, and emergency regulations, which curtail the rule of law, were enforced almost exclusively on
Palestinians between 1949 and 1966. The number of emergency regulations varied as, over the years, Israel made some additions to and subtrac-
tions from the Mandatory ones. While there were some 150 regulations

operating during the first two decades of the state (Segev, 1998), Mishal Shoham, who headed the Military Government, stated in 1958 that his apparatus relied mostly – though not exclusively – on just six regulations: 108, 109, 110, and 111, which were "used against individuals and make possible their placement under police supervision," as well as 124 and 125, which related to "territories and crowd" (Arab Affairs Committee, January 30, 1958a, pp. 5–6). Together, these regulations imposed many restrictions, including allowing the military commander to imprison Palestinians without charge for up to six months, to keep evidence secret, to impose curfews, and to restrict communication or ban organizations. Most importantly from the perspective of surveillance, the regulations granted the military commander extensive control over the movement of people, including the ability to restrict Palestinians to certain areas and require them to register with police, as well as an extensive pass system enabling systematic surveillance (Jiryis, 1976; Kretzmer, 1990).

By the end of 1949, the Military Government was composed of some 1,000 employees (Segev, 1998); over the next decade, however, its staff steadily declined and by 1958 they numbered just 116 persons, 87 of whom were assigned a variety of administrative and operational duties (such as liaising with the local population), while the remainder comprised three squads for escorting and patrolling (Arab Affairs Committee, August 14, 1958b). This shrinking staff was responsible for a rapidly growing population. For example, in 1958, the 116 staff members ruled over 180,000 Palestinians and had to fulfil the formidable duties that were entrusted to the government. In addition to its main task of stopping the return of Palestinian refugees, the Military Government was responsible for a number of key assignments, which included imposing various emergency regulations such as closing certain areas, overseeing curfews, and confining movement, as well as documenting and gathering information on Palestinians, appointing *mukhtars* (heads of villages or neighbourhoods), advising on the appointment of teachers and civil servants, and more (Arab Department, n.d., pp. 2–3).

Perhaps the most salient feature of the Military Government was its projection of state power, which was occasionally no less significant than its deployment. In this regard, Uri Lubrani, the third Advisor to the Prime Minister on Arab Affairs, summed up the significance of the Military Government at the end of its first decade as follows:

1. It represented, to a frightened, segmented, and distressed population, the new regime.
2. It presented to this population the military power which this regime has built.

3. It comprised the only address for all state branches which were active in the Arab sector. As such, every Arab citizen felt dependent in his everyday life on the military governor of his area.
4. Through Mukhtar, Sheikhs, and heads of Hamulas (extended families or clans), it had been able to rule over an entire population by a very small staff. (quoted in Bauml, 2007, p. 224)

Variances between the Panoptic and the Military Government

Lubrani's assessment might create the impression that the Israeli administration was reminiscent of Bentham's panopticon, where the projection of power induces compliance among a large and segmented population through the establishment of extensive surveillance networks. Moreover, the small staff of the Military Government and the popular characterization of the areas under its rule as a prison might have amplified this image (Eyal, 2006; Ozacky-Lazar, 2002). The metaphor of the panopticon, however, merits further exploration. In fact, in various ways, the Military Government was, I argue, strikingly different from the panopticon, especially with regard to its goals. The objectives of the Military Government were not confined to surveillance and normalization as in the case of Bentham's panoptic prison; rather, they stemmed from a generalized conception of state security. In 1958, Mishal Shoham, the head of the Military Government, made a distinction between two conceptions of state security: overt/direct and covert/accumulative. The former includes the aims of preventing the return of refugees, smuggling, and espionage, as well as preventing the establishment of Palestinian organizations deemed hostile to the state, and the seizure of Palestinian lands for military training (both when it was necessary and when it was not, Shoham stated) (Arab Affairs Committee, August 14, 1958b).

The covert/accumulative conception of security, on the other hand, encompassed five goals. These did not have an immediate effect on security, but their accumulation, according to Shoham, bolstered it:

A) To prevent the rehabilitation of deserted villages by their inhabitants who became internal refugees and who lived in nearby localities. For example, in the Galilee there were ninety-seven such villages whose 20,000 residents were scattered in fifty-four villages and the city of Nazareth.
B) To stop the Palestinian workforce from reaching the labour market in the cities and Jewish settlements in order to keep the available jobs for Jewish migrants.

C) To prevent Palestinians from moving in "security-sensitive areas." These areas were composed of the country's major part: from Benyamena (to the south of Haifa) to the south of the Negev.

D) To limit the seizure of state-declared lands. These lands were frequently declared as closed areas, "[assigned] for military training or the disguise of training." Moreover, it was intended to facilitate land settlements with regard to the confiscation, registration, and purchase of Palestinian lands.

E) To protect newly established Jewish settlements that were physically and organizationally weak by preventing Palestinians from passing through their lands. (Arab Affairs Committee, August 14, 1958b, pp. 7–8)

Another goal to which Shoham alluded but did not elaborate upon was the transfer of Palestinians should an opportunity for such a move emerge: "They know that we shall not act like the Mandatory government in the 1937 events or similar to the way the French act in Algeria. Our way will be either us or them, therefore they are very cautious" (Arab Affairs Committee, August 14, 1958b, p. 4). Indeed, in line with the comprehensive plan that was laid down in 1958 on how to surveil and manage the Palestinians, an internal memorandum of the Military Government specified that in the case of war, it should "encourage and make it possible for certain parts of the population to move to neighbouring countries" (quoted in Eyal, 2006, p. 154). Such an atmosphere trickled down to the level of the soldiers and policemen who were stationed in Arab localities. Over the course of the trial of the military personnel who took part in the Kafr Qasim massacre, which was carried out by the Israeli army in an Arab village located on the border of Israel and the Jordanian-controlled West Bank on the eve of the Suez War in 1956, it was revealed that in line with the oral tradition – Torah Shi-Ba'al Peh – a leaflet distributed in the battalion stated: "[From here the Arabs ought to be] going to Jordan [i.e., the West Bank], maybe there would be a need to give a punch here and a slap there, to let them escape beyond the borders, [there] they can do whatever they like" (Rosenthal, 2000, p. 18).

Shoham used a canonized conception of security, described here in the words of Kretzmer (1990):

[S]ecurity of the state is synonymous with security of the Jewish collective, and that is often seen as being dependent on promoting "Jewish national goals." Acts that strengthen the Jewish collective are perceived as acts that promote security. On the other hand, acts that tend to strengthen Arab national aspirations among Israeli Arabs are regarded threatening to the Jewish collective. (p. 136)

It is thus instructive to view surveillance as a tool that aims to achieve these so-called Jewish national goals.

Another key difference between the Military Government and the panoptic metaphor relates to the way in which power was practised. The Military Government was not based on routinized procedures or sets of rules and rituals as in the institutions described by Foucault; rather, in representing a state of exception, it was based on unrestricted arbitrary power. Indeed, the emergency regulations gave the military and its governors unlimited authority that was subject neither to administrative nor to judicial reviews. For example, the first head of the Military Government, Colonel Elimelech Avner, thought that these powers would make each governor an "absolute monarch" in his small domain (Pappe, 1995, p. 639). Later Avner's main job would become protecting Palestinians from acts of revenge and looting by his own staff (Robinson, 2005). The head of the Military Government, along with the Prime Minister and his Advisor on Arab Affairs, however, would not be bothered as much by the abusive behaviours towards Palestinians as by the collapse of discipline within the organization. Given the quality of the soldiers and the nature of their work, disorder and corruption were inevitable. The soldiers mostly came from the human surplus of the army: they were either unfit due to age or health or had been injured in battles (Robinson, 2005; Segev, 1998). The only body that could have imposed restrictions, the High Court of Justice, ruled that "it cannot interfere in the military governor's absolute discretion when he is driven by security considerations, and that the military governors are not to be interrogated regarding their reasoning as this might endanger state's security" (Jiryis, 1976, p. 20).

A final difference between the Military Government and the panopticon model has to do with Bentham's main concern, namely, utilitarianism. The Military Government did not rule effectively over the Palestinians; instead, it was the outer layer of multiple control and surveillance apparatuses. It has drawn the attention of researchers, however, because of both its visibility and the legal powers awarded to it by the emergency regulations. As Lubrani put it earlier, it was the symbol of the occupying army. In governing the Palestinians, the Military Government was aided by various bodies that were in charge of surveillance and security directly as well as with organizations employing subtle forms of power. These organizations include first and foremost the Shin Bet (General Security Services – Sherut Bitachon Klali). Established in the summer of 1950, the Shin Bet's main assignment was the prevention of sabotage and espionage activities, yet it engaged in wide-ranging surveillance of various aspects of Palestinian lives, including monitoring them in classes, offices, mosques, public spaces, and social gatherings to learn about their political attitudes (regarding Shin Bet's more recent political surveillance, see Khoury & Yoaz, 2007). Such activities were conducted in addition to the more common surveillance tactics of wiretapping, intercepting mail, and bugging communication systems. The Shin Bet also screened, and in many cases continues to screen, Palestinian candidates for positions in state and public sectors such as teachers, headmasters, inspectors, bureaucrats in state and Histadrut-related bodies, and functionaries in Islamic religious institutions. Its decisive recommendations were passed to the Office of the Advisor on Arab Affairs. Further, the Shin Bet gave advice to policymaking bodies regarding the policy options towards the Palestinians (Benziman & Mansour, 1992).

In addition to the political power of Shin Bet, the police played, and continues to play, an important role in the surveillance of Palestinians. Besides its duty of maintaining law and order, police had other assignments in the Palestinian-populated areas, including political surveillance and control. Police services, particularly "the department for special assignments" (*Matam*), were entrusted with surveillance over the Palestinians as well as coordinating police activities with the Shin Bet and the Military Government. The *Matam* operated sections at the district and the regional levels, known as *Latam*.

Together, the Military Government, the Shin Bet, and the police (*Matam*), along with the Prime Minister's Advisor, coordinated the running of Palestinians' everyday lives. This coordination was carried out at two levels: the Central Committee (*Hava'ada HaMerkazit*), through which the overall policies, conduct, and activities of these bodies were coordinated; and the district committees (*Va'adot Mirchaviot*). The Central Committee was headed by the Advisor on Arab Affairs and included the head of the Military Government and representatives of the Shin Bet and the police. Its responsibilities revolved around, among other things, screening of candidates for teaching and dismissal of teachers on political grounds and deciding on the awarding of licences for taxis, trucks, the opening of businesses, and much more (Arab Affairs Committee, August 14, 1958b; Avivi, 2007; H. Cohen, 2006). Meanwhile, the three district committees (corresponding to the areas under the Military Government) were composed of representatives of the security agencies along with the head of the Regional Bureau of the Advisor's Office, and were led by the Regional Military Government Commander. The district committees were in charge of running the day-to-day activities of the

Palestinians at the micro-level as well as composing recommendations to the Central Committee and to state offices.

While the functions of these organizations were to supervise, punish, inhibit, disallow, restrict, suppress, and expropriate, the role of the Histadrut, the umbrella labour organization, was to incorporate Palestinians into state structures and the economy as second-class citizens. Formalized in August 1949, it was agreed with the Advisor on Arab Affairs that the Histadrut would be in charge of, among others things, banking, marketing organizations, transportation, local cooperatives, and the awarding of credit, with the aim of serving political ends: "The development of the Arab economy has to contribute to the struggle against forces in the Arab community that oppose de facto or de jure the Israeli state, its security or development" (Mol & Palmon, August 21 and 25, 1949, n.p.). While the Histadrut promoted Arab agricultural products in Jewish cities and settlements by setting up open markets, it was aided by the Military Government in establishing shops in Arab villages where Israeli products were sold in order to "circulate" Palestinian money to the Jewish economy (Histadrut's Arab Department, n.d.), as the following excerpt from a 1956 letter from Histadrut's Arab Department exemplifies:

> 15.2.56
> To Atta Ltd. [textile company]
> The northern district
> Haifa
> Permission [was given for the opening of an] Arab cooperative store in Deir Hanna, Nazareth district ... [It] was organized by our department and it comprises a tool for *the introduction of Israeli industrial products to Arab villages ...*
>   Sincerely, Yaakov Cohen (February 15, 1956, emphasis added)

The Histadrut had other roles of social significance as well. It was the main supplier of vital services, providing health insurance and health services through its nationwide clinics, called *Kupat Holim*, and training for paramedical personnel. As a workers' union, it provided protection for Palestinian employees after their acceptance into the labour federation in 1960 (Histadrut's Arab Department, n.d.). And it established sports clubs in Palestinian villages, particularly soccer teams, soccer being the most popular game among Palestinians.

Alongside these activities, the Histadrut aimed to influence Palestinian consciousness through wide-ranging educational and cultural activities. For example, it set up in some localities "clubs [that] included a library, a reading hall, games and newspapers" (Histadrut's Arab Department, n.d., p. 4). These clubs also screened films, showed plays, and hosted public lectures. As early as 1961, special attention was paid to Palestinian women because, as one Histadrut report revealed: "From our activity in this field, we learnt that Arab women are susceptible to our *Hasbara* [propaganda] and are ready to be incorporated in the Histadrut's and state's life" (Histadrut's Arab Department, n.d., p. 6). Therefore, various courses for women were created – modelled on colonial education for Native women – and focused on teaching house management, handicrafts, and Hebrew.

In order to more widely disseminate Israeli propaganda (*Hasbara*), the Histadrut's Arab Department assumed in 1960 the management of *Al-Yom*, the semi-official Arabic daily. It also published a number of other newspapers and magazines designated for various audiences, including several geared to schools and teachers. The Histadrut further published books in Arabic, mostly translations from Hebrew, as well as calendars that emphasized Israeli dates and celebrations. Some of these publications were meant to impress foreign audiences in particular. The Histadrut tried to influence the Palestinian population (not only the literate) in other ways as well. Films, for example, were screened in Palestinian villages either for the purposes of distraction or to transmit hidden messages,[2] and a theatre group of Iraqi Jews (Ohel group) was established and performed plays written by Arab playwrights, such as the classical romance *Majnon Lila*, to which the cultural attachés of foreign delegations were invited, including from France, Russia, America, and Britain (see Barkatt, September 4, 1956a and September 5, 1956b).

The Histadrut's scope of influence, therefore, was not limited to local communities, but extended into state and government politics. Indeed, it worked to enlist Arab support for the centre-left Mapai, the dominant party in Israeli politics prior to its merger into the Labour Party in 1968. Given this range and diversity of activities, a report by the Arab Affairs Committee (1968) stated:

> The Histadrut is the main public body which materializes Israel's presence in the Arab villages year-long. In Arab villages there are almost no branches of governmental ministries or [Mapai] party's branches. The Histadrut is the only body that occupies buildings and centers of activities which show in practice, through signs, flags etc. the presence of Israel in the Arab villages, small as large, and this is important. I would say that the Histadrut became hegemonic in the social, cultural and political aspects. (p. 12)

Amnon Linn, who served as director of Mapai's Arab Department in the Northern District between 1951 and 1965 and nationally between 1965 and 1969, was similarly unambiguous in describing the Histadrut's social and political role:

> I remember ... [in the 1950s and early 1960s] Mapai's activity among the Arab population was carried out by the Histadrut's Arab department ... Although the Histadrut was a general [union], but the Arab department was more or less homogenous ... I am telling this because the Histadrut's Arab department executive was controlled by us; it was possible to use it as Mapai's instrument in the villages. Our delegates and the Histadrut institutions in the villages were our operative arm. Then we could do in the Histadrut all what we desired. (Arab Affairs Committee, June 6, 1968, pp. 3–4)

Despite this, Histadrut activities did not bring about the anticipated success. Yaakov Cohen of the Histadrut's Arab Department had to admit that the organization's activities

> don't affect wide audiences in the Arab localities ... [It] ought to be considered, whether we are content with these dimensions (of participation) or require wider circles to collaborate and get immersed in cultural, social, sport or artistic activities of the Histadrut as a sign of identification with the state's values and orientation. (Secretariat of the Arab Affairs Committee, June 24, 1964, p. 1)

It seems that these state values were the problem. Despite the persecutions of the founders of independent Palestinian clubs established in some villages in the Triangle area, they managed with meagre resources to compete with the Histadrut's clubs.

Israeli Panoptic Practices

While the model of the panopticon does not fully capture the dynamics of the Military Government's surveillance practices – including but not limited to the activities of the Shin Bet, the police (*Matam*), and the Histadrut – in other ways, the model does reflect the activities of the State of Israel. This section will thus explore three interconnected examples of the Military Government's panoptic surveillance and control: its fixing of the Palestinian population to specific geographic spaces and localities, focusing in particular on the Negev Bedouins; its binary classification of Jew and non-Jew, which resulted in both a legal duality and the establishment of military courts alongside civilian ones; and its intimate and continual surveillance and registration, in part through Palestinian collaborators and their operators.

The three regions under the Military Government's control – Galilee, the Triangle, and Negev – were divided and subdivided into smaller units, which in many cases formed the boundaries of a single locality; for example, until 1954, Galilee was divided into 46 areas, and passes were required to move between them (Kafkafi, 1998). This restriction and regulation of movement was a common dimension of many colonial regimes, as we can see in a number of other chapters in this book (see Jacob Steere-Williams, chapter 2; Ian Warren & Darren Palmer, chapter 5; and Uma Dhupelia-Mesthrie & Margaret Allen, chapter 6). Even after the relaxation of restrictions, the areas under military rule were divided into 16 units. These spatial divisions were used as the criteria according to which the military commanders made decisions with regard to the allocations of permits, supplies, transportation, and services. Social communications and relationships between residents were consequently confined to their areas of residence, thus giving rise to localism (Eyal, 2006). Moreover, such divisions made it easier to control Palestinians through state-sponsored clubs and programs, such as those noted above. In this regard, Reuven Barkatt, who headed the Histadrut's Arab Department during the 1950s and was general secretary of Mapai from 1962 to 1966 stated: "clubs should be established and not only for youth and young people ... where they can play their games and drink coffee etc. In this way it becomes possible to concentrate, and [consequently] influence them" (Confined Secretariat, March 19, 1964, p. 9).

In line with this objective, various measures were taken to concentrate Palestinians in small areas. For example, on Prime Minister Ben-Gurion's request to settle Palestinian commuter workers in Haifa, the city's mayor, Abba Hushi, set up a housing company with Archbishop Maximos V Hakim to build and market apartments, yet not much came out of it. Palestinian workers preferred to commute rather than be packed up in dense neighbourhoods. More generally, Aharon Becker, secretary of the Histadrut from 1961 to 1969, stated: "we should have an interest in the thinning-out of Arab villages. Otherwise, by our deeds, we would ensure the continual concentration of Arabs in one place [i.e., region]" (Confined Secretariat, March 19, 1964, p. 6).

Perhaps the clearest example of the aim to control Palestinians through geographic concentration can be found in the attempts to forcibly resettle the Negev Bedouins. Although they were confined to the *Seig* (an infertile area in the northeastern part of the Negev; the name was given to this area by the Israeli authorities and it means "the enclosure") after the 1948 War (which amounted to less than 10 per cent of the areas in which they had previously lived), two plans were developed to restrict them to much smaller areas. There is no doubt that the plans for the Bedouins stemmed from the Zionist principle of "liberating the land," yet they also reflect the principle of fixing the population in small fragmented zones. The first plan in 1960, outlined by Moshe Dayan – a revered Israeli commander of the 1948 War who later became Chief of Staff, Knesset member, and, between 1959 and 1964, Minister of Agriculture – aimed to settle the Bedouins in working-class neighbourhoods in the mixed cities of Ramle and Jaffa and in the town of Beersheba. Meanwhile, the second plan in 1962, devised by Yigal Allon – a 1948 War military commander and Minister of Labour during the 1960s – aimed at concentrating them in a small number of townships. Both plans were premised on the spatial confinement of the Bedouins (see, for example, Bauml, 2007). Meanwhile, Bechor Sheetrit – the Minister of Police between 1949 and 1967 – thought that such plans ought to take into account the Bedouin's social structure. Thus, regarding the first plan by Dayan, Sheetrit maintained: "I was against this. If you don't uproot the whole tribe you achieve nothing. If [only] part of the tribe leaves, the tribe will grow again and nothing is achieved ... you achieve results if you move the [whole] tribe from one place to another" (Confined Secretariat, March 19, 1964, p. 4).

A further similarity between the Military Government and the panopticon is the employment of a binary classification in which surveillance enabled practices of social sorting (see Lyon, 2003). It is obvious from above that a binary division of the population into Jews and non-Jews constituted a cornerstone in Israeli policy, which, as I have argued elsewhere (Sa'di, 2014), also involved policies to remake the Palestinian population into a divided collection of minorities. The imposition of the Military Government on Palestinian-populated areas meant, in the legal sphere, the establishment of two legal systems: one for Palestinians and another largely for Jews. Although the emergency regulations were stated in universalistic terms, their application was almost always confined to Palestinians. For example, military governor Shoham stated in 1958 that the areas under the Military Government were drawn in such a way as to be imposed on Palestinian localities only: "Understandably exempted from this [area are] the Jewish settlements, the Circassians (Kafr Kama) and every person who served in the Israeli Defence Force and carries with him this certificate or he is in the reserve service or during a compulsory service" (Arab Affairs Committee, August 14, 1958b, p. 6). The discrimination in the application of the law was not only on spatial criteria but on ethnic grounds as well. This is noted by the state comptroller in his 1957/1958 report:

> An order from the military governor declaring an area closed is, in theory, applicable to all citizens without exception, whether living in the area or

outside it. Thus anyone who enters or leaves a closed area without a permit from the military governor is in fact committing a criminal offense. In practice, however, Jews are not expected to carry such permits and in general are not prosecuted for breaking the regulations. (Quoted in Jiryis, 1976, p. 26)

Another dimension of the legal duality was the establishment of military courts alongside civilian ones. While the emergency regulations specified the nature of offences that were tried in either of these courts, military commanders were given discretion in deciding the choice of the court that would deal with any case. Military courts, which epitomized the arbitrary judicial system that was the essence of the state of emergency, were of two types. The first was composed of three officers (who did not necessarily have legal education), who were mandated to deal with any breach of emergency regulations and to pass any verdict the officers deemed appropriate. The second type of military court, of a lower rank, was composed of a single officer who could pass sentences of up to two years' imprisonment and impose fines. Until 1963, the verdicts of these courts were final. Many Palestinians passed through this legal system; from March through December 1951, some 2,028 Palestinians stood in these courts (Korn, 1995). The tribunal for the prevention of infiltration was another body established on the basis of the Prevention of Infiltration Law (1954). It consisted of a one-officer court that was authorized to deal with all offences of this law, though an appeal could be filed to request a tribunal of three officers. This type of tribunal operated until 1959, when offences under this law were transferred to civilian courts.

To make these arrangements more effective, the police force acting in Military Government zones was put under military authority (Korn, 1995). The implication of this duality, according to Korn (1995), was the criminalization of Palestinians on political grounds:

Many categories of crime are a clear "outcome" of the political character of the law and its selective implementation on the Arab population. During the military government ... crime in the Arab population was, to a large extent, a result of political control over it ... the political use made of the criminal law, both in respect of its content and the methods of its enforcement, played a central role in "creating" crime and delinquency among Arabs ... [Thus] a very broad area of social, economic and political activity was defined as "crime" and was dealt with by the rhetoric and practices of crime control. (p. 659)

Finally, intimate surveillance was sought not only through Palestinian collaborators – who passed on information made public in social gatherings or while travelling on public transportation (for example H. Cohen, 2006) – but also through their operators. The regional representatives of the Military Government were required to live in the area under their supervision in order to obtain first-hand and unfiltered information when necessary and to be in reach of the *mukhtar* and collaborators. They preserved what might be considered a primitive archive, a "record of sins" in which the names and addresses of offenders and their punishments were recorded (Eyal, 2006, p. 155). More generally, Cohen (2006) maintained that the security agencies paid special attention to the method of face-to-face interview with Palestinians, assuming that the balance of power in such encounters was in their favour as the majority of their interviewees would be anxious and shaken, thus enabling them to use ways and means of hearsay, promises, or intimidation. The desire for intimate knowledge took two forms: attempts to get in-depth knowledge, which took the shape of psychologism such as the construction of pseudo-psychological profiles for "leaders and collaborators by the security agencies" (H. Cohen, 2006, p. 21), and the opening of a file by the security agencies for any Palestinian who approached any of the state's institutions for any reason: work, licence for business, a pass, permit, etc. (Bauml, 2007).

The State of Exception and the Suspension of the Normative Law

I have discussed thus far the structures of power that characterized the Military Government and the ways in which power was deployed to control the Palestinian population. It is important, however, to also consider this form of power through the premise of exceptionalism. What characterized the state of exception was not only the suspension of the normative law but the awarding of power to people who would render the law irrelevant and whose behaviour would turn any appeal to justice or rule of law a mockery. The lack of legal foundation or recourse enabled Israeli officials to bypass accountability, with Prime Minister Shimon Peres even claiming that "the military government is a small apparatus that does not oppress the Arabs ... [it even] helps them" (quoted in Kafkafi, 1998, p. 360). The state of exception was not devoid of laws and regulations, but emergency laws possess different motivations and ends than normal ones. In this regard, it is worth distinguishing *rule by law*, which is characteristic of emergency laws, from the normative *rule of law*.

In the first instance, rules and regulations are not universally applied as they are considered another tool of domination. Therefore, the law loses its objective and universal aura, and its stature is diminished in the eyes of both the dominant and the dominated. In this regard, Jiryis (1981) described Israel's rule in the discussed period as domination by law. Indeed, the Military Government as the embodiment of this regime had the vices that were characteristic of it, mainly the pervasiveness of large-scale abuses. Such abuses took many forms, including spectacular punishment and the pleasure of control, both of which I discuss below. In addition, the suspension of normative law needed to be legitimated, requiring what I call an aestheticization of power that obscures the suspension of the normative rule of law.

Spectacular Punishment

There are few Palestinian communities in which stories of spectacular punishment do not exist. These forms of punishment demonstrated the non-panoptic dimensions of surveillance especially well. Police and military surveillance was key, but this also involved putting people themselves on display, the spectacle here reminiscent in some respects of the pre-panoptic forms of surveillance discussed by Foucault, and that Thomas Mathiesen (1997) argues have in fact remained central to contemporary surveillance practices. The story in one village revolved around Commander Blume:

> He used to patrol the village and whenever he encountered a man he would ask him: "Are you married?" If the answer was positive, he would beat him up while roaring: "Do you want to increase this wicked nation?" If the answer was negative, he would say: "What is a donkey like you lacking? Do you think you are still young?" Then he would beat him up. One day he encountered an elderly man and ordered him to draw a circle and stand inside it. Then he threatened him: "If you step outside this loop, I'll kill you." The man stayed inside the circle from the morning till the evening. The commander returned in the evening to check if the man obeyed the order. When he found that the man was standing in the same place, he punched him while yelling: "What a stupid donkey. Why didn't you run away?" (Ghanim, 2009, p. 11)

Jiryis (1976) likewise provides several similar stories. Probably the one that has stuck in the public imagination more than others is that of Ahmad Hasan, a man from a tribe located close to the village of Arraba:

> [I]n August 1958 ... the military governor ordered him to sit every day for six months, from sunrise to sunset, under a large carob tree which stands to the west of the village of Deir Hanna. The purpose was to prevent him from contacting smugglers. (Jiryis, 1976, pp. 28–9)

While these forms of spectacular punishment might have been intended to frighten the Palestinian population and break its resolve or resistance, there were other forms that represented direct assaults on the Palestinians' fundamental beliefs, dignity, and what they considered the essence of their humanity. Among these were the desecration of holy sites or scriptures and the violation of fundamental moral values. For example, a Military Government officer named Avraham Yarkoni and his assistant were accused by the residents of the village of Deir Hanna of extortion, theft, and severely beating some of its residents. More ferocious behaviours included "urinating on residents in public places and taking Avraham's dog to defecate inside the mosque" (Robinson, 2005, p. 152).[3] Although such abuses were not exceptional, the manner in which the Military Government managed the daily life of Palestinians was by its nature abusive, with even the military governor of Jaffa surprised by the brutality of his soldiers, complaining that "They do not stop beating people" (cited in Pappe, 2006, p. 205).

The manner of getting passes was also a humbling experience (see, for example, Ozacky-Lazar, 2002), and it was often used as a means of exerting reward or punishment. Palestinians who were engaged in independent political or social activities were quite often denied passes, which meant that they could not access the labour market or reach educational, medical, or bureaucratic institutions. Moreover, they could not visit relatives or access any business they might have outside their locality. Violation of the pass system led to trial in a military court and possible imprisonment (Jiryis, 1976). Exceptionalism was also characterized by the absence of relevant knowledge by those who were under surveillance, a state which guaranteed their precarious position. The passes were written in Hebrew, a language that the vast majority of the Palestinian population could not read (Ozacky-Lazar, 2002), and the boundaries of the closed areas were not known to the population.

The Military Government never published the extent of the areas under its control and very rarely disclosed anything about its activities. Anyone wanting to find out which areas they could visit without a permit had to go to one of the few Military Government offices or to a police station, which could rarely provide the information. Anyone entering or leaving a closed area without a permit was liable to prosecution for breaking the emergency regulations, regardless as to whether or not they knew the boundaries. Ignorance was not a valid excuse before a military court (Jiryis, 1976).

The Pleasures of Control

In contrast to the above-mentioned abusive and restrictive acts, Military Government officials and Jewish employees in Palestinian communities made a habit of inviting themselves to the houses of Palestinian citizens or making sure that Palestinians understood that hospitality was part of the dues they had to pay (Benziman & Mansour, 1992; Robinson, 2005). In late 1949, the Military Government sought to tackle this habit by reminding low-level officers and clerks that everyone but the governors themselves had to follow "strict orders on gatherings and meals in Arab villages" and should "undertake visits ... without promise to take care of anything" (Robinson, 2005, p. 156).

Such lunches and celebrations were often explained by reference to Palestinians' cultural values of generosity and hospitability. In practice, however, they violated the essence of these values, as hospitability and generosity rested on underlying perceptions of mutuality, reciprocity, goodwill, and voluntarism. Conversely, these lunches reflected the existing hierarchy of power while reinforcing power relationships; in fact, they were a sort of extortion. Beyond the pleasures of eating and controlling the natives and their impact on the morality and commitment of the Military Government's employees, these invitations, I argue, had a panoptic dimension. The invitees were able to scrutinize the local population's customs, behaviours, and social relations and to use the acquired knowledge in controlling the Palestinians. Thus, in these invitations both the panoptic and the exceptional dimensions of surveillance converged.

In the end, the pursuit of pleasure and the desire to accumulate wealth seem to have largely overweighed the commitment to the the organization's and the state's objectives. Yehoshua Palmon, the first advisor on Arab affairs who in 1950 expressed apprehension regarding Palestinians' ability to corrupt his staff, had to admit within a year and a half that corruption had become endemic in the state bureaucracy that dealt with Palestinians, not because of the Palestinians but because of the abuse of authority by state officials:

In the field there is wide-ranging corruption by most of the staff. Bribery is taken in exchange for doing things. This finds expression in ... joint ventures of military government officials in areas of their responsibility like quarries, the custodian [for absentees' property] etc. ... [Moreover] in most cases the difference [in the prices paid to Palestinian farmers and the prices in the market] reached 60 per cent instead of the official rate of 25 per cent ...

There is corruption among Jews, however, it has limited political implications, the corruption in the relationships between Jews and Arabs gives the Arabs the feeling that this is not corruption but *"Khawa"* (extortion). This means an Arab has to pay it because he is an Arab; and to pay for a Jew because he is a Jew. (Political Committee, January 24, 1952, pp. 5–6/3)

Aestheticizing Power

Exceptionalism is usually imposed to confront imminent real or imaginary danger, and it is supported by the covert or overt promise of overcoming the danger and establishing a better state of affairs. But how can those who reject exceptionalism and its false promise impose it themselves? This contradiction has been at the heart of the Israeli leadership's discussions around the aesthetics of power; namely, how can the representation of the exception conceal its nature? Before 1948, Zionist politicians, lawyers, and jurists condemned the Mandatory emergency regulations in the strongest language. As declared by Yaakov Shimshon Shapira, who after 1948 became the legal advisor to the Israeli government:

> Even in Nazi Germany there were no such laws ... It is mere euphemism to call the military courts "courts." To use the Nazi title, they are no better than "Military Judicial Committees Advising the Generals" ... No government has the right to draw up such laws. (Jiryis, 1976, p. 12)

When the State of Israel reintroduced these regulations after 1948 to rule the Palestinians, however, there seems to have been a need to justify this change. In 1953, Prime Minister Ben-Gurion explained this alteration as follows:

> We opposed this law of the Mandate government because a foreign government, neither elected by us, nor responsible to us, had given itself the right to detain any one of us without trial. In the present instance the law is being applied by the state of Israel, through a government chosen by the people and responsible to them. (Quoted in Peretz, 1991, p. 91)

Ben-Gurion's argument is problematic as his juxtaposition between elected and non-elected government and the right to impose a state of exception is spurious. Imposition of a state of exception had to do with sovereignty rather than legitimacy. This is clear from the opening sentence of Schmitt's (2005) *Political Theology*, where he declared: "[S]overeign is he who decides on the exception" (p. 5).

Democratic and non-democratic countries alike, as recent history has shown, have imposed decrees of emergency that legitimated and extended a range of surveillance practices. As Matthew Ferguson, Justin Piché, and Kevin Walby demonstrate in this book's final chapter, in many contexts police seek to frame their surveillance practices through legitimizing narratives in which exceptional practices and forms of social sorting are rendered invisible. Indeed, we might argue that the effective operation of systems of surveillance *requires* these forms of legitimization.

The state of emergency in Israel should be compared to the juridico-formal point of view (in Agamben's words), or, as Schmitt postulated, the state of exception represents the sovereignty of men and should be juxtaposed against normalcy as the sovereignty of law. Given this contradiction in the Israeli imposition of the Mandatory emergency regulations, various sections of the Israeli elite had been uncomfortable with the Military Government, though not with its goals. For example, the left-Zionist party Mapam publicly stood at the forefront of the struggle for the abolition of the Military Government. Yet its leaders did not hesitate to pressure military governors and commanders to confiscate lands of Arab villages – on the bases of these regulations – and transfer them to their Jewish settlements (Arab Affairs Committee, January 30, 1958a). They wanted to keep the Military Government but in a different form.

The main question that bothered the opponents of the Military Government was its representation, rather than what it did. For example, in the discussion on the first official plan to control the Palestinians that was introduced in 1958 and discussed by the main figures in charge of surveillance and control of the Palestinians, Michael Assaf, an orientalist and one of Mapai's leaders, suggested playing down this contradiction through obfuscation by "maintaining the Military Government but changing the name – the Military Government – which has become a monster" (Arab Affairs Committee, January 30, 1958a, p. 12). Assaf's idea was to be repeated on several occasions by other Israeli leaders. For example, Mordechai Namir, who filled many leading positions including Mapai Knesset member, general secretary of the Histadrut, and later mayor of Tel Aviv, stated in the discussion on the Military Government:

> In general I also claim that we can achieve the same social and economic goals, which adjoin the security issue after ten years of Military Government rule, not under this awful title "Military Government." Maybe a change of the name under existing complex circumstances plays a positive role. (Arab Affairs Committee, August 14, 1958b, p. 18)

He went on to suggest: "I don't see that others have better methods ... and I don't absolve our party and those who work [on this issue] from looking for ways to add 'lipstick, powder and rouge'" (Arab Affairs Committee, August 14, 1958b, p. 21). Even the head of the Military Government itself in 1958, Mishal Shoham, was well aware that his organization was not viewed favourably:

> I want to repeat some of the things I have already said in the ministerial committee ... Maybe the Military Government is not an aesthetic pot however it contains good wine ... it can be crushed ... however with one

stipulation, that the wine be transferred to another appropriate container. (Arab Affairs Committee, August 14, 1958b, p. 24)

Finally, in the 1960s the severity of the Military Government was eased, not due to theories of justice or moral considerations but to two other reasons: first, the fast economic growth that was triggered by the import of capital, particularly from West Germany, following the Reparations Agreement and consequently the growing demands for workers in the labour market (Sa'di, 1995); and, second, the political pressure by other Jewish parties from the right and left, who argued that the Military Government was used to coerce Palestinians to vote for Mapai (see Bauml, 2007; Jiryis, 1976; Sa'di, 2003).

In 1963, Prime Minister Levi Eshkol, who succeeded Ben-Gurion, expressed his wish, in line with the colonial vision, that the Military Government would, like Bentham's prison inspector, "see without being seen" (quoted in Bauml, 2007, p. 238). As of that year, most Palestinians no longer needed to acquire specific passes for movements outside the areas of residence, although they were still not allowed to enter closed areas. While I have argued that the panoptic model does not always capture how Israeli surveillance practices functioned, in this case the invisibility of the surveillance authority matches a key characteristic foregrounded by Foucault. It is telling, though, that it is only with the relative *relaxation* of surveillance that this element moves to the fore, displacing the more spectacular forms of surveillance.

Conclusion

This chapter has attempted to analyse the structures of power and forms of surveillance through which Israel's Military Government ruled the Palestinians. Two contradictory paradigms were introduced to explain the functioning of these structures: the panopticon and the state of exception. The paradigm of the panopticon, at least as it was employed by Foucault (1991), was meant both to achieve a low-cost (illusion of) total surveillance and normalization of those deemed as deviants. Politically, this would mean an administered integration of unfit citizens to the mainstream of society. This could be achieved through the use of some of the techniques that Bentham (1995) and Foucault (1991) described, including the fixing of the subjects, the illusion of their subjection to continuous supervision, their quarantining, the documentation of their offences and the application of reward and punishment on the basis of binary divisions and branding, and their division and subdivision to governable units. Indeed, most of these techniques were used by the Military Government. Yet, according to the panopticon paradigm, normalization is carried out by a clear and universal set of rules. In this regard, the Military Government was closer to the premises of the state of exception, where the governing body was not confined by a set of normative laws and their universal application. Rather, decisionism or the arbitrary exercise of authority was the driving force, where the subjects led a precarious life and lacked control over their environment, and their ability to predict events relating to their lives was radically decreased.

By 1965, the heads of the security apparatuses in the Central Committee reached the conclusion that the Military Government had run its course. On December 1, 1966, it was abolished and its responsibilities and authority transferred to the police and the Shin Bet.[4] The emergency regulations, however, remained unchanged, and many of the restrictions imposed on Palestinians were not lifted. Moreover, the new system that was introduced was no less oppressive than its predecessor; the Israeli system of surveillance over the Palestinians moved on the continuum toward the pole of the panopticon in order to invigorate the economy and to mobilize the much-needed Palestinian labour force in the construction sector. The abolition of restrictions on the freedom of movement of Palestinian citizens finally took place on October 3, 1967 (Bauml, 2007, pp. 226–45), although this was also the year in which the occupation of Gaza and the West Bank brought non-citizen Palestinians under Israeli control. The following year, in 1968, Shmouel Tolidano, Advisor to the Prime Minister on Arab Affairs, presented another new plan to control the Palestinians, which was not significantly different from the 1958 plan. According to this new plan, which continued until the 1990s, surveillance and the regulation of Palestinians' social relations were to be conducted in a more sophisticated, subtle, and largely indirect way. To this day, the continued state hostility towards the Palestinians is deemed to require extralegal measures of surveillance and therefore the emergency regulations have never been revoked.

NOTES

1 This chapter is an updated and revised version of a previous publication: Sa'di, A.H. (2014). Chapter 3: Legal frameworks, institutions and approaches to power. *Thorough surveillance: The genesis of Israeli politics of population management, surveillance and political control towards the Palestinian minority* (pp. 49–68). Manchester, UK: Manchester University Press.

2 Among these films was *King Solomon's Mines*, which is based on a colonialist motif. Eliahu Agasi, from the Histadrut's Arab Department, asked the cinema department to lend him this film as well as the film *Children of the Prairie* (see Agassi, August 29, 1956).

3 This information was confirmed by personal communication with an elderly
   resident of Deir Hanna, who recalled many such stories, several of which
   were on the desecration of places of worship.
4 The Military Government was reinstituted for two weeks in 1967 during and
   following the Six-Day War.

REFERENCES

*Primary Sources from the Pinchas Lavon Archive of the Histadrut and
Beit Berl Archive of the Labor Party, Israel*

Agassi, E. (1956, August 29). Letter to the cinema department: To Hahaver Arie
   Brzam. IV-208-1-8559. (Beit Berl)
Arab Affairs Committee. (1958a, January 30). The protocol of the meeting.
   7/32. (Beit Berl)
Arab Affairs Committee. (1958b, August 14). The protocol of the meeting –
   The Military Government. 7/32. (Beit Berl)
Arab Affairs Committee. (1968, June 6). The protocol of the meeting. 7/32/68.
   (Beit Berl)
Arab Department. (n.d.). The Problem with the Military Government [a
   discussion paper]. 26/11. (Beit Berl)
Barkatt, R. (1956a, September 4). Letters of invitation to: Cultural Secretary,
   Russian Embassy; Dr Thomas H. McGrail Cultural Attache, American
   embassy; and Robert E. Gramble, Second Secretary, British Embassy. IV-208-
   1-8559. (Lavon Archive)
Barkatt, R. (1956b, September 5). Letter of invitation to: Mlle E. Fischer,
   Attaché Culturel, Ambassade de France. IV-208-1-8559. (Lavon Archive)
Cohen, Y. (1956, February 15). Letter to Ata Company Ltd. Document No.
   IV-208-8559. (Lavon Archive)
Confined Secretariat, Arab Affairs Committee. (1964, March 19). The protocol
   of the meeting. 26/14/11. (Beit Berl)
Histadrut's Arab Department. (n.d.). [Report]: The activities of the economic
   section of the Arab department acting beside the Histadrut's executive,
   submitted to the subcommittee of the Knesset's Arab Affairs Committee.
   Document No. IV-208-1-5814. (Lavon Archive)
Mol, M. 'An., & Palmon, Y. (1949, August 21–5). An outline of a meetings that
   was carried out between and on 21 and 25 of August, 1949. Document No.
   IV-208-1: 5815. (Lavon Archive)
Political Committee [of Mapai]. (1952, January 24). Protocol of the meeting.
   2-026-1952-10. (Beit Berl)
Review of the Histadrut's Arab Department. (n.d.). (In the Political Committee's
   files). (Beit Berl)

Secretariat of the Arab Affairs Committee. (1964, June 24). Protocol and a Summary of a Coordination meeting. 26/14/11. (Beit Berl)

Secondary Sources

Agamben, G. (2005). *State of exception.* Chicago, IL: The University of Chicago Press.

Avivi, S. (2007). *Copper plate: Israeli policy towards the Druze 1948–1967.* Jerusalem, IL: Yad Ben-Zvi.

Bauml, Y. (2007). *A blue and white shadow: The Israeli establishment's policy and actions among its Arab citizens; The formative years 1958–1968.* Haifa, IL: Pardes Publishing House.

Bentham, J. (1995). *The panopticon writings.* London, UK: Verso.

Benziman, U., & Atallah, M. (1992). *Subtenants.* Jerusalem, IL: Keter Publishing House.

Cohen, H. (2006). *Good Arabs: The Israeli Security Services and the Israeli Arabs.* Jerusalem, IL: Ivrit Publishing House.

Eyal, G. (2006). *The disenchantment of the Orient: Expertise in Arab affairs and the Israeli state.* Stanford, CA: Stanford University Press.

Foucault, M. (1991). *Discipline and punish: The birth of the prison.* London, UK: Penguin.

Ghanim, H. (2009). *Reinventing the nation: Palestinian intellectuals in Israel.* Jerusalem, IL: Hebrew University Magnes Press.

Jiryis, S. (1976). *The Arabs in Israel.* London, UK: Monthly Review Press.

Jiryis, S. (1981). Domination by the law. *Journal of Palestine Studies, 11*(1), 67–92.

Kafkafi, E. (1998). Segregation or integration of the Israeli Arabs: Two concepts in Mapai. *Journal of Middle East Studies, 30,* 347–67.

Khoury, J., & Yoaz, Y. (2007, May 20). Shin Bet: Citizens subverting Israel key values to be probed, *Haaretz.* Retrieved from www.haaretz.com/news/shin-bet-citizens-subverting-israel-key-values-to-be-probed-1.220965

Korn, A. (1995). Crime and law enforcement in the Israeli Arab population under the Military Government, 1948–1966. In S. I. Troen & N. Lucas (Eds.), *Israel: The first decade of independence* (pp. 659–79). New York, NY: SUNY Press.

Kretzmer, D. (1990). *The legal status of the Arabs in Israel.* Boulder, CO: Westview.

Lyon, D. (Ed.). (2003). *Surveillance as social sorting: Privacy, risk, and digital discrimination.* New York, NY: Routledge.

Mathiesen, T. (1997). The viewer society: Michel Foucault's panopticon revisited. *Theoretical Criminology, 1*(2), 215–34.

Ozacky-Lazar, S. (2002). The Military Government as an apparatus of control of Arab citizens in Israel: The first decade 1948–1958. *Hamizrah Hehadash, 43,* 103–32.

Pappe, I. (1995). An uneasy coexistence: Arabs and Jews in the first decade of statehood. In S. I. Troen & N. Lucas (Eds.). *Israel: The first decade of independence* (pp. 617–58). New York, NY: SUNY Press.

Pappe, I. (2006). *The ethnic cleansing of Palestine.* Oxford, UK: Oneworld.

Peretz, D. (1991). Early state policy towards the Arab population, 1948–1955. In L. J. Silberstein (Ed.), *New perspectives on Israeli history* (pp. 82–102). New York, NY: New York University Press.

Robinson, S. (2005). *Occupied citizens in a liberal state: Palestinians under military rule and the colonial formation of Israeli society, 1948–1966.* Unpublished doctoral dissertation. Stanford University.

Rosenthal, R. (2000). Who killed Fatma Sursor: The background, the motivations and the unfolding of the Kafr Qassem Massacre. In R. Rosenthal (Ed.), *Kafr Kassem: Myth and history.* Tel-Aviv, IL: Hakibbutz Hameuchad.

Sa'di, A. H. (1995). Incorporation without integration: Palestinian citizens in Israel's labour market. *Sociology, 29*(3), 429–51.

Sa'di, A. H. (2003). The incorporation of the Palestinian minority by the Israeli state, 1948–1970: On the nature, transformation and constraints of collaboration. *Social Text, 21*(2), 75–94.

Sa'di, A. H. (2014). *Thorough surveillance: The genesis of Israeli politics of population management, surveillance and political control towards the Palestinian minority.* Manchester, UK: Manchester University Press.

Schmitt, C. (2005). *Political theology.* Chicago, IL: The University of Chicago Press.

Segev, T. (1998). *1949: The first Israelis.* New York, NY: An Owl Book.

Legislation Cited

Law and Administration Ordinance (1948). The Provisional Council of State, May 19, 1948. No. 13 of 5708. Retrieved from www.geocities.ws/savepalestinenow/israellaws/fulltext/lawandadministrat480707.htm

Mandatory Emergency (Defence) Regulations of 1945. Retrieved from http://nolegalfrontiers.org/military-orders/mil029ed2.html

SECTION THREE

State Security, Policing, and Dissent

State security is a theme that runs through much of this book, whether in relation to medical surveillance, eugenic conceptions of the nation and/or race, systems of identification, the securitization of borders, or a host of other practices. In this final section contributors turn more directly to policing and state security, focusing in particular on the surveillance of dissidents and dissent. In many respects the suppression of dissent represents one of the oldest and most ubiquitous uses of state surveillance, but what constitutes dissent is less clear. In some instances it refers to oppositional political movements such as the anti-colonial and anti-capitalist Ghadar Party we discussed in chapter 1. In other cases dissent is somewhat less organized, reflected for example in the relatively common designation of non-normative sexualities as "dissident sexualities" (see, for example, Epprecht, 2004; Holden & Ruppel, 2003). When we examine how agents of state surveillance have conceptualized dissent, we find that their interpretation is often exceptionally broad, encompassing a wide range of people and activities that may include those actively challenging state power, but are also frequently collective subjects designated as intrinsically threatening (for example, "dissident sexualities" or the "criminal tribes" noted in the introduction to section 2), and who may not be engaged in any deliberate political action. In Canada, for example, the targets of state political policing have ranged from Fenians in the nineteenth century, to South Asian migrants, communists, gay and lesbian government workers, Québec separatists, Indigenous peoples and movements, environmental groups, and Muslims in the twentieth and twenty-first centuries (Crosby & Monaghan, 2018; Kinsman & Gentile, 2010; Whitaker, Kealey, & Parnaby, 2012). Clearly, in the eyes of agencies of state surveillance, dissent can encompass a wide range of perceived threats.

The chapters in this section look primarily at policing, state security, and dissent in the second half of the twentieth century. The Cold War provided one of the key contexts for these developments, with states on both sides of the Iron Curtain seeing the growth of extensive surveillance and security apparatuses. In chapter 9 Christina Plamadeala explores this history through an examination of communist Romania's Securitate, and the ways in which this notorious security agency both surveilled dissidents and facilitated networks of collaboration with the state. As with so many surveillance practices, this was based on systems of identification and documentation that she calls "dossierveillance." Plamadeala offers a powerful insight into the systems of terror and cooptation by which the state cemented its power, and the specific ways in which Soviet bloc systems of surveillance functioned.

Counterposing this discussion of Romania with two chapters on the United States allows readers to consider the ways in which the surveillance practices on opposing sides of the Iron Curtain mirrored and/or diverged from each other. The Church Committee (1975–1976) revealed the assassination programs and destabilization of foreign governments undertaken abroad by the United States, but also, prefiguring Edward Snowden's leaks, exposed the extent of information sharing between telephone companies and the National Security Agency (NSA). The FBI's COINTELPRO program famously targeted suspected communists and members of the Black Panther Party, but J. Edgar Hoover's agency went after others as well, for example in the so-called Sex Deviates Program aimed at gays and lesbians (Charles, 2015), or in the surveillance of academics, writers, and artists (Culleton & Leick, 2008; Price, 2004). In chapter 10 Kathryn Montalbano looks at another of the FBI's surveillance targets, the Quaker-affiliated American Friends Service Committee (AFSC). The AFSC was caught up in the broader anti-communist hysteria, but Montalbano paints a complex picture of FBI surveillance, foregrounding the contested ways in which notions of "ideal citizenship" and of "true" religion, derived from a Protestant hegemonic framework, informed these practices, including shaping the day-to-day actions of agents in the field. This history has added significance given the role played by William C. Davidon and Ann Morrissett Davidon, who worked with Quaker organizations, in exposing the workings of COINTELPRO.

Chapter 11, by Elisabetta Ferrari and John Remensperger, likewise considers FBI surveillance of dissidents but in addition examines the contested politics of representation, the subject of the book's final chapter as well. Ferrari and Remensperger focus on how two underground newspapers that were themselves the targets of FBI surveillance – the

Los Angeles Free Press and the *San Francisco Good Times* – engaged with and reported on state surveillance. They uncover a rich and evolving understanding on the part of radicals of how surveillance worked, as well as the development of activist responses to that surveillance, which has profound resonance in the contemporary context. The final chapter, by Matthew Ferguson, Justin Piché, and Kevin Walby, also focuses on representations of surveillance, but shifts the analysis from the surveilled to how those *doing* surveillance represented and memorialized their practices, in this case police forces in Ontario, Canada. Their chapter takes the concerns of this book up to the present, but also returns us in some respects to the archival questions posed in our introduction. In their examination of police museums, Ferguson, Piché, and Walby trace how historical documentation of surveillance is used by police to construct narratives that serve to buttress their authority and legitimacy in the present, illuminating the challenge of doing historical research on surveillance in a very different way. In tandem with the others, chapter 12 works to develop a critical history challenging the often violent and repressive dimensions of state surveillance, while remaining attentive to its many complexities and changes over time.

REFERENCES

Charles, D. (2015). *Hoover's war on gays: Exposing the FBI's "sex deviates" program*. Lawrence, KS: University Press of Kansas.
Crosby, A., & Monaghan, J. (2018). *Policing Indigenous movements: Dissent and the security state*. Winnipeg, MB: Fernwood Publishing.
Culleton, C. A., & Leick, K. (Eds.). (2008). *Modernism on file: Writers, artists, and the FBI, 1920–1950*. New York, NY: Palgrave Macmillan.
Epprecht, M. (2004). *Hungochani: The history of a dissident sexuality*. Montréal, PQ: McGill-Queen's University Press.
Holden, P., & Ruppel, R. J. (Eds.). (2003). *Imperial desire: Dissident sexualities and colonial literature*. Minneapolis, MN: University of Minnesota Press.
Kinsman, G., & Gentile, P. (2010). *The Canadian war on queers: National security as sexual regulation*. Vancouver, BC: University of British Columbia Press.
Price, D. H. (2004). *Threatening anthropology: McCarthyism and the FBI's surveillance of anthropologists*. Durham, NC: Duke University Press.
Whitaker, R., Kealey, G. S., & Parnaby, A. (2012). *Secret service: Political policing in Canada from the Fenians to fortress America*. Toronto, ON: University of Toronto Press.

9 Dossierveillance in Communist Romania: Collaboration with the Securitate, 1945–1989

CRISTINA PLAMADEALA

Despite being closely connected with the lives of many Romanians who lived under communism, collaboration with the Securitate during Romania's communist era is still a taboo subject, even now, more than a quarter of a century since the fall of communism. In Romania, as of 2008, someone's collaboration with the surveillance apparatus of the secret police and/or the communist state is determined solely by the courts, by consulting the Securitate files of, as well as conducting interviews with, the accused party. In order to be considered a collaborator, those so suspected must have in their Securitate files written proof that they were remunerated, often financially, from collaboration. Secondly, the documents must indicate that suspected persons had agreed to collaborate without being forced to do so (Horne, 2015). Collaboration, in this case, is treated similarly to an employment position, amounting to a win-win situation of some sort, a mere exchange wherein one reaped the benefits of the service(s) provided while the other obtained financial remuneration or some social or professional benefit for oneself or one's loved ones. In many instances, however, the reality behind collaboration with the Securitate was considerably more complex.

The term *collaboration*, as used in this chapter, includes the activities consciously undertaken by the Securitate's entire surveillance network (*rețeaua informativă*), composed of its collaborators (*colaboratori*), informers (*informatori*), residents (*rezidenți*), and support persons (*persoane de sprijin*).[1] All of the individuals within this espionage system collected information to target and incarcerate its imagined or real opponents. Indeed, collaborators were recruited to provide information needed in various Securitate investigations, and depending on the quality and efficiency of the work performed, they could be promoted within the hierarchy of this surveillance network. The camouflaged presence of informers and collaborators in the daily lives of Romanians, some of whom may have been members of one's immediate family, teachers or university professors, next-door neighbours, or even one's childhood friends, helped to form an atmosphere of "morbid fear" (Stan, 2013, p. 9), suspicion, and distrust. The word *collaboration*, as used here, stands for the totality of actions of all the individuals belonging to this surveillance network, actions connected to their *active* and *conscious* engagement, carried out in a coerced or voluntary capacity, or a mix of both.

This chapter examines the changing nature of, and motives behind, people's collaboration with the Securitate. In seeking to quell real and imagined dissent, the terror tactics used by the state changed over time, and across two national leaders. The overt brutality, violence, and mass incarceration of political prisoners of the Gheorghe Gheorghiu Dej era (1947–1965) shifted to a more general and less directly violent instilling of fear and surveillance by the Securitate under the rule of Nicolae Ceauşescu (1965–1989) (Deletant, 1995; Stan, 2013). While collaboration during the Dej era was driven primarily by one's need for survival, it became in later decades more a means for social and economic advancement. This shift in tactics correspondingly influenced the motives behind individuals' reasons to collaborate with the Securitate.

At the heart of the surveillance apparatus was the file or dossier that contained the information collected by the Securitate. I call this process "dossierveillance," which I discuss by drawing parallels between this phenomenon and the concept of "dossier society" put forth by Kenneth Laudon (1986). Hannah Arendt's work (1977) on the nature of political evil is also helpful, in particular for exploring the overt terror in Dej's era, the culmination of which was arguably embodied by the Piteşti experiment, as well as for examining the Securitate's widespread surveillance of the population in the Ceauşescu era. The latter regime's surveillance mechanisms, including the different forms of individuals' collaboration with the secret police, kept its targets under its omnipresent "gaze," to use Michel Foucault's language (Foucault, 1973; Galic, Timan, & Koops, 2016). Those who did the watching and those who were watched met in the Securitate file(s), and indeed it was to these files that many of the members of the former group, including collaborators, dedicated their professional careers. In time, the file, or the fear of having a file, or the fear that doing something may bring one to the attention of those who were responsible for file writing, was one of the Ceauşescu regime's most effective "disciplinary" tools (Foucault, 1991).

The Piteşti Re-education Experiment in Prison and Beyond[2]

In Romania, the communist regime was established in the winter of 1947, shortly after the abdication of King Michael, as a result of the adoption of the Law 363, which abolished the country's constitutional monarchy (Roper, 2000). Following closely the Soviet Muscovite system, particularly in the usage of terror (Vatulescu, 2010), the overt brutality of the regime was especially evident in the Dej era, a period which may be described as "extreme evil," as coined by Hannah Arendt in a letter to Gershom Gerhard Scholem (cited in Morgan, 2001, p. 22). This was especially evident in the notorious re-education experiment that was carried out in the 1940s and early 1950s in the male prisons at Piteşti and other localities (including Târgu Ocna, Ocnele Mari, Târgşor, Baia Sprie, and Aiud) (Tismăneanu et al., 2006). The key objective of this experiment was to "erase the difference between victims and perpetrators," as Cristina Petrescu and Dragos Petrescu put it (2014, p. 65), to modify or recreate the human essence of its participants, torturers and their victims alike, all partaking in a vicious cycle of never-ending suffering of the mind, body, and soul (Muresan, 2011; Stănescu, 2010a, 2010b, 2010c).

Sometime in 1948, prison authorities in Piteşti allowed a few prisoners to traumatize other fellow prisoners, and thus gain lower sentences or some meagre benefits. The program sought "to re-educate" its targets, as its name suggests, through physical and psychological violence, so that they would embrace Marxist and Soviet ideology. A great number of them were former Legionaries, a name used to refer to members of the Iron Guard, the notorious anti-communist and anti-Semitic paramilitary organization that was active primarily during the interwar period (Clark, 2015; Livezeanu, 1990). Prisoners Alexandru Bogdanovici and Eugen Ţurcanu, two of this experiment's key leaders who were also former Iron Guard members, created the Organization of the Detainees with Communist Convictions (ODCC), under whose aegis the experiment was carried out in four stages, each more grotesque than the next (Ciobanu, 2015; Tismăneanu et al., 2006).

During the first stage, called "external unmasking" (*demascare exterioara*), the prisoner had to divulge anything he may have withheld from the Securitate interrogators to the ODCC leaders. This was done so that the victim would show full allegiance to the ODCC. During the second stage, called "internal unmasking" (*demascare interioara*), inmates were asked to report to ODDC the individuals who were most kind to them in prison. This was so that these inmates would be assigned to become these individuals' torturers, thereby wiping out in themselves and their targets any remaining sense of humanity, decency, or camaraderie (C. Plamadeala, 2019). In the third stage, called "public moral unmasking," they were asked to publicly reject their own past and insult or denounce their family members and loved ones, including their faith in God and country (Tismăneanu et al., 2006). Everything that they held dear had to be publicly rejected, including their own former selves. This was done not only oratorically but also in writing. It was in the final stage that the actual "metamorphosis" occurred: the former victims, now themselves vicious abusers, subjected other prisoners to the violence and torture they had experienced throughout this brutal experiment. Starved, sleep deprived, beaten to the point that some contemplated suicide, these inmates were also made to feel pain and humiliation when eating and drinking (Deletant, 1995). Sometimes they were forced to eat while kneeling on the floor with their hands behind their backs, and other times they were made to eat their own excrement along with their food. As Dennis Deletant (1995) explains, "eating became a source of humiliation as well as of pain and the sense of taste, smell and touch were repeatedly associated with pain" (p. 35). Other diabolical and brutal acts were inflicted as well, all of which treated prisoners as creatures able of being reformed or reshaped to serve the needs of the regime. Similar to the Soviet gulag or the concentration camp, the Romanian prisons where these experiments took place served as settings where the terrorized victims, to quote Valerie Hartouni (2012), "perform[ed] their ... part in a cacophonous symphony of death" (p. 43).

The idea that humans could be reshaped, that a "new man" could be created, is also evident in other forms of surveillance discussed in this book, most notably medical and eugenic surveillance that sought to produce new colonial (Jacob Steere-Williams, chapter 2) and national (Holly Caldwell, chapter 3; B Camminga, chapter 4) subjects. What distinguished the Pitești experiment was its extreme violence, but this was also part of the broader violence enacted in society at large (C. Petrescu & D. Petrescu, 2010). Commencing in the spring of 1948 and continuing until summer 1964, Romanians experienced widespread violence unknown in the later Ceaușescu years (1965–1989). The police investigations and mass arrests of political elites, peasants, and former Legionaries, as well as members of the anti-communist resistance movement (mostly students), ceased only in 1964 (Tismăneanu et al., 2006). During this era, the villages, like the urban areas, were not left unaffected. Thousands of peasants and landowners were murdered, confined for years to forced labour in labour camps, resettled to other parts of the country, and/or forced to leave their homes, lands, and belongings as a result of the regime's collectivization program (*colectivizare*), also known as the nationalization of the country's agriculture (Kligman & Verdery, 2011). Thousands were deported to the Soviet Union shortly after the end of the war, most of whom were ethnic Romanian refugees from Bessarabia, which had been part of Greater Romania in the interwar period, and who sought refuge in Romania after fleeing the

Soviet Army's invasion (Tismăneanu et al., 2006). This widespread overt violence against Romanians slowed down in the 1960s, with the coming to power of Nicolae Ceaușescu.

While there was a significant shift in the Ceaușescu era, the goal of redesigning the inner identity of the targets of surveillance, their psyches or souls, remained. The prison experiments aimed at transforming mainly former Legionaries into adherents of communist ideals, but the wide surveillance carried out during the Ceaușescu years sought to monitor and ensure adherence to communist ideals among the Romanian population more broadly, and discourage and inhibit any act of rebelliousness against the regime. Both policies aimed at turning victims into perpetrators: in the earlier period prisoners physically tortured their fellow inmates, while under Ceaușescu victims wrote reports on their neighbours, acquaintances, family members, and friends. Furthermore, both policies were made to look as if they began from "within": seemingly initiated by the ODCC, in the case of the Pitești experiment, and by one's neighbour now turned a Securitate informer, for example, in late communism. In both cases, however, these actions were carried out under the real or imagined gaze of some sort of authority figure – the prison guard or the Securitate officer. Without this gaze and the human apparatus that stood behind it, and, most importantly, without the oppressive regime that gave it legitimacy to operate, there would have probably been neither a prison experiment of such level of cruelty in Romania, nor the same level of collaboration that took place during the Ceaușescu period. Both phenomena required essentially similar circumstances from which to emerge. Such circumstances had to be able to instil immense fear in targets, sufficient, at times, to make them even turn against those they cherished or held dear – or, to put it another way, to turn against themselves. In the Ceaușescu decades, during which the change in terror tactics came about primarily due to the structural modifications within the Securitate, the Securitate file, as will be discussed later in this chapter, became one of the regime's primary means through which this level of fear was effectively disseminated.

The Securitate and Its Surveillance Network (*rețeaua informativă*)

Nicolae Ceaușescu came to power shortly after Dej's death from cancer in 1964. Despite his later infamy, Ceaușescu enjoyed great support and even admiration from many hopeful Romanians, especially intellectuals, due primarily to his bold and vociferous opposition to the 1968 Soviet invasion of Czechoslovakia. The Ceaușescu "thaw," however, was short-lived. Soon, the man who had once convincingly promised to return his country to normalcy and prosperity by investigating the human rights abuses of his predecessor would go on to be infamously known, both domestically and abroad, for the human rights abuses inflicted upon Romanians, abuses carried out under his watch, and, at times, under his direct orders, with the assistance of the Securitate. The same man would go on contribute to terror's proliferation and not to its extinction, as he had once promised, by personally ordering an internal investigation of the Securitate and its human rights abuses in 1968 (C. Petrescu, 2014). The terror employed by the Securitate under Ceaușescu's reign, embodied in the wide surveillance of the population, was generally of a different kind, targeting, this time, not the body but, as mentioned earlier, the soul.

Surveillance, in this case, was carried out to create a new and more elevated society by kneading the "human material" into a "new man," who, to use Peter Holquist's (1997) description of the Soviet project, was to become a "more emancipated, conscious, and superior individual" (p. 417). This new man, in fact, was merely a platonic form, and society as a whole, as the regime intended it, was to aspire to attain the qualities attributed to this human-shaped chimera, whose villains were often referred to in Securitate files as "elements" (*elemente*) (ACNSAS, 1969, p. 23; ACNSAS, 1988, pp. 39, 57). Holquist (2001) describes these individuals, as depicted in Bolshevik files for example, as somehow undesirable or dangerous to the regime, and they were targeted in order to have their real or potential rebelliousness tamed. In Securitate files, the information collected on these "elements" renders them dangerous to the State due their religious (ACNSAS, 1966, p. 96), ethnic, class, and/ or political affiliations, via a process resembling "social sorting" (Lyon, 2003). Some of the Securitate's targets were at times employed in various surveillance operations (ACNSAS, 1988, pp. 74–111), being subject to what Oscar Gandy referred to as the "panoptic sort" (cited in Gates, 2011, p. 112), with the information gathered on these targets being employed not only to monitor them, but also to serve the Securitate's interests and goals. The tangible, it seems, was no longer as important as the non-tangible in the crafting of the "new Soviet man," the chiselling of which was carried out primarily through instilling fear. Alluding to ancient Greek philosophy on the indestructible and unchanging nature of the soul and the ephemeral nature of the body that houses it, destroying the human spirit or redesigning it meant, in this context, gaining full and ultimate control of a person's essence. In a world wherein anyone and everyone could be an informer, as it was the case of Romania in this period, the necessity to be constantly vigilant of one's surroundings became a universal survival mechanism that undoubtedly predisposed many to social alienation, and even paranoia and internal torment. This sense that anyone could be either an informer or a target was arguably the primary difference from the surveillance enacted in the United States during the same time period, which tended to be more focused (see, for example, the discussion in the following pages by Kathryn Montalbano in chapter 10 and Elisabetta Ferrari and John Remensperger in chapter 11). In Romania in such circumstances, while the human body remained intact, the soul was bound to be shattered.

One of the changes in the Securitate, imposed shortly after Ceauşescu's arrival in power, consisted in a significant increase in the calibre of its informers and collaborators, many of whom joined the Securitate following Ceauşescu's "Khrushchev-style public acknowledgments" pronounced in the late 1960s (C. Petrescu, 2014, p. 389). These individuals were employed to gain access to the most intimate aspects of people's lives, including friendships, family dynamics, and life in the workplace. The Piteşti experiment moved outside the cold and dark cells of the prison in the years following Ceauşescu's coming to power, an argument developed by Horia Roman Patapievici (C. Petrescu, 2014; C. Petrescu & D. Petrescu, 2010) and further developed in the works of C. Petrescu and D. Petrescu, who coined the term *Pitesti syndrome* (2014, p. 132) in order to describe life under communism in Romania during the Ceauşescu era (1965–1989). The physical limitations, hardships, psychological torment, and brainwashing measures to which the Romanian communist regime subjected its population for the sake of adherence to Marxist and Soviet ideals echo the human rights abuses and brainwashing techniques inflicted on the victims of the Piteşti experiment, although to a much lesser extent. Furthermore, I argue, as part of the greater Piteşti syndrome phenomenon, as in the case of the Piteşti experiment that was overseen by prison personnel who were members of the Securitate, former victims were lured into collaboration with the Securitate and, thus, turned perpetrators against those who were once closest to them.

Like the prisoner targeted as a potential torturer in the Piteşti experiment, the Securitate's candidate for recruitment, as one of the manuals employed to train Securitate agents on how to attract people into collaboration indicates, was subject to a process of "unmasking" of some sort. Only by knowing the very core or sense of self of the respective human beings, in addition to their vulnerable points (*puncte vulnerabile*), were the Securitate officers able to provide strong enough incentives to collaborate and/or coerce them to submit to collaboration and the agency's demands, guidelines, and requests (ACNSAS, 1976). Hence, as part of the greater Piteşti syndrome phenomenon, as in the case of the Piteşti experiment, people may have ended up harming those close to them, thereby subjecting themselves to the same "former-victim-turned-perpetrator" paradigm (C. Plamadeala, 2019). In the Ceauşescu decades, this paradigm was evident on a nationwide level, under the umbrella of dossierveillance.

As will be described below, the dossier became the key technology in this system of surveillance, with the fear that someone may be watching you, or writing a file on you, central to the regime's disciplinary measures. The result was to transform even the most rebellious creature into a docile one, obliged to embrace some sort of "self-surveillance" (Vaz & Bruno, 2003) when engaging with others. Cristina Petrescu (2014) describes the outcome of this phenomenon: the "idea that the secret police was the most powerful organization in the country acting on its own and in control of everything, induced the majority of the population into submission" (p. 319).

The new officers brought into the Securitate's network of surveillance beginning with the year 1968 represented "young, intelligent, ambitious and highly educated individuals" (C. Petrescu, 2014, p. 392). According to Elis Neagoe-Pleşa (2008), the ideal candidate was "[ethnically] Romanian, from urban settings (*orşşean*), preferably an intellectual [holding at least a bachelor's degree], knowledgeable of at least one foreign language and motivated by patriotic sentiments in relation with the Securitate" (pp. 12–13).[3] These individuals first had to undergo a thorough investigation of their character, for to be recruited, one had to "present a guarantee that s/he would collaborate honestly with the Securitate" (Neagoe-Pleşa, 2008, p. 12).

In order to hire new individuals, many former collaborators had to be let go or abandoned (*a abandona*). This occurred when their relationship with the Securitate was accidentally or deliberately revealed, or when they lacked the access to the information sought out by the Securitate. At times, the abandonment occurred when the informers themselves decided to discontinue their collaboration. In total, the number of those involved in the surveillance network at the end of 1968 – informers, residents, hosts of so-called meeting houses (*gazde de çase de întâlniri*) and conspiring houses (*gazde de case conspirative*)[4] – dropped by roughly 30 per cent from the number registered at the beginning of that same year. By the end of 1968, the surveillance network consisted of roughly 85,042 people, almost half of them informers (43,498). The network had also 29,761 collaborators, 2,296 residents, 320 meeting house hosts, and 167 conspiring house hosts (Neagoe-Pleşa, 2008, p. 11).

It is around this time, Neagoe-Pleşa (2008) points out, that the collaborator emerged as a special and "newly established category" (p. 11) within the notorious surveillance network of the Securitate. The "official" or "formal" collaborators were to be given "specific tasks" (*sarcini specifice*), assignments that most likely required a lot more than just transcribing word by word one's mundane conversation with one's neighbour, a job that often went to the informer or the support person. I borrow the term *formal* from the terminology employed by the Stasi police to differentiate between the formal and less formal collaboration, the latter being referred to as IMs informants (*Inoffizielle Mitarbeiter*). Unlike their more official counterparts, the IMs informants assisted the secret police on a less frequent basis (Fulbrook, 2005; Garton Ash, 1997; Willis, 2013). In the context of communist Romania, the IMs' counterparts would qualify as the Securitate's informers.

Most of the Securitate's informers (85 per cent) were from an urban background. In the cities, as the Securitate's report points out, lived most of society's "hostile elements" (*elemente ostile*) (Neagoe-Pleşa, 2008, p. 13), and thus urban areas were in the greatest need of informers and collaborators for surveillance purposes. With respect to the ethnic origins of those newly recruited into the surveillance network in 1968, 81 per cent were of Romanian origins, and the other 19 per cent were of primarily German, Hungarian, Jewish, or Ukrainian backgrounds. In comparison to the Dej era, the percentage of Hungarian Securitate agents increased significantly while those of German origin decreased at a similar rate. By the early 1970s, Romania's Nazi past and its collaboration with the Germans was no longer as pressing a national concern as the "Hungarian irredentist" threat from the West, an issue that preoccupied Ceauşescu till the end (ACNSAS, 1988, pp. 80–3).

Ceauşescu's Community Party sought to systematically destroy bourgeois culture, and thus the secret police targeted elites, intellectuals, representatives of the arts and culture, men and women who spoke foreign languages, those who read books – including Western ones in the hidden corners of their homes – and probably even those who listened to classical music. Stated differently, the regime turned to those few who managed to survive Dej's repression, the torturous treatment in prisons or labour camps, or the generation that came after them and who in some ways strove to attain these political prisoners' cultural acumen. To become simultaneously efficient and deceitful, the regime needed the help of those it had attempted to eliminate; it needed the assistance of the intellectuals and of the sophisticated polyglots. For terror to become more "subtle," to quote Stan (2013, p. 7), it needed, as the Securitate realized, more sophisticated minds behind it to mastermind what may have turned out to be, in retrospect, a senseless and vicious plot against the nation, a plot that would keep people in distress. Surveillance, in particular the practice of dossierveillance, became a central task of collaborators in furthering the work of the Securitate.

Dossierveillance and Collaboration: Examining the Securitate Files

Dossierveillance is a type of surveillance morphologically and seman-
tically similar to the concept of "dossier society" coined by Kenneth
Laudon in 1986. The "dossier" in dossierveillance, like the one in the
"dossier society," represents "thousands of officially selected moments
in your past to confront you with the threads of an intricate web, revealing your 'official life,' the one you must live with and explain to whatever authority chooses to demand an explanation" (Laudon, 1986, p. 4).
Unlike its seemingly more benign counterpart in the "dossier society,"
the file in dossierveillance accumulates information on a given person
with deliberate intent to cause harm, incriminate, and even punish.
Its targets, often referred to as "elements," as discussed above, tend to
be depicted as needing to be reformed, penalized, and monitored for
their potential to cause real or imagined harm to the regime that stands
behind this type of surveillance activity. In this dossier, victims can gradually turn into perpetrators and vice versa, or, to use the words of Stan and
Turcescu (2007), "a person's relationship with the Securitate could easily
change from torturer to tortured, then back again" (p. 66).

Collaboration is, in most cases, this file's key necessary ingredient.
Without it, this type of dossier can't possibly attain the level of importance and power it can have in carrying out this type of surveillance
activity, a power arising from the quality of information it can collect.
A symbiotic relationship arises, I argue, or possibly even one of mutual
dependency, between collaboration and the dossier that officiates it. It
grants legitimacy and, ultimately, represents the tangible proof of something that, without it, remains a rather abstract phenomenon, embodied at best by fleeting human interactions and handshakes. Similarly,
without the "human interactions" that are recorded and analysed in the
dossier, it is merely paper, simply put. For this paper to become a dossier in the dossierveillance operation, it needs to serve as the recording
tool of these interactions. The result, I argue, is a chicken-or-egg causal
dilemma in respect to the Securitate dossier and collaboration work carried out around it, the story of which this dossier may in turn narrate.

Page by page, if one peruses Securitate files that are currently stored
at the Council for the Study of Securitate Archives in Bucharest and
Popești-Leordeni, one gets the opportunity to learn about the life stories
of those who had contributed to the writing of these files. These dossiers
can also help one understand why people acquiesced, or felt compelled
or coerced to collaborate. They lived in a world of state and police file
making, especially in the Ceaușescu era, a world of dossierveillance in
which the very thought of a Securitate file could paralyse every other

thought that may have instigated one to take action against the communist regime.

Reading a Securitate file is a lot like embarking on a pilgrimage through unexplored terrain. The file represents people who once lived and some who no longer do, and in this sense, reading these "paper cadavers," to quote Kirsten Weld (2014, p. 3), may be their last hope at a symbolic resurrection. The closer our interpretation of these files is to the truth, the closer we are to factually resuscitating them. Studying these archives calls for what Weld refers to as a special "archival thinking" whose purpose is twofold: to provide "a method of historical analysis" and "a frame for political analysis" (2014, p. 13). As Timothy Garton Ash (1997) further argues, reading these secret police files "may even teach us something about history and memory" (p. 23). And, if we are to treat these files as a *peregrinatio* into the unknown, the journey a researcher embarks on in reading such a file can turn into a pilgrimage into a past that begs not to be forgotten, and whose last chance at being remembered is embodied in the dusty pages of which it is composed.[5]

Yet, as I have suggested elsewhere (see C. Plamadeala, 2015), reading these files is not without challenges, one of which is their lack of reliability in telling the truth. In some cases, secret police files may have been destroyed or modified by government agents in order to further incriminate someone or to cover up someone else's crime. In Romania, for example, 100,000 files were "lost, 'misplaced,' or destroyed by the case officers" (Stan & Nedelsky, 2013, p. 3). Furthermore, some secret agents likely provided faulty reports in a deliberate manner (Stan & Nedelsky, 2013, p. 4), while others may have falsified information in these files in order to reach the "required quota of recruitments and information activity" (Ursache, 2013, p. 114).

Lastly, although gaining physical access to these files may require time, usually a month or two in order to be accredited as a researcher, acquiring an understanding or a fluency in reading them can be significantly harder to attain, something which Cristina Vatulescu (2010) rightfully referred to as "patiently (re)learn[ing how] to read" (p. 13). Loaded with the political jargon of communist-era wooden language (*limbaj de lemn*), as it is referred to in academic literature, and similar to the "newspeak" in George Orwell's *1984* or Adolf Eichmann's "officialese" (Arendt, 1977; Hartouni, 2012), readers (myself included) are often left perplexed in the process of file analysis and must construct their own timeline of events discussed, as pages are not necessarily organized in a chronological manner. The files might as well be seen as a discordant chorus of "voices mingling in the surveillance file" (Vatulescu, 2010, p. 38), voices that are at times contradictory to each other: those of the Securitate officers; of conflicted, and possibly sometimes guilt-stricken informers; and of the nefarious system in which they all operated. The Securitate file can be both a precious historical relic for a detached scholar, and possibly a nightmarish reminder of one's past, buried deeply in one's memory, if it concerns one's own file or the file one helped write as an officer, collaborator, or a simple informer – a tiny cog in a dictatorial regime one tried to make sense of, adjust to, by tacitly acquiescing with and conforming to its demands.

With the exception of those who agreed to work for the Securitate due to what Neagoe-Pleşa (2008) calls "patriotic sentiments" (p. 13) or honest admiration for communist ideals, many collaborators and informers joined the Securitate's espionage network while serving time in prison during both the Dej and the Ceauşescu eras. For these individuals, collaboration falls closer to what Kokoška (2013) describes as forced participation as opposed to voluntary (Turcescu & Stan, 2017). Ceauşescu's condemnation of the USSR's invasion of Czechoslovakia in 1968 may have inspired in some a genuine eagerness to get involved with the internal affairs of the state in order to protect their nation from the "threat coming from the East" (C. Petrescu, 2014). This motive, needless to say, was not applicable for everyone. Roughly 8.3 per cent were brought in based on blackmail, either because of their criminal record or former ties to the fascist Iron Guard (Neagoe-Pleşa, 2008). After 1947, while many former fascist Legionaries were incarcerated and murdered, many were also set free. Among those granted this new-found liberty, some were even allowed to climb up the ladder of the Orthodox Church's hierarchy (Turcescu & Stan, 2015).

The case of the Vladimireşti Monastery (interchangeably referred to in Securitate files as Vladimirescu) serves as a good example in this case, as Securitate files on this monastery portray some of those within the inner circle of the then Orthodox Patriarch Justinian Marina as former Legionaries (ACNSAS, 1958–1959, p. 50). These former Legionaries were also helping the Patriarch quell the heretical tendencies within this monastery (ACNSAS, 1958–1959, p. 129). But most importantly, the case of Vladimireşti merits noting here because it brings into question the ethical and moral grounds behind the acts of resistance its leadership carried out against the communist state. According to Securitate files, the monastery promoted vandalism, anti-Semitism, and social unrest in the surrounding areas (ACNSAS, 1954, pp. 9–10, 272, 281; ACNSAS, 1958–1959, pp. 63, 102–3), all carried out under the facade of anti-communist resistance against "the red dragon" (balaurul roşu). This is the way in which its leadership referred to the communist regime (ACNSAS, 1966, p. 238). Other places of worship, such as the Ciolanu Monastery, for example, attempted to follow Vladimireşti's example in mimicking its strategies of revolt against the state (ACNSAS, 1958–1959, p. 137).

The leadership of the Vladimireşti Monastery was composed of the priest Ioan Iovan and the nuns Veronica Gurău and Mihaela Iordache, the latter being the only fascist Legionary member of the two, with the other two being sympathizers of the Legion (ACNSAS, 1962, pp. 13–14; C. Plamadeala, 2016). A pilgrimage site for faithful believers who, until its shutdown in 1955, gathered here periodically in protest against the communist regime (ACNSAS, 1954, p. 26), the monastery was also a target of the Orthodox Christian Patriarchate. The then Patriarch declared its leadership heretical (ACNSAS, 1958–1959, p. 129) due to its teachings and practice of mass confession (*spovedanie în bloc*) (ACNSAS, 1969, p. 23), a ritual forbidden in the Orthodox Church. Vladimireşti was also a target of the Patriarchate because of the great popularity of the religious visions (*vedenii*) of the monastery's nun Veronica, in which she described her encounters with Jesus and Virgin Mary (ACNSAS, 1954, p. 107). These visions, disseminated to visitors in the form of brochures produced by the monastery's own printing press (ACNSAS, 1966, p. 238), attracted a great number of pilgrims from all corners of the country (ACNSAS, 1966, p. 35), who came to hear Ioan's anti-communist sermons. As Securitate files on this monastery suggest, the popularity of Veronica's visions and Ioan's anti-communist preaching represented a dangerous yet powerful combination to both the Church and the regime. As a hotbed for anti-communist resistance and a place of worship that was not fully abiding by the Orthodox traditions and ideals, Vladimireşti was threatening the very foundation of the communist system and the stability of the Romanian Orthodox Church both parties sought out to actively protect (C. Plamadeala, 2016).

In the case of this monastery, as suggested in the files pertaining to it, resistance under communism in Romania was not at all morally and ethically impeccable. Indeed, current literature on collaboration and resistance under communism *may* tend to depict this subject in a rather dichotomous manner, with collaboration bad and resistance good. The story of the Vladimireşti Monastery shrinks this gap, demonstrating that resistance could also involve the deployment of an anti-Semitic and reactionary politics in opposition to the state.

But there is something else worth mentioning here that the Vladimireşti files and the other files mentioned here may reveal, namely, how these officers would recruit new informers for its espionage network. One of their strategies, to put it simply, was to focus on these targets' vulnerable points (*puncte vulnerabile*). The Securitate officers had a special name for this: they called it "compromising materials" (*materiale compromiţătoare*) (ACNSAS,

1958–1959, p. 105) or a "hostile past" (*trecut duşmănos*) (see ACNSAS, 1958–1959, pp. 104–5), such as one's Legionary past, for example – a strategy practised by Securitate to recruit informers in many other monasteries, and not just Vladimireşti.

In a system constructed on fear, violence, tribute, exchange of favours, and pursuit of power at all costs, it is almost impossible to pinpoint a specific case wherein collaboration with the Securitate may have been driven *solely* by pure and noble ideals and innocent desire to cause positive social change, although it may have been *initiated* by such sentiments or longings. Instead, collaboration was driven by many other factors. Instances of collaboration motived by financial and economic gain increased significantly in the Ceauşescu era as overt terror subsided (Albu, 2008). Some collaborators, however, were motivated by what Stan (2013) refers to as "misplaced patriotism" or a desire to seek revenge (p. 64). Others may have been "attracted by the compensations," not only financial in nature (Stan, 2013, p. 66). Such benefits included, but were not limited to, the chance to relocate from rural to urban settings or to travel outside of the country, as travel passports in communist Romania were issued under the condition of collaboration with the secret police (Deletant, 1995).

Some of these cases of collaboration were most likely formalized "through convincing" (*prin convingere*), as Securitate files refer to it (ACNSAS, 1958–1959, pp. 114–15). The term *participatory dictatorship* (Fulbrook, 2005) comes to mind when attempting to contextualize this phenomenon of collaboration "through convincing" within the greater socio-political milieu within which it was carried out. The term "participatory," as used here, does not necessarily mean either free or fully willing to engage in the respective agreement (Bruce, 2010). Yet, it also does not suggest being completely deprived of agency and some freedom of choice. Being "convinced" in this case most likely meant being given something that may have looked advantageous, like the cheese in the mousetrap: tantalizing, enticing, and even necessary for survival, yet potentially deleterious nevertheless. Similar to the mouse's incapacity to comprehend the peril that comes with attempting to get the much-desired bait, those convinced to collaborate could not possibly comprehend the intricacies of the dynamic they were entering, or the potential repercussions such a scheme would have in their lives, or the lives of those it involved. For that, one must be endowed with a bird's-eye view on a given situation, something that we historians cannot fully attain by contemplating the recent past.

Securitate files dating from 1988, only months before the outbreak of the bloody 1989 December revolution that took the lives of more than a thousand people and culminated with the execution of Ceauşescu and his wife by a firing squad after a brief televised show trial, offer further insight into the work of collaboration. One 300-page file (ACNSAS, 1988), for example, describes a "plan of action" to tackle the many "hostile elements" (ACNSAS, 1988, pp. 74–111) that, as discussed earlier, were seen as threats to Romanian society. With its various clichés and "Party talk," as if unaware of the grim reality outside of their windows, the Securitate officers who wrote this file spoke in a language out of touch with the world in which they lived. The "sheer thoughtlessness" (cited in Hartouni, 2012, p. 77) and "banality" of the wrongdoing depicted in these files, to quote Arendt (1977), served to label as criminal, hostile, and dangerous innocent men and women, extending even to peaceful practitioners of Transcendental Meditation (ACNSAS, 1988, p. 157). The writing and editing of the file, and all the long meetings and red tape that may have been needed to discuss and approve it, was in the service of what turned out to be a "lie," to quote Daniel Chirot (1999, p. 36), in the spirit of keeping alive a defunct system and its faded ideology that may once have offered many what Vaclav Havel (1979) called an "illusion of an identity, of dignity, and of morality."

Another example worth highlighting here is the case of the Romanian philosopher and former political prisoner Constantin Noica (1909–1987), not only because of his leadership of the Paltiniş School but also because of the ambiguous status of his relation to the Securitate. As Securitate documents reveal, Noica served as a collaborator while overtly professing cultural resistance by being the leader of this school, composed of some of the finest minds of Romania of that time, who were convening periodically in the 1980s in a reclusive dwelling in the Romanian mountains of Transylvania to read and discuss various philosophical works by Plato, Kant, Hegel, and Heidegger. Having signed an agreement to collaborate with the Securitate shortly before his release from prison in the 1960s, some scholars believe that it was that very agreement that permitted Noica to organize such intellectual activities, under the facade of apparent tolerance on the part of the Securitate for such literary gatherings (Turcescu & Stan, 2017).

Noica's case evokes the notion of "Ketman" in Czesław Miłosz's *The Captive Mind*, a term Miłosz appropriated from the Islamic legislation (from *kitman*) that permits one to hide one's support for a religious minority group subjected to discriminatory treatment. In a communist context, the actions of Miłosz's practitioners of Ketman did not reflect their thoughts and feelings, as they had adjusted to the regime's requirements while simultaneously maintaining within what Miłosz describes as an "autonomy of a free thinker – or at any rate a thinker who has freely chosen to subordinate himself to the ideas and dictates of others" (cited in Turcescu & Stan, 2017, p. 36). It is possible that the Ketman in Noica may have worked on both fronts, with "Lady Philosophy," to use Boethius's (2008) language in his *Consolations of Philosophy*, serving as the artful mediator between the two.

Needless to say, not all collaboration implied carrying on passionate intellectual discussions about the nature of the human soul, nor were all instances of collaborations initiated on relatively peaceful terms. For example, some informers, as mentioned earlier, were blackmailed about their Legionary pasts (Stan & Turcescu, 2007). While in prison, many political prisoners who had later turned to collaboration were starved and/or deprived of medical help or visiting rights from family and loved ones. Others were subjected to, or threatened with, solitary confinement or physical torture. These prisoners, for obvious reasons (i.e., anti-communist sentiments, rhetoric, and activity), were the most difficult to convince or lure to collaborate with a regime towards which they had little to no allegiance. Their recruitment was often launched only days before they were to be set free – information that only the prison personnel and the Securitate knew and that was not divulged to the prisoners. Weeks before the due date to be liberated from prison, these individuals were told that their prison stay would be significantly increased if they refused to collaborate. At that time, however, these prisoners were most often *not aware* that they would soon be set free. As a result of the withholding of such crucial information, many were becoming desperate and were agreeing to collaborate due to the psychological pressure exerted on them. Following their signing of a "document of commitment to collaborate" (*angajament de colaborare*) they were promised a "normal life," without the stigma of having served time in prison, unaware that by placing their signature on that document many were acquiescing to the continuation of their suffering and psychological pain (Tismşneanu et al., 2006).

What seemed like an optimal solution to the torment experienced in those prison conditions was, in reality, the commencement of a long agonizing journey towards what may have possibly felt for many as a living hell: hence the name of the title of the semi-autobiographical book written in the 1960s about this era by the Romanian theologian Antonie Plămădeală, entitled *Trei Ceasuri în Iad* (*Three Hours in Hell*). Here we return to the Dej period, with its more overt violence that set the stage for the normalization of Securitate surveillance under Ceaușescu. In *Three Hours in Hell*, the author offers an allegorical analysis of his experience as a political prisoner in communist Romania in the 1950s, alluding to the brainwashing techniques inmates were subjected to during this period.

The outcome, the re-educated man, is a foreigner to himself and those who once knew him. Adam-Ghast, the book's main character, is both Adam, his baptismal name, and Ghast, the name given to him by the regime that re-educated him. In search for the old Adam, Adam-Ghast loses his hope and trust in life and in himself, leaving behind everything and everyone he once knew for an unknown world that will never know or understand who he really was or is (A. Plămădeală, 2013a; C. Plama-deala, 2016).

But of what kind of hell am I talking about here? Definitely not one found in the works of Dante Alighieri, but more as a metaphor in line with the political evil described in the works of Arendt, a hell that is more connected to one's inner being and soul. Devoid of any religious mean-ing, this hell would be more of an allegory for what may have been felt by many living under the Securitate's surveillance network. In a letter to his friends dating from September 29, 1964, A. Plămădeală (2013b) writes the following about what he hopes to transmit in his then unfinished manuscript of *Three Hours in Hell*:

> I want to capture the tragedy of the loneliness felt by the human being nowadays … who is obliged to simulate to the maximum, until he gets to think that he no longer is subject to simulation, but he is authentic … What you at first consider as violence inflicted unto yourself, you discover that it is freedom. In reality, it is still violence, but it is so extreme that you lose your capacity to understand. (p. 234)

In the communist era, many Romanians may have experienced this camouflaged infliction of violence perceived as freedom, living in a society wherein one's capacity to understand may have been somewhat destroyed. This may explain, albeit partially, the widespread collabora-tion of individuals in Romania, who, as in the actual Pitești experiment, may have been former victims now turned perpetrators, all playing a role of some sort in maintaining this hell. For some, paradoxically, this hell may have even looked like a mirage of freedom and, possibly, even of prestige, as it may have been the case for at least a few of the Securitate officers mentioned above.

Conclusion

In the 1980s, Romanians made up an anecdote about the Securitate, most probably to deal with the overwhelming fear that they felt because of its ubiquitous presence. The joke goes as follows: "Around midnight, a man hears loud knocks on his door. Awakened from his sleep, he asks: who's

there? From the other side of the door, a scary voice answers: 'Death!'
Breathing relievededly, the man responds: 'Thank God! I thought it was
the Securitate!'" (cited in Banu, 2008, p. 204). This anecdote humor-
ously highlights the grim reality of many Romanians in the communist
era, and especially in the Ceauşescu decades. In late communism, suspicion and fear of being constantly watched or under the surveillance of
an informer disguised as a friend, lover, brother/sister, or neighbour was
for many a habit ensuring one's survival in a world of which few could
probably fully make sense.

In this chapter, I referred to this world as that of dossierveillance,
whose primary symptom was fear. Many, as the joke told above illustrates,
may have felt this fear very strongly, to the point that, at least for some,
it may have superseded their fear of death itself. The Securitate file was
the phantom that haunted many Romanians of that era. Some took part
in the writing of files. Others were the reason why they existed to begin
with, being the targets of the Securitate. As this chapter sought to show,
the Securitate couldn't possibly have been that effective in instilling fear
as it was, if it wasn't for the widespread collaboration. That is because
collaboration relied, ultimately, on the Securitate file for its legitimacy.
Similarly, the file depended on collaboration for it to exist and justify its
existence. Together, as in the case of communist Romania, they turned
out to be a deleterious combination, yet a fertile ground for dossierveillance to be successfully carried out.

NOTES

1 Working under the guidance of the assigned Securitate officer or the
   resident, the *informer* gathered information deemed important for the
   Securitate. The *informer* was hired after a thorough investigation of his/her
   character and background. In the first decade following the establishment
   of the Securitate, informers were also classified as qualified and unqualified,
   with the qualified informers considered of a higher level of preparation
   and ability to carry out the tasks assigned by the Securitate personnel.
   The *residents*, often members of the Communist Party, were not involved
   in the recruiting process. They were in charge of managing informants,
   support persons, and lower-level collaborators. Sometimes, the residents
   were undertaking some of the responsibilities assigned to the Securitate
   liaison officer. (Index de termeni şi abrevieri cu utilizarea frecventă în
   documentele Securității, pp. 1–11).
2 I thank Lavinia Stan for her assistance in crystalizing the argument put forth
   in this section of the chapter.

3 Translation from Romanian into English was performed by the author.
4 A meeting house (*casă de întâlnire*) was a house, office, or space made
   available to the Securitate via a written contract with the owner, landlord, or
   renter of the respective facility. It was used to arrange meetings between the
   Securitate officers and informers and residents belonging to the surveillance
   network. Those who acquiesced to this kind of arrangement were called
   "hosts of the meeting houses" (*gazde de case de întâlniri*). A conspiring house
   (*casă conspirativă*) was an apartment or a house used by the Securitate
   officers to meet informers (Index de termeni și abrevieri cu utilizarea
   frecventă în documentele Securității, pp. 1–11).
5 I thank Sara Terreault for the discussions we had throughout these past two
   years on what constitutes a pilgrimage, as they contributed to stretching this
   concept into the realm of archival research.

REFERENCES

*Primary Sources from the Archives of the National Council for the Study of
Securitate Archives, Romania (hereafter ACNSAS)*

ACNSAS. (1954). Fond informative, 160128, vol. 1.
ACNSAS. (1958–1959). Fond documentar, "Proiect Referitor la Evacuarea
   Manastirei Tudor Vladimirescu si Altele, 1955–1959," [Project to Evacuate
   the Monastery Vladimirescu and other ones, 1955–1959], D00066.
ACNSAS. (1962). Fond operativ, "Iordache Maria și altii" [Iordache Maria and
   others], 160128, vol. 5.
ACNSAS. (1966). Fond operativ, "Ioan Silviu I. Cornel," I1211014, vol. 1.
ACNSAS. (1969). Fond operativ, "Iordache Maria," I234104.
ACNSAS. (1976). Fond documentar. "Criterii Privind Recrutarea de Informatori
   si Colaboratori Pentru Munca de Securitate" [Criteria for Recruitment of
   Informers and Collaborators for Securitate Work]. D08712, vol. 1P19.
ACNSAS. (1988). Fond operativ, "Documente de Organizare si Planificare a
   Muncii de Informații" [Documents for the Organization and Planification of
   the Work Pertaining to Information Gathering], D000003, vol. 5.

Secondary Sources

Albu, M. (2008). *Informatoru: Studiu asupra colaborării cu Securitate* [*The secret
   informer: A study of collaboration with the Securitate*]. Iasi, RO: Polirom.
Arendt, H. (1977). *Eichman in Jerusalem. A report on the banality of evil.* New York,
   NY: Penguin Press.

Banu, F. (2008). Cateva consideratii priind istoriografia Securitatii [Some considerations pertaining to Securitate's historiography]. Editors: Florin Banu, George Enache, Silviu B. Moldovanu, Liviu Taranu. In *Caietele CNSAS* (pp. 187–220). Bucharest, RO: Editura CNSAS.

Boethius. (2008). *The consolation of philosophy* (D. R. Slavitt, Trans.). Cambridge, MA: Harvard University Press.

Bruce, G. (2010). The people's state: East German society from Hitler to Honecker [Review of *The people's state: East German society from Hitler to Honecker*, by M. Fulbrook]. *Journal of Cold War Studies, 12*(3), 137–40.

Chirot, D. (1999). What happened in Eastern Europe in 1989? In V. Tismăneanu (Ed.), *The revolutions of 1989* (pp. 19–50). London, UK: Routledge.

Ciobanu, M. (2015). Piteşti: A project in reeducation and its post-1989 interpretation in Romania. *Nationalities Papers: The Journal of Nationalism and Ethnicity, 42*(5), 615–33.

Clark, R. (2015). *Holy legionary youth: Fascist activism in interwar Romania*. Ithaca, NY: Cornell University Press.

Deletant, D. (1995). *Ceauşescu and the Securitate: Coercion and dissent in Romania, 1965–1989*. Armonk, NY: M. E. Sharpe.

Foucault, M. (1973). *The birth of the clinic: An archaeology of the medical perception* (A. M. Sheridan Smith, Trans.). New York, NY: Pantheon Books.

Foucault, M. (1991). *Discipline and punish: The birth of the prison*. London, UK: Penguin.

Fulbrook, M. (2005). *The people's state: East German society from Hitler to Honecker*. New Haven, CT: Yale University Press.

Galic, M., Timan, T., & Koops, B. J. (2016). Bentham, Deleuze and beyond: An overview of surveillance theories from the panopticon to participation. *Philosophy and Technology, 30*(1), 9–37.

Garton Ash, T. (1997). *The file: A personal history*. New York, NY: Random House.

Gates, K. A. (2011). *Our biometric future: Facial recognition technology and the culture of surveillance*. New York, NY: New York University Press.

Hartouni, V. (2012). *Arendt, evil, and the optics of thoughtlessness*. New York, NY: New York University Press.

Havel, V. (1979). The power of the powerless (Paul Wilson, Trans.), available at http://chnm.gmu.edu/1989/archive/files/havel-power-of-the-powerless_be62e5917d.pdf. Last accessed on April 9, 2019.

Holquist, P. (1997). "Information is the alpha and omega of our work": Bolshevik surveillance in its Pan-European perspective. *Journal of Modern History, 69*(3), 415–50.

Holquist, P. (2001). To count, to extract, and to exterminate: Population statistics and population politics in late imperial and Soviet Russia. In R. G. Suny & T. Martin (Eds.), *A state of nations: Empire and nation-making in the age of Lenin and Stalin* (pp. 110–43). Oxford, UK: Oxford University Press.

Horne, C. (2015). "Silent lustration": Public disclosures as informal lustration mechanisms in Bulgaria and Romania. *Problems of Post-Communism, 62,* 131–44.

Index de termeni și abrevieri cu utilizarea frecventă în documentele Securității [Index of the terms and abbreviations that are frequently used in Securitate documents], retrieved from www.cnsas.ro/documente/arhiva/Dictionar%20 termeni.pdf. Last accessed on December 9, 2017, pp. 1–11.

Kligman, G., & Verdery, K. (2011). *The collectivization of Romanian agriculture, 1949–1962.* Princeton, NJ: Princeton University Press.

Kokoška, S. (2013). Resistance, collaboration, adaptation: Some notes on the research of the Czech society in the protectorate. *Czech Journal of Contemporary History, 1,* 54–76.

Laudon, K. (1986). *Dossier society: Value choices in the design of national information systems.* New York, NY: Columbia University Press.

Livezeanu, I. (1990). Fascists and conservatives in Romania: Two generations of nationalists. In M. Blinkhorn (Ed.), *Fascists and conservatives: The radical right and the establishment in twentieth century Europe* (pp. 218–39). London, UK: Unwin Hyman.

Lyon, D. (Ed.). (2003). *Surveillance as social sorting: Privacy, risk and digital discrimination.* London, UK: Routledge.

Morgan, M. L. (2001). *Beyond Auschwitz: Post-Holocaust Jewish thought in America.* New York, NY: Oxford University Press.

Muresan, A. (2011). *Pitești: Cronica unei sinucideri asistate* (Trans. *Pitești: The chronicle of an assisted suicide,* 2nd ed.). Iași, RO: Polirom.

Neagoe-Pleșa, Elis. (2008). Securitatea: Metode și acțiuni. 1968 – Anul reformării agenturii Securității [The Securitate: Methods and activity. Year 1968 – The year of reforms of the Securitate agency]. Editors: Florin Banu, George Enache, Silviu B. Moldovanu, Liviu Taranu. In *Caietele CNSAS* (pp. 9–22). Bucharest, RO: Editura CNSAS.

Petrescu, C. (2014). The afterlife of the Securitate: On moral correctness in postcommunist Romania. In M. Todorova, A. Dimou, & S. Troebst (Eds.), *Remembering communism: Private and public recollections of lived experience in southeast Europe* (pp. 385–416). New York, NY: Central European University Press.

Petrescu, C., & Petrescu, D. (2010). The Pitești syndrome: A Romanian vergangenheitsbewaeltigung. In S. Troebst & S. Baumgartl (Eds.), *Postdiktatorische Geschichtskulturen im Süden und Osten Europas: Bestandsaufnahme und Forschungsperspektiven* (pp. 502–618). Wallstein Verlag.

Plămădeală, A. (2013a). *Trei ceasuri in iad* [*Three hours in hell*]. Bucharest, RO: Editura Sophia.

Plămădeală, A. (2013b). Letter dating from 20 Septembrie 1964 to Petru and Angela Ciobanu. T. (pp. 233–5). Aioanei, Mitropolitul Antonie Plămădeală și

Aminitirea unei Prietenii: Scrisori Inedite [Metropolitan Antonie Plămădeală and the recollation of a friendship: Unpublished letters], vol. I.

Plamadeala, C. (2015). The use and misuse of information in the Securitate's Files: The case of Plămădeala. *Eurostudia*, *1*(1), 125–41.

Plamadeala, C. (2016). Antonie Plămădeală and the Securitate in the years 1940s–1950s. *Archiva Moldaviae*, *8*, 215–51.

Plamadeala, C. (2019, forthcoming). The Securitate file as a record of psuchegraphy. *Biography*, *42*(3), special issue on "Biographic Mediation: The Uses of Disclosure in Bureaucracy and Politics."

Roper, S. D. (2000). *The unfinished revolution*. Amsterdam, Netherlands: Hardwood Academic.

Stan, L. (2013). *Transitional justice in post-communist Romania: The politics of memory*. New York, NY: Cambridge University Press.

Stan, L., & Nedelsky, N. (2013). Access to secret files. In L. Stan & N. Nedelsky (Eds.), *Encyclopedia of transitional justice*, vol. 1 (pp. 1–11). New York, NY: Cambridge University Press.

Stan, L., & Turcescu, L. (2007). *Religion and politics in post-communist Romania*. New York, NY: Oxford University Press.

Stănescu, M. (2010a). *Reeducarea în Romania comunistă (1945–1952): Aiud, Suceava, Pitești, Brașov* [*Re-education in communist Romania (1945–1952): Aiud, Suceava, Pitești, Brasov*]. Iași, RO: Polirom.

Stănescu, M. (2010b). *Reeducarea în Romania comunistă (1948–1955): Târgșor, Gherla* [*Reeducation in communist Romania (1945–1952): Târgșor, Gherla*]. Iași, RO: Polirom.

Stănescu, M. (2010c). *Reeducarea în Romania comunistă (1949–1955): Targu-Ocna, Ocnele Mari, Canalul Dunare-Marea Neagra* [*Re-education in communist Romania (1945–1952): Targu-Ocna, Ocnele Mari, Canalul Dunare-Marea Neagra*]. Iași, RO: Polirom.

Tismăneanu, V., Dobrincu, D., Vasile, C., et al. (Eds.). (2006). *Raport final: Comisia Prezidențială Pentru Analiza Dictaturii Comuniste din Romania* [*Final report: The Presidential Commission for the Analysis of Communist Dictatorship in Romania*]. Retrieved from http://old.presidency.ro/static/rapoarte/Raport _final_CPADCR.pdf

Turcescu, L., & Stan, L. (2015). Church collaboration and resistance under communism revisited: The case of Patriarch Justinian Marina (1948–1977). *Eurostudia*, *10*(1), 75–103.

Turcescu, L., & Stan, L. (2017). Collaboration and resistance: Some definitional difficulties. In L. Stan & L. Turcescu (Eds.), *Justice, memory and redress in Romania: New insights* (pp. 24–44). Newcastle upon Tyne, UK: Cambridge Scholars Publishing.

Ursache, R. (2013). Archival records as evidence. In L. Stan & N. Nedelsky (Eds.), *Encyclopedia of transitional justice*, vol. 1 (pp. 112–17). New York, NY: Cambridge University Press.

Vatulescu, C. (2010). *Police aesthetics: Literature, film, and the secret police in Soviet times.* Stanford, CA: Stanford University Press.

Vaz, P., & Bruno, F. (2003). Types of self-surveillance: From abnormality to individuals at risk. *Surveillance & Society, 1*(3), 272–91.

Weld, K. (2014). *Paper cadavers: The archives of dictatorship in Guatemala.* Durham, NC: Duke University Press.

Willis, J. (2013). *Daily life behind the Iron Curtain.* Santa Barbara, CA: Greenwood.

10 The FBI and the American Friends Service Committee: Surveilling United States Religious Expression in the Cold War Era

KATHRYN MONTALBANO

The American Friends Service Committee (AFSC) was formed in 1917 by Quakers in Philadelphia, Pennsylvania, in the throes of the First World War. Its goal was to provide conscientious objectors with service opportunities and to assist with the war-related struggles of European nations. The organization soon expanded to provide relief to communities in war or conflict zones in the United States and abroad, including by feeding the hungry and supporting immigrant and refugee communities. From the early to the mid-twentieth century, the Federal Bureau of Information (FBI), led by J. Edgar Hoover, suspected AFSC members of associating with communists through humanitarian projects, especially during the Cold War period. Beginning in 1956, Hoover's secret FBI Counterintelligence Program (COINTELPRO) instructed FBI agents to surveil subversive individuals and groups, including the AFSC, in order to dismantle the United States Communist Party. In 1971, a group of then anonymous activists publicly exposed COINTELPRO by breaking into the FBI office in Media, Pennsylvania, and releasing incriminating files. William C. Davidon, a Haverford College professor, led the break-in. His wife, Ann Morrissett Davidon, a writer and AFSC volunteer, archived and annotated many of the FBI files on the AFSC. Neither of the Davidons held a religious affiliation, yet they worked closely with Quaker and Catholic activist organizations committed to peace and social justice ("Historical Background," n.d.). Ann Morrissett Davidon (1978) succinctly relayed the revelation that AFSC Quakers were under surveillance with her headline in *The Nation*, "Watching for Cominfil: Even the Quakers Scared the FBI."

These FBI files, ranging from AFSC news releases to allegations of communist infiltration, compose the backbone of this chapter, which examines FBI surveillance of the AFSC from the early 1940s to the early 1960s. As an alumna of Haverford College, where William C. Davidon taught from 1961 to 1991, I was aware of the vast collection of Quaker material in the archives of Haverford and Swarthmore Colleges. Through advice from the curators of the Swarthmore College Peace Collection and Friends Historical Library, which I visited first, I turned my attention to the AFSC Archives in Philadelphia. After preliminarily examining all of the materials in the archives pertaining to FBI surveillance of the AFSC, I focused on the following five groups of files due to the relevance of their contents to Quakerism and communism: (1) "Government Surveillance," which contained AFSC-official material released after they discovered the surveillance program; (2) "Criticisms and Investigations," which included assorted FBI files from the 1950s and 1960s, including AFSC correspondence, congressional reports featuring AFSC members, and annotations of AFSC mass media publications; (3) "FOIA Box #1," which stored FBI files received through FOIA in 1976 dating from 1943 to the early 1970s; (4) "FBI Material Rec. on Appeal, 4/80," which held additional files dating back to 1921 that the AFSC received in 1980 after appealing the attempt of the FBI to withhold files from the original 1976 FOIA release; and (5) "CIA/Air Force/Army/Navy," containing files the FBI retrieved from these other government agencies.

Methodologically, archival research expands the depth of surveillance and communication scholarship by foregrounding the central role of information systems in surveillance practices. Harold Innis (1950/2007) innovatively showed how efficient communication systems, organizing spheres of government such as administration and law, have historically formed the backbone of the political, economic, and geographic power of empires. As such, historical analyses of government surveillance might trace agents' production of information within the surveillance systems that in turn control populations, which is analogous to Innis's (1951/2008) attention to how systems of writing historically both coordinated and controlled human activities. With Innis's analyses of control through communication systems in mind, this chapter considers government agents who, by observing and categorizing religious expression, produced information within their own records about the AFSC that could have effected real-world consequences had the surveillance program not been uncovered.

From 1976 to 1980, through Freedom of Information requests, the AFSC received its FBI files dating from 1921, when the FBI observed conscientious objection and connections to the Soviet Union among AFSC members, to the early 1970s. In 1921, an unidentified Philadelphia FBI agent characterized the AFSC as a "religious and philanthropic society," composed of officers who were "of high standing in commercial and social circles, many of whom are devoting all of their time, without monetary gain, in the Committee's undertaking to help the famine stricken people of Soviet Russia, and in their other philanthropic undertakings" (AFSC, 1976b, p. 1). The Philadelphia FBI office, in particular, would come to the defence of the AFSC more than once during Hoover's surveillance campaign.

The FBI received some information on the AFSC from other government agencies, but it predominantly gathered its own intelligence through FBI agents and informants. Hoover positioned FBI special agents in charge around the country, agents who had enjoyed relative autonomy prior to Hoover's role as Director of the FBI from 1924 until 1972 (Cecil, 2014). The special agents served as information gatherers and communicated with influential members of their respective communities (Cecil, 2014). Within this centralized government agency, Hoover sought unlimited power to identify and prosecute political dissenters. He would have to wait until the Cold War, however, as the FBI only acquired the capacity, and not the authority, to detect radicalism in the 1920s. When Hoover was put in charge in 1924 of the agency that would later become the FBI, he was told by Harlan Fiske Stone, the attorney general and former Columbia Law School dean who had appointed him, to investigate people's actions rather than their opinions (Schwarz, 2013). The rise of civil liberties following the First World War shaped and constrained the environment in which J. Edgar Hoover assumed his leadership position.

Before Hoover consolidated power within the FBI in the Cold War period, many requests for information about the AFSC from the 1920s to the early 1950s came from local citizens and FBI agents. Hoover often denied requests for information from citizens. He reassured them that the AFSC was a "committee of the Quaker faith," built to engage in peace and relief efforts at home and abroad (quoted in AFSC, 1976a, p. 12). In the late 1950s and the 1960s, FBI agents began to systematically surveil the AFSC by reporting on meetings, demonstrations, and vigils against Vietnam that AFSC members attended or organized. However, Hoover encountered dissent from his own FBI agents about surveillance of the AFSC.

Despite the rise of civil liberties following the First World War (Foner, 1998) and culture of transparency (Schudson, 2016) that accompanied his tenure as Director of the FBI, Hoover concealed that the FBI was spying on the AFSC and other organizations marked as infiltrated by communists. Informants provided intelligence on the AFSC to local FBI offices throughout the country, while undercover agents observed and reported on AFSC-related gatherings (AFSC, 1976a). FBI agents struggled to determine the legitimacy of surveilling the AFSC, trying to distinguish between religious expression and threatening action, and to determine the extent to which Quaker beliefs shaped the actions of the organization. Comparing FBI surveillance of the AFSC to that of the underground newspapers discussed by Elisabetta Ferrari and John Remensperger in chapter 11 of this book, both parallels and contrasts are striking. The AFSC was similarly caught up in the dense networks of Cold War FBI surveillance, but in their case, unlike with many other radical groups, there was significant debate and many reservations expressed within the FBI over the scope of this surveillance.

This chapter first probes surveillance theories to assess what kind of framework is most useful for understanding FBI surveillance of the AFSC, as well as other historical case studies in which religious groups have been surveilled. It then turns to the shifting institutional identity of the AFSC, including the role of Quakerism, from its formation in the early twentieth century. The relationship between the AFSC and Quakerism shaped FBI surveillance, but it was complicated by the fact that neither the AFSC nor the Religious Society of Friends (the collective name for Quaker groups) fit comfortably within the cultural hegemony of Cold War America. The chapter then moves to a discussion of the FBI files themselves, which reveal the nuanced approach of the FBI in ascertaining the religiosity of the AFSC and the extent to which Quakerism shaped AFSC actions in the public sphere. The FBI surveillance program, in turn, suggests that panoptic systems are not always cohesive or infallible. The chapter concludes with an analysis of the conditions under which the FBI willingly compromised religious expression and, through surveillance, aligned ideal citizenship with only certain kinds of religious expression.

Histories of Government Surveillance of Religion: A New Theoretical Framework

Many scholars have ruminated on the usefulness of the panopticon as a theoretical construct for more contemporary examples of surveillance (Giddens, 1985; Haggerty & Ericson, 2000; Lianos, 2003; Lyon, 1994, 2006). This chapter contends that, while FBI surveillance of the AFSC was panoptic in some respects, relying solely on the panopticon in analysing historical examples of surveillance is unduly limiting. David Lyon (2007) wrote that focusing on the panopticon can lead to the mischaracterization of surveillance as a "total, homogeneous situation" (p. 25), rather than one that distinctively affects specific groups, including, as this case study emphasizes, religious organizations in particular historical periods. Michel Foucault's (1977) panoptic model, based on Jeremy

Bentham's proposed prison, emphasizes centralized, repressive, powerful, disciplinary mechanisms of control, whereas this case study tells a different story: one of a decentralized and at times incoherent surveillance program composed of conflicting elements and agents, thereby challenging the notion that surveillance encompasses a singular objective or can be detected at a singular locus. Some FBI agents disagreed with one another and with J. Edgar Hoover on whether the AFSC posed a viable threat and, stemming from varying levels of familiarity with the Service Committee, had to repeatedly clarify to other agents and to inquiring citizens that they did not, at that time, hold evidence of communist infiltration of the AFSC. A more neutral, less panoptic surveillance framework also allows for the broader, nuanced, and more significant part of this particular story to emerge: how the FBI probed, defined, and categorized religion, and how, in turn, the dissonance between government agents within the FBI surveillance program shed light on the "practices through which states are formed" (van der Meulen & Heynen, chapter 1).

Kevin Haggerty (2006) has argued that the panoptic model does not allow room for a consideration of the attitudes, biases, and prejudices of those surveilling. This perspective is important in this chapter, which examines FBI agents' own cultural and legal frameworks for understanding legitimate versus non-legitimate religious expression. Several agents noted either explicitly or indirectly the respectability of the AFSC within United States society as an obstruction to surveilling them. Nevertheless, FBI concern that the AFSC was a Unitarian organization in disguise demonstrated that a hierarchy of respectable religions also existed, even within the Protestant umbrella. FBI surveillance of the AFSC offered a glimpse into how the fear that communism had permeated the Quaker group led the "surveillant assemblage" (Haggerty & Ericson, 2000), comprising the Air Force, Army, Internal Revenue Service, Navy, National Security Agency, Secret Service, and State Department, to surveil even a highly respected, socially acclimated, religiously affiliated organization.

As Emily van der Meulen and Robert Heynen likewise note in the book's introduction, Anthony Giddens (1985) has argued that the storage of information entails forms of surveillance central to the formation of the nation state. FBI files on the AFSC helped to solidify a cultural hegemonic framework for acceptable religious expression, indirectly contributing to a nation-building process that monitored identity. Also significant for this chapter, scholars such as John Fiske (1998) and Simone Browne (2015) have demonstrated the ways in which surveillance operates differently upon Black and racialized bodies, dating back, in the case of Browne, to the surveillance of blackness in the transatlantic slave trade. Her theorization of "racializing surveillance," described as a "technology of social control" that defines "boundaries, borders, and bodies along racial lines" (p. 16), is particularly illustrative. The overwhelming whiteness of the AFSC in the mid-twentieth century, juxtaposed with its overarching (alternative) Protestant identity, likely granted the religious organization a certain level of protection. Their surveillance was less systematic than that of their contemporary subversive peers, thereby raising questions of how the white, middle-class, educated Protestant identity of the AFSC at this time inhibited the reach of FBI surveillance. Browne's emphasis on race and surveillance parallels the concern of this chapter on how surveillance defines boundaries along racial, socio-economic, and religious lines.

AFSC members had always looked outward from their own networks on the question of conscientious objection. Yet it was not until the AFSC confirmed it was being surveilled by the federal government, based on information that emerged from a lawsuit including the Chicago AFSC called the Chicago 8 trial, that the organization examined how surveillance affected the wider United States public. As an organizational response to COINTELPRO, the AFSC launched the Board Level Task Force on Government Surveillance and Citizens' Rights in 1975, which later prompted the Campaign to Stop Government Spying and raised awareness about government surveillance of groups that were deemed dissident or subversive. The AFSC members who worked for Government Surveillance and Citizens' Rights emphasized that the intimidation, dossier keeping, and surveillance tactics of the FBI particularly affected the poor, as well as Blacks, Hispanics, and other minorities. As Browne (2015) has extended the term *sousveillance*, or surveillance-from-below (Mann, Nolan, & Wellman, 2003), to demonstrate how "dark sousveillance" operated as resistance in the transatlantic slave trade, the AFSC, instilled with societal and political power that silenced slaves could not access themselves, became even more radical and politically active than it had been prior to the exposure of the FBI surveillance program through its own form of sousveillance. In this way, government surveillance of the AFSC shaped its evolving organizational structure.

FBI surveillance targeted religious groups that did not reside comfortably within the legal parameters for religious expression or the hegemonic norms of the religious and cultural majority. Surveillance of religious groups was often fraught. As we saw in the previous chapter by Cristina Plamadeala, even in officially atheist communist Romania, it was primarily religious figures and institutions that were subject to repression, in that case particularly those, like the Vladimireşti Monastery, that had fascist ties. Although the focus of the FBI was communist influence in the AFSC, FBI agents consistently worried over how to determine the relationship between the religious and the political, examining specifically the role of Quakerism in shaping the alleged communism of the AFSC. This intermingling of religion and communism in FBI discourse suggests that government surveillance can exhibit nuanced approaches that aim not only to impose immediate sanctions, but to mould and categorize publics – who is included and who is excluded in certain spaces – through the collection of information. In the case of the AFSC, the FBI considered how they fared as citizens, and indirectly, whether the boundaries of religious expression were adequate.

The American Friends Service Committee: An Evolving Organization

Over the course of its surveillance program, the FBI observed the evolving organizational structure of the AFSC, and to what extent that structure was congruent with Quaker ideology. Ivan Greenberg (2012) has argued, "In order to understand surveillance in America, it is necessary to study its ideological roots: the beliefs, attitudes, and worldview that framed leadership decisions governing political monitoring" (p. 189). I further suggest that to understand United States government surveillance, it is additionally necessary to study the tension between ideologies of the surveilling agent(s) and the surveilled, as well as how the surveilling agent(s) interpret the ideology of the surveilled subject.

The Free Exercise Clause, which protects religious practice from government intrusion, was already legally defined when the FBI began surveilling the AFSC in 1921. Yet the FBI struggled with how to craft a policy for surveilling the organization using that definition. Although the 1921 Philadelphia FBI agent mentioned at the outset of the chapter might have disagreed, some have noted that the AFSC did not fit comfortably within high-status society. Sociologist Lester M. Jones (1929) described the original members of the AFSC: "It was not a group distinguished for wealth, political power, or social prestige. It was, however, a strong group, and particularly so after the committee was enlarged and made representative of the various groups of American Friends" (p. 19). The AFSC would, nonetheless, amass social prestige through the twentieth century, gradually recruiting college-educated Quakers (or Friends, as they were more commonly known) to increase its numbers. AFSC members made important contacts with government officials, including Herbert Hoover, a Quaker-raised "birthright Friend" (Jones, 1929, p. 22), who asked the AFSC to carry out his feeding program in Germany after the First World War (Sutters, 2010).

Such contacts strengthened the AFSC network and forged bipartisan connections in the executive branch, Congress, and abroad. Bronson Clark, Executive Secretary of the AFSC from 1968 to 1974, during the increasing involvement of the United States in Vietnam, once remarked that it could maintain amicable relations with the United States government while also criticizing some government policies (Jones, 1971). J. William Frost (1992) wrote on the topic of Quaker conscientious objection, "Friends presented to the government a problem to be solved: how to avoid persecution, respect religious freedom, and, if it could not directly enlist pacifists in the fighting, then to make their activities seem not subversive and even helpful to the war effort" (p. 3). Quakers who appeased influential members of society while simultaneously objecting to war mirrored the simultaneous insider and outsider status of the AFSC, which understood the need to adhere to its goals with minimal government criticism.

While it is difficult to generalize about the Society of Friends due to the diversity within the group, Thomas D. Hamm (2003) has argued that certain religious beliefs have historically united Quakers in America, allowing them to maintain a cohesive identity across geographic and political differences. The first Quakers were English, the movement emerging in the 1640s and 1650s and developing a political, religious, and social world view that put its members at odds with their fellow citizens (Hamm, 2003). Friends thus migrated to various locations, including North America, with the first arriving in the Chesapeake Bay region in 1655 or 1656. By 1662, approximately sixty Friends had visited Virginia and Maryland, establishing several Quaker meetings (Hamm, 2003). The Friends and the AFSC were not interchangeable, but they shared institutional links and members. That the FBI struggled to differentiate the two, as well as differentiate Quakers from other Christian religions, moulded their surveillance strategy into one that was cautious and reflective. FBI officials pondered whether the religious beliefs of AFSC members bolstered their actions. Quaker ideology, and its overall congruence with good citizenship (Giddens, 1985), threatened the boundary between communism and non-communism through an aggressive pacifism that distinguished Quakerism from mainstream Protestant culture.

Consequently, FBI agents first tried to deconstruct Quaker theology and determine the extent to which it shaped the AFSC. As early as the 1940s, the FBI (1943) published a one-hundred-page report on the AFSC, reproduced in correspondence between Hoover and then Assistant Secretary of State Adolf A. Berle. The report itself was fairly objective, comprising a history of the American Friends, an explanation of Quaker religious beliefs, and a summary of the history and activities of various subsections of the AFSC. One section of the report, entitled "The Tenets of the Quaker Religion," is particularly fascinating in that it suggests the FBI was trying to understand the relationship between Quakerism and the AFSC. It appears as though the FBI accepted how the AFSC intertwined religious and political beliefs, despite later reports insisting on distinguishing between the two, observing that the AFSC "actually practices, in a modern fashion, the theories and teachings of the Quaker religion. It can be seen from this that some knowledge of the history and tenets of Quakerism should be had to more completely understand the purpose of this organization" (FBI, 1943, p. 1). The agent who produced the report also cited a summary of Quaker theology entitled "The Faith of Friends" from a 1916 booklet, *The Five Years Meeting of the Friends in America, Year Book*, which was "said" to provide a comprehensive outline of the Orthodox portion of the Friends, or the segment in the "ranks of Evangelical Churches" (FBI, 1943, p. 2). The booklet pointed to the primary tenet of Quakerism as the relationship between the individual soul and God. The overall neutral tone of this section of the report tried to organize the core principles of Quakerism, demonstrating the early attempt of the FBI to both classify and surveil the AFSC.

Much as the AFSC did not seamlessly blend with the Society of Friends or with mainstream United States culture, Quakerism itself did not fit comfortably within Protestant hegemony. As early as the 1960s, some Friends became uncomfortable with what they perceived as leftist politics driving AFSC actions and alliances with people whom they deemed questionable (Hamm, 2003). H. Larry Ingle (1998) has argued that Friends believed the AFSC should remain "a community of holy individuals" committed to peace and opposed to hostile aggression (p. 34). The anti-communist campaign sweeping the nation during the Cold War caused such Quakers, who already disagreed with these leftist, aggressive inclinations, to reconsider the AFSC. In its surveillance of the AFSC, the FBI observed this growing divide in the broader Quaker community between those who advocated stronger radical action at home and abroad, and those who insisted that passivity best reflected Quaker pacifism. This divide was made evident to FBI agents through citizen letters from Quakers throughout the nation, questioning whether the AFSC was in fact free from communist influence (AFSC, 1980a).

The AFSC increasingly distanced itself from a purely Quaker identity in order to execute radical projects. The inclusion of non-Quaker employees in the AFSC during the post-war period "gradually imperiled the Quakerism at its base and recast it as just one more pressure group within the secular political community" (Ingle, 1998, p. 28). But it did not conceal its Quaker inspirations for activism in a 1979 news release following the FBI surveillance revelation: "Its work is based on a profound Quaker belief in the dignity and worth of every person, and a faith in the power of love and nonviolence to bring about change" (AFSC, 1979, p. 3). The FBI, in turn, continually struggled to determine whether the AFSC based its actions on Quaker principles.

Searching for Subversion: FBI Surveillance of the American Friends Service Committee

From late April to December of 1975, the AFSC completed Freedom of Information requests to solicit files from the FBI, Air Force, Army, Internal Revenue Service, and Navy, as well as the National Security Agency, Secret Service, and State Department in order to determine the scope of the surveillance of their activities. Thereafter, they received additional files from various government agencies on appeal (AFSC, 1976a). A total of thirteen thousand pages of files were received from sixteen different federal government agencies between 1975 and 1979 (AFSC, 1979). The documents and, relatedly, the scope of the FBI surveillance program were quite comprehensive, probing large-scale official AFSC gatherings, meetings, demonstrations, and marches; citizen letters written to the FBI asking about the AFSC; special reports and documents; and minor events that were two or three degrees removed from the AFSC, such as the suspicion that the AFSC had paid a Haverford College student, Russell Stetler, to show a movie allegedly produced by the National Liberation Front of South Vietnam to students of Pennsylvania State University (FBI, 1965a). Chronologically, the documents shifted from a focus on general investigations of the AFSC in the 1920s and 1930s – often induced by inquiring citizen letters – to conscientious objection and relief efforts in Europe and for Japanese and Japanese American civilians during the Second World War, to the Cold War and Vietnam War communist investigations.

Though there was a copious amount of information in the files, occasionally names were redacted. For instance, AFSC volunteer Helena Michie observed that one file deleted the name of someone writing an article who supposedly, due to FBI influence, "stated he would change it immediately in as much as that was not his intent at all," to whom the FBI responded that it was in fact "an investigative agency and was not in any way editing his article" (AFSC, 1980c, p. 1). Although Michie does not note what kind of article was in jeopardy, it quite possibly was a newspaper article, raising the question as to whether the FBI was concealing its tenuous relationship with the press to prevent future accusations of free expression censorship, as well as whether it was intentionally bolstering its image as an agency of investigation rather than one of intervention, which, in its surveillance of religion, was perhaps more accurate. In addition to periodic redactions, sometimes the FBI files, such as assorted summaries of intercepted AFSC correspondence, were also simply illegible.

On March 30, 1979, the AFSC released a report to its members that police spying existed on a large scale, well beyond the parameters of the AFSC. The FBI files confirmed AFSC leaders' suspicions about surveillance of its members, which they documented in an eighteen-page analysis by Ann Davidon, whose husband had led the Media, Pennsylvania, break-in that unveiled COINTELPRO (AFSC, 1976a), mentioned at the outset. The exposure of COINTELPRO prompted the FBI to refine its objective for surveillance (Greenberg, 2010), but it did not terminate its questionable and unjustified surveillance practices. The unveiling of the files did reveal an interesting dynamic about how the FBI collected and processed intelligence, however, and AFSC archivists acknowledged in their notes on the files that many FBI evaluations provided fair assessments of the Quaker organization. Nonetheless, letters from non-AFSC Friends, concerned about a communist AFSC but lacking substantive evidence to their claims, had flooded the FBI. Hoover offered vague responses to these letters, explaining that FBI files were confidential and the AFSC did not require evaluation, but concealing that the AFSC was in fact under the radar of the FBI.

In contrast to these ambiguous citizen letters, many files that were received by the AFSC did claim to offer evidence that communism had infiltrated the AFSC. Some accounts produced by government agents observing the AFSC feared the otherwise peaceful Quaker organization was compromised, perhaps because the group was, from their perspective, centrally engrained in the white, mainstream Protestant culture of United States society. For example, the Inspector General of the United States Air Force (1958) cited information from the United States Army to imply that the AFSC had been "unwittingly" used by the Communist Party as a "semi-front organization" (p. 1). Other accounts placed agency more readily within the AFSC itself. One informant, for example, wrote to the FBI in 1966 that the AFSC was "definitely a red front" used by the Communist Party to resist the United States military (FBI, 1967a, p. 2). The informant denounced the AFSC as a leftist group that used "Biblical and Quaker quotations" to justify anti-United States government and pro-communist views and accused the AFSC of collecting the majority of its funds from "red supporters" rather than Quakers (FBI, 1967a, p. 2). Without providing any evidence, the informant accused AFSC figureheads of being duped into supporting communism.

This informant was a Quaker who claimed to speak on behalf of the Five Years Meeting of the Friends, established by the Orthodox Yearly Meeting in 1902 and, as reported by the FBI agent in the 100-page AFSC report noted earlier, that aimed "to unite the entire Society of Friends in carrying on the religious, evangelical, educational, missionary, charitable, and benevolent work" (FBI, 1943, p. 2). In contrast to the Orthodox Quakers, the agent noted in the report that the AFSC was said to have had been formed by the "Liberal" Quakers, but that all Yearly Meetings in the United States thereafter united in the work of the organization "so that today there seems to be greater unity amongst all members of the Society than there has been in 1827" (FBI, 1943, p. 2). While AFSC leaders did not admit to espousing communist views, they might not have been offended by the leftist label, as there was certainly a range of both conservative and liberal Quakers in the United States at this time.

J. Edgar Hoover himself partially constructed the dichotomy of communism and religion. In a 1947 testimony before the House Un-American Activities Committee, he remarked, "I confess to a real apprehension so long as Communists are able to secure ministers of the gospel to promote their evil work and espouse a cause that is alien to the religion of Christ and Judaism" (Investigation of Un-American Propaganda, 1947, n.p.), claiming not only that some ministers themselves supported communism, but that communism was incompatible with Judaism and Christianity. Hoover argued that communist ideology and a national tolerance for religious freedom threatened the religious core of the nation and of Western civilization. Hoover (1958) tried to distinguish between true and false religions beyond the organizational boundaries of the FBI in an article in which he claimed communism was a false religion. He appealed to the religious commitment of each United States citizen to stand up to the reality of communism in the world in the name of democracy, uniting the two principles rhetorically in the phrase "democratic faith" (Hoover, 1958, p. 1). Communists threatened this faith by infiltrating religious groups to subtly disseminate their propaganda, gain respectability, and introduce a false form of peace to church communities: "Every possible deceptive device is being used to link the party's 'peace' program with the church" (Hoover, 1958, p. 3). Although he never explicitly named Quakers or individual groups in the article, his focus on peace as a link between communist and church communities was his focal point for detecting subversion within religious networks: "He cannot be a Marxist and adhere to a religion" (Hoover, 1958, p. 4). Hoover's dichotomous view of religion and atheistic communism helped to explain his troubled approach to determining whether or not the AFSC had in fact been infiltrated, and if so, what that meant for the religious nature of the organization. For in Hoover's mind, while the ultimate goal of communism was "the utter elimination of all religion" (Hoover, 1958, p. 4), it would never defeat religion, the core source of strength of the United States.

In the years leading up to the Cold War, the FBI had already been concerned about the Quaker identity of the AFSC. Some FBI agents even conflated the terms AFSC and Quaker in their reports to head-quarters, suggesting that they struggled to distinguish the religious beliefs and actions of the AFSC from its official endeavours. FBI head-quarters in Washington, D.C., found no evidence that the AFSC had engaged in "purely political matters" (FBI, 1941a, p. 4) or subversive activities in 1941. That the FBI distinguished between what was or was not purely political action foreshadowed its central concern with defin-ing the AFSC. That year, the FBI had also surveilled the Peace Caravan in Osborne, Kansas, which hosted a series of meetings addressing peace and conscientious objection in the community. According to the FBI, its observations of the Peace Caravan suggested the AFSC represented the Society of Friends "in the fields of social action" (FBI, 1941b, p. 2). In this case, and unlike in the post-war period when such exemptions were not extended to the AFSC, the incorporation or representation of religious beliefs in AFSC actions protected the Peace Caravan from further investigation.

That the FBI linked the AFSC closely with the Society of Friends rela-tively early on in its investigation is most salient in its one-hundred-page report from 1943. In particular, the introduction of the report described the AFSC under the heading "The Society of Friends," indicating its con-flation of the two groups. Indeed, the FBI agent who wrote the report accepted how the AFSC intertwined religious and political beliefs, despite other reports insisting on distinguishing between the two:

> The American Friends Service Committee is an organization formed by Quakers, or Friends as they are also called, and actually practices, in a mod-ern fashion, the theories and teachings of the Quaker religion. It can be seen from this that some knowledge of the history and tenets of Quakerism should be had to more completely understand the purpose of this organiza-tion. (FBI, 1943, p. 1)

Here, the agent explicitly wrote that to understand the AFSC, the FBI needed to understand the Quakers. The agent purposefully character-ized the movement as exhibiting a certain form of Christianity that also gave women an equal place with men in the church, all of which helped the agent argue that the AFSC was not entirely dangerous to the United

States. The agent characterized the organization of the Quaker church as "essentially democratic" (FBI, 1943, p. 1), and then provided a chart delineating the organization as well as a listing of how and where various Society of Friends meetings occurred, thus suggesting the FBI was trying to construct a clear-cut framework for surveilling the Quaker group beyond the official meetings of the AFSC.

Confusion over the identity of the AFSC subtly showed that the FBI was interested in the religious nature of the organization to determine whether and how it could be surveilled, as well as how to define it. An Air Force memorandum from 1947 reported that an informant had claimed that a member of the AFSC, whose name was withheld from the file, was

> one of the powerful influences in Socialism and Communism. It was stated that [deleted] was at one time the shining light for religious inspiration of the American Friends Service Committee. This informer said that the American Friends Service Committee was alleged to be a Quaker organization but was "actually a Unitarian group." (AFSC, 1976c, p. 9)

This memorandum excerpt shows that the religious identity of the AFSC baffled government agencies, in this case the Air Force and FBI, and that government agencies speculated that the AFSC was perhaps only nominally committed to Quaker pacifism, thereby revealing the biases of the agents (Haggerty, 2006) about these two distinctive religious groups. The biases of surveilling agents perpetuate ideologies that shape the sorting of populations and, this chapter adds, work to distinguish authentic from inauthentic religion. FBI agents' biases derived from varying levels of familiarity with both the AFSC and the Society of Friends, as well as from divergent normative frameworks for understanding religion within mainstream Protestant culture. In her AFSC analysis (1976a) of the FBI files, Davidon corroborated the Unitarian claim by adding: "It was stated that the organization was about 15% Quakers and 85% non-Quakers" (p. 11). The attempt of the FBI to categorize the religiosity of the AFSC by storing this report showed the extent to which it was worried about the communist threat. Moreover, by aligning the AFSC with Unitarianism, the FBI was liberated from the burden that pacifism is a Quaker belief since pacifism is not a doctrine of Unitarianism. Thus, the FBI legitimized the surveillance of the AFSC by questioning whether Quaker pacifism truly influenced its service activities or was merely a facade.

The investigation of the Unitarian association, nonetheless, also demonstrated the earnestness of the FBI to objectively analyse the goals and theological motivation of the AFSC, as well as how FBI surveillance of the AFSC was comparatively less severe than that of the plethora of other groups targeted in COINTELPRO (Lyon, 2007). This characterization resonated with J. Edgar Hoover's union of Judaeo-Christian and democratic ideals, as parallel framings of democratic versus non-democratic religious ideals. Years later, on behalf of the AFSC, Davidon (1976a) would reflect on the FBI allegations that the AFSC was a Unitarian organization by highlighting its all-Quaker leadership, despite welcoming non-Quaker members, and by noting its connections to 21 of 31 Quaker Yearly Meetings: "It is true that only 20% of the staff is Quaker, but the rest are sympathetic to Quaker beliefs" (p. 11). The AFSC reiterated that it included non-Quaker employees to send overseas based on their merit, rather than on religious dedication. The AFSC also claimed it was creating a more diverse, multicultural environment by hiring minority workers (Austin, 2012).

Hoover in one instance deviated from his typical neutral response maintaining that FBI surveillance was confidential when faced with questions about the AFSC from members of the inquiring public. He proactively defended the AFSC to a college professor who wrote to Hoover in 1952 that he had heard the organization was "practically Communistic." Hoover replied that the AFSC is "a committee of the Quaker faith ... engaged in projects designed to promote peace and to afford young people the opportunity for constructive patriotic service, and to provide relief assistance in this country and abroad" (AFSC, 1976b, p. 13). This unusual response was perhaps Hoover's attempt to rebut public accusations that the FBI should be more actively surveilling the AFSC or was insufficiently doing so. While members of the public feared a communist-infiltrated AFSC, dissonance about surveilling it resounded within the FBI. In March of 1954, an FBI Philadelphia report implored Hoover to cease his campaign against the organization. Philadelphia agents likely experienced the benefits that the Philadelphia-based AFSC offered to their local communities. They affirmed the sincerity of pacifism in the AFSC while acknowledging its parallels with communism. But they insisted that their informants, who were familiar with AFSC members, had no knowledge of any communist infiltration.

Some Philadelphia-based informants came from within the Religious Society of Friends itself. One informant noted it would not be out of character for an AFSC member to maintain contact with an individual or group affiliated with the Communist Party, but the Philadelphia FBI office continued to defend the AFSC. The in-depth Philadelphia SAC report, "Communist Infiltration of the American Friends Service Committee," dated February 25, 1955, outlined the structure, activities, and philosophy of the AFSC, and included copies of literature the organization produced (AFSC, 1980e). Sections on the philosophy and structure of the AFSC drew from its own material, while other sections constituted assessments by FBI agents. The report reiterated that while the AFSC limited its corporate membership to Quakers, it was not an official organization of the Society of Friends (FBI, 1955a). Hoover instructed the Philadelphia office that the 1955 Philadelphia SAC report demonstrated "there are no facts to establish that the Communist Party is attempting to penetrate" the AFSC and, as such, they should bring the case to a "logical conclusion" (FBI, 1955b, n.p.). The majority of FBI reports from other cities, including Chicago, San Francisco, and Houston, supported the Philadelphia claim that no evidence could prove Communist Party infiltration of the AFSC (see FBI, 1955c, 1955d, 1955e). Some local FBI offices believed that the AFSC was not a mouthpiece for the Society of Friends and therefore fair grounds for investigation. Yet another 1965 report claimed the AFSC was in fact the "social outreach arm of the religious group known as the Society of Friends (Quakers)" (FBI, 1965b, p. 1). Discord within the FBI demonstrated the diversity within the bureau.

Despite the attempt by some local FBI offices to mitigate Hoover's concerns about the AFSC as an organizational whole, the FBI constantly opened and closed cases on individual AFSC members in 1955. Internal FBI defiance, such as that pioneered by the Philadelphia office beginning in the mid-1950s, remained secretive, and therefore had no impact on public suspicion of the AFSC. One 1966 letter from a citizen queried the FBI about its official opinion of the AFSC (FBI, 1966a), with Hoover (FBI, 1966b) providing the standard response that he offered to many such letters: that the FBI was an investigative agency and did not make evaluations or conclusions about the integrity of any organization, publication, or individual. However, he continued that the citizen's concern about communism was "understandable" and that a general knowledge of the communist movement was essential for all United States citizens. In his response, he cited a few of his own texts, *Masters of Deceit* and *A Study of Communism*, asserting that the AFSC was a pacifist, Quaker organization. Occasionally, Hoover included informative enclosures about the AFSC, or referenced the Senate Internal Security Committee and the House Un-American Activities Committee in more detailed responses.

Despite Hoover's increased efforts in public communication about the AFSC, cracks in his surveillance policy surfaced during the 1960s. He repeatedly claimed that the FBI never investigated the AFSC in his correspondences with concerned citizens. However, Hoover surprisingly deviated from his formulaic response by admitting to one citizen in a letter dated June 24, 1958, that, back in 1942, the FBI had formally investigated the AFSC but found it was not engaged in subversive activities (AFSC, 1980b). Although Hoover's acknowledgment did not extend back to 1921 when the FBI first started investigating the AFSC, he contradicted his prior standardized responses to citizens throughout the late 1940s and 1950s, in which he testified that the FBI never investigated the AFSC. Hoover also began reconsidering the accuracy of the Communist Party label. He suggested the term was employed too loosely in discussions of non-Communist Party groups: "An increasing number of communications are received under the 'Communist Party, USA' where the subject matter does not pertain to the CPUSA or only indirectly pertains to the CPUSA" (FBI, 1962a, p. 1). Hoover was strategically starting to reconsider his AFSC surveillance campaign. By the 1970s, as Davidon noted in her 1978 article for *The Nation*, Hoover began to realize that he might not prove that the AFSC was infiltrated by communism. Instead, the FBI added new groups to one of its other surveillance programs, Cominfil, under a different name, the New Left, which included the Vietnam Veterans Against the War, Students for a Democratic Society, and the AFSC.

Despite these three points of contention complicating the integrity of FBI surveillance of dissent – dissent from FBI local offices, Hoover's contradictory public communication about the existence of FBI surveillance of the AFSC, and the evolving identities of the AFSC and Society of Friends, not to mention of the Communist Party itself – Hoover restored the AFSC file from closed to "pending" in 1962 due to new information that allegedly incriminated the AFSC as infiltrated by communists. A 1962 report on a Communist Party meeting quoted a speaker who referred to the Quakers as a "key force in establishing united action in various phases of peace activities" and claiming that they worked "consistently" with the Communist Party in private but not in public (AFSC, 1980d, pp. 1–2). The FBI Philadelphia office, fairly consistent in its defence of the AFSC by this point, responded that there was far too much material to sort through and that a report would be futile since the AFSC was a pacifist organization that did not merit investigation (FBI, 1962b). As Michie noted in a summary of the files, the campaign raised "the question of whether the FBI sees itself as an instrument of change as well as an investigatory agency" (AFSC, 1980d, pp. 1–2). Whether the FBI saw itself as an instrument of change as a whole is perhaps less ascertainable than the fact that local branches exerted influence. It is clear, for instance, that the Philadelphia FBI office led the resistance to Hoover's surveillance of the AFSC, supported as well as contested by other local FBI offices around the nation. It follows that local FBI offices clashed with each other just as members of the Society of Friends and of the

AFSC sometimes quarrelled about their respective organizational identities and goals. In this case, then, the panoptic power of the FBI was quite nuanced and not as complete as we tend to assume (Lyon, 1994, 2007), as demonstrated by the contestations that took place not only between and with those groups being surveilled, but also among the surveillers and their agencies as well.

Conclusion: Deconstructing FBI Secrecy

Both the FBI and the AFSC underwent organizational transformations in the mid-twentieth century that affected the outcome of the FBI surveillance program. J. Edgar Hoover tried to restructure the FBI by centralizing his authority within the organization. His intention backfired, however, as agents from offices around the country challenged Hoover's stance on the threat that the AFSC posed. The Philadelphia FBI office in particular resisted Hoover's surveillance of the AFSC Quakers. Their resistance can be attributed to the insider status of both Quakers and AFSC members in United States society, which would have been particularly evident to FBI agents based in Philadelphia who maintained personal or undercover contacts with the leading figures of both the AFSC and the Society of Friends. The efforts of Quaker and AFSC activists would ultimately unveil surveillance of numerous citizen groups that the FBI worked so hard to conceal.

The AFSC, in turn, was shaped by the rise of civil liberties in the postwar period, as well as by its Quaker-rooted commitment to social justice. The organization consciously hired non-Quaker members who identified with Quaker ideals in order to professionalize the organization through long-term employees, and to promote the radicalism necessary to execute projects abroad, sometimes at the expense of unrelenting pacifism. The AFSC underwent a second major reordering to focus on its own legal defence, as well as the defence of other surveilled organizations, following the exposure by a few of its own members of mass United States government surveillance. It not only continued to engage in radical activities, but it began to mobilize and partake in legal defences against unjustified governmental spying of citizen organizations.

Despite the bewilderment articulated by many AFSC members about their surveillance, the FBI files revealed that its agents did not all blindly follow Hoover's prescription to cure the alleged communism of the AFSC. Many FBI agents, in particular those of Philadelphia and at times Hoover himself, displayed nuanced, thoughtful attempts to deconstruct Quakerism and trace the relationship between the Society of Friends and the AFSC. Prudent agents, aware of the rising demand for civil liberties by the United States public, possibly considered that, due to the opening and closing of the AFSC case in the mid-1950s, disaster could ensue should it later be revealed that the FBI was wrongfully surveilling a respectable, lawful organization. That disaster would have multiplied if the boundary between the Friends and the AFSC was found to be less than clear-cut, given the legal and cultural protections of religious expression in the United States.

Hoover did not live to see the unwinding of much of the secrecy of his once protected organization. But the legacy of the AFSC would persist through its campaign to end unconstitutional government spying and to protect those who were less willing or able to protect themselves. The FBI files that William C. Davidon and others dispatched to the nation in the 1970s did not end government surveillance of United States citizens. Yet AFSC retaliation against localized FBI surveillance foreshadowed further sousveillance of government surveillance and secrecy. That sousveillance was shaped by the nature of the surveillance directed at the AFSC. The hostility of the FBI to radical media (see Elisabetta Ferrari and John Remesperger's discussion in the next chapter), meant that the sousveillance they practised and promoted involved much greater degrees of secrecy. More broadly, the chapters in this book show how different surveillance practices generate varying forms of resistance. The forms of sousveillance practised by those negotiating their criminalization under the law of attainder in Australia (Ian Warren and Darren Palmer, chapter 5), or those resisting the policing of gender expression in South Africa (B Camminga, chapter 4), for example, were quite distinctive. In the case of the AFSC, then, not only the specific contexts of the United States during the Cold War, but also the range of possibilities afforded the AFSC by virtue of their religious affiliation, shaped the ways in which they could resist surveillance, including how they could support other radical groups who were being targeted.

This chapter has shown that the FBI relied on its centralized, powerful position within the federal government to conduct its surveillance of the AFSC. Yet FBI agents deliberated religious expression and national identity, disagreeing with Hoover's rationale for surveilling the AFSC and demarcating a boundary between religious expression and dissent. Future historical and contemporary case studies of government surveillance of religion should consider a new theoretical framework for surveillance. Scholars might consider the kinds of data that government agents amass to evaluate the religiosity of surveilled subjects across space and time, including observations of public demonstrations and marches; inquiring letters from members of the public who are, consequently, entering into the surveillant assemblage themselves; and surveillance of individuals even remotely associated with the original target. These government documents can indirectly or directly affect legal and cultural norms that impact how the surveilled religious group operates in the public sphere. Therefore, close evaluation of the discourse produced about religious groups through surveillance is a necessary step toward understanding how surveillance both enables and inherits certain kinds of religious communication and expression.

REFERENCES

Sources from the American Friends Service Committee Archives, Philadelphia, PA

AFSC. (1976a). *Freedom for Americans: 1976* [Unpublished draft]. A. M. Davidon.
AFSC. (1976b, March 12). *Report to AFSC board executive meeting on FBI files* [Unpublished report]. A. M. Davidon. ("Government Surveillance" folder).
AFSC. (1976c). *The war at home: An AFSC report.* Philadelphia, PA: A. M. Davidon.
AFSC. (1979). *Major report concludes that political spying by police exists on large scale* [Press release]. Philadelphia, PA: P. E. Brink.
AFSC. (1980a, February 20). *Summary of files* [Unpublished document]. Philadelphia, PA: H. Michie. ("FBI Material Rec. on Appeal, 4/80" box #4, section #1, folder #1).
AFSC. (1980b, April). *Summary report of FBI file* [Unpublished report]. Philadelphia, PA: H. Michie. ("FBI Material Rec. on Appeal, 4/80" box #4, section #8, folder #4).
AFSC. (1980c, April). *Summary report of FBI file* [Unpublished report]. Philadelphia, PA: H. Michie. ("FBI Material Rec. on Appeal, 4/80" box #4, section #9, folder #4).
AFSC. (1980d, April). *Summary report of FBI file* [Unpublished report]. Philadelphia, PA: H. Michie. ("FBI Material Rec. on Appeal, 4/80" box #4, section #10, folder #4).
AFSC. (1980e, May 5). *Narrative summary of materials received by AFSC from FBI on appeal, March 24, 1980* [Unpublished document]. Philadelphia, PA: H. Michie. ("FBI Material Rec. on Appeal, 4/80" box #4, section #5, folder #4).
Davidon, A. M. (1978, March 11). Watching for Cominfil: Even the Quakers scared the FBI. *The Nation,* 266–8.
FBI. (1941a, February 13). *American Friends Service Committee* [Unpublished memorandum]. Washington, DC. ("FBI Material Rec. on Appeal, 4/80" box #4, section #1, folder #1).

FBI. (1941b, August 18). *Peace caravan: American Friends Service Committee* [Unpublished report]. Washington, DC. ("FBI Material Rec. on Appeal, 4/80" box #4, section #1, folder #1).

FBI. (1943, November 24). Introduction. In *The American Friends Service Committee* [Unpublished report]. Washington, DC. ("FBI Material Rec. on Appeal, 4/80" box #4, section #5, folder #2, file #100-11392).

FBI. (1947, November 17). *American Friends Service Committee, Internal security* [Unpublished report]. Washington, DC: SAC Baltimore. ("FBI Material Rec. on Appeal, 4/80" box #4, section #3, folder #3).

FBI. (1955a, February 25). *Communist infiltration of the American Friends Service Committee* [Unpublished report]. Washington, DC: SAC Philadelphia. ("FBI Material Rec. on Appeal, 4/80" box #4, section #5, folder #4).

FBI. (1955b, March 9). *Communist infiltration of the American Friends Service Committee* [Unpublished report]. Washington, DC: SAC Philadelphia. ("FBI Material Rec. on Appeal, 4/80" box #4, section #5, folder #4).

FBI. (1955c, May 26). *Communist infiltration of the American Friends Service Committee* [Unpublished report]. Washington, DC: SAC Omaha. ("FBI Material Rec. on Appeal, 4/80" box #4, section #5, folder #4).

FBI. (1955d, June 20). *Communist infiltration of the American Friends Service Committee* [Unpublished report]. Washington, DC: SAC Houston. (FBI Material Rec. on Appeal, 4/80" box #4, section #6, folder #4).

FBI. (1955e, October). *Communist infiltration of the American Friends Service Committee* [Unpublished report]. Washington, DC: SAC San Francisco. ("FBI Material Rec. on Appeal, 4/80" box #4, section #6, folder #4).

FBI. (1962a, March 21). *Proposed peace march, Seattle Washington, April 21, 1962, information concerning (internal security)* [Unpublished memorandum to SAC Seattle]. Washington, DC: J. E. Hoover. ("FBI Material Rec. on Appeal, 4/80" box #4, section #10, folder #4).

FBI. (1962b, March 22). *American Friends Service Committee, incorporated information concerning (internal security)* [Unpublished memorandum to John Edgar Hoover]. Washington, DC: SAC Philadelphia. ("FBI Material Rec. on Appeal, 4/80" box #4, section #10, folder #4).

FBI. (1965a, May). May 2 movement. Philadelphia FBI report (pp. 1–4). [Unpublished document]. Washington, DC. ("FOIA box #1, "FBI materials" folder #1, 1-1 through 50-2, folder #2, 51-1 through 85-19, folder #3, 86-1 through 98-2).

FBI. (1965b, October 1). Folk rock & all cause protest music festival. Seattle FBI report (pp. 1–3). [Unpublished document]. Washington, DC. ("FOIA" box #1, "FBI materials" folder #1, 1-1 through 50-2, folder #2, 51-1 through 85-19, folder #3, 86-1 through 98-2).

FBI. (1966a, March 4). Letter from citizen to J. E. Hoover [Unpublished letter]. Washington, DC. ("FOIA" box #1, folder #1, item #19).

FBI. (1966b, March 6). Letter from J. E. Hoover to citizen [Unpublished letter]. Washington, DC. ("FOIA" box #1, folder #1, item #19).

FBI. (1967a, March 17). *Memorandum to FBI director from SAC Philadelphia, re. AFSC* [Unpublished document]. Washington, DC: SAC Philadelphia ("FOIA" box #1, folder #3, item #48).

Investigation of Un-American Propaganda Activities in the United States: Hearings on H. R. 1884 and H. R. 2122, Bills to Curb or Outlaw the Communist Party of the United States, Part 2, Before the Committee on Un-American Activities, House, 80th Cong. 1 (1947) (Testimony of John Edgar Hoover).

United States Air Force, Office of Special Investigations. (1958, October 31). *Report of investigation: Propaganda concerning missile base, Francis E. Warren AFB, Wyoming* [Unpublished report]. Washington, DC. ("CIA/Air Force/Army/ Navy" box #5, "Air Force" folder).

United States Navy, Fourth Naval District, Philadelphia, District Intelligence Office. (1943, October 28). Section C, Statements issued by the AFSC, of Section III, Activities (pp. 13–37). Section B, Characterizations of organizations, of Section VI, Association with organizations designated by the United States Attorney, general pursuant to Executive Order 10450 (pp. 45–50). In *American Friends Service Committee and Kindred Organizations* [Unpublished report]. Philadelphia, PA. ("FBI Material Rec. on Appeal, 4/80" box #4, section #5, folder #4).

Secondary Sources

Austin, A. W. (2012). *Quaker brotherhood: Interracial activism and the American Friends Service Committee, 1917–1950.* Urbana, IL: University of Illinois Press.

Browne, S. (2015). *Dark matters: On the book surveillance of blackness.* Durham, NC: Duke University Press.

Cecil, M. (2014). *Hoover's FBI and the fourth estate: The campaign to control the press and the Bureau's image.* Lawrence, KS: University Press of Kansas.

Fiske, J. (1998). Surveilling the city: Whiteness, the Black man and democratic totalitarianism. *Theory Culture Society, 15*(2), 67–88.

Foner, E. (1998). *The story of American freedom.* New York, NY: W. W. Norton & Company.

Foucault, M. (1977). *Discipline and punish: The birth of the prison* (A. Sheridan, Trans.). New York, NY: Vintage Books.

Frost, J. W. (1992). "Our deeds carry our message": The early history of the American Friends Service Committee. *Quaker History, 81*(1), 1–51.

Giddens, A. (1985). *The nation-state and violence: A contemporary critique of historical materialism* (Vol. 2). Cambridge, UK: Polity Press.

Greenberg, I. (2010). *The dangers of dissent: The FBI and civil liberties since 1965.* New York, NY: Lexington Books.

Greenberg, I. (2012). *Surveillance in America: Critical analysis of the FBI, 1920 to the present.* New York, NY: Lexington Books.

Haggerty, K. (2006). Tear down the walls: On demolishing the Panopticon. In D. Lyon (Ed.), *Theorizing surveillance: The Panopticon and beyond* (pp. 23–45). Cullompton, UK: Willan.

Haggerty, K. D. & Ericson, R. V. (2000). The surveillant assemblage. *British Journal of Sociology, 51*(4), 605–22.

Hamm, T. D. (2003). *The Quakers in America.* New York, NY: Columbia University Press.

Hoover, J. E. (1958, April 21). Communism: A false religion. *Human Events, 15*(16), 1–4.

Ingle, H. L. (1998). The American Friends Service Committee, 1947–49: The Cold War's effect. *Peace & Change, 23*(1), 27–48.

Innis, H. A. (1950/2007). *Empire and communications.* Toronto, ON: Dundurn Press.

Innis, H. A. (1951/2008). *The bias of communication* (2nd ed.). Toronto, ON: University of Toronto Press.

Jones, G. (1971). *On doing good.* New York, NY: Charles Scribner's Sons.

Jones, L. M. (1929). *Quakers in action: Recent humanitarian and reform activities of the American Quakers.* New York, NY: The Macmillan Company.

Lianos, M. (2003). Social control after Foucault. *Surveillance & Society, 1*(3), 412–30.

Lyon, D. (1994). *The electronic eye: The rise of surveillance society.* Minneapolis, MN: University of Minnesota Press.

Lyon, D. (2006). The search for surveillance theories. In D. Lyon (Ed.), *Theorizing surveillance: The Panopticon and beyond* (pp. 3–20). Cullompton, UK: Willan.

Lyon, D. (2007). *Surveillance studies: An overview.* Cambridge, UK: Polity Press.

Mann, S., Nolan, J., & Wellman, B. (2003). Sousveillance: Inventing and using wearable computing devices for data collection in surveillance environments. *Surveillance & Society, 1*(3), 331–55.

Schudson, M. (2016). *The rise of the right to know: Politics and the culture of transparency, 1945–1975.* Cambridge, MA: The Belknap Press of Harvard University Press.

Schwarz, Jr, F. A. O. (2013). An historic perspective of intelligence gathering. In F. Patel (Ed.), *Domestic intelligence: Our rights and our safety.* New York, NY: Brennan Center for Justice at New York University School of Law.

Sutters, J. (2010, March 29). *Warmth and sweetness: The beginnings of a postwar feeding program in Germany.* Retrieved from https://afsc.org/story/warmth-and-sweetness-beginnings-postwar-feeding-program-germany

11 "When Under Surveillance, Always Put on a Good Show": Representations of Surveillance in the United States Underground Press, 1968–1972

ELISABETTA FERRARI AND JOHN REMENSPERGER

In May 1970, the *Los Angeles Free Press*, an alternative newspaper connected to the American radical and countercultural movements of the time, reported that individuals calling the paper's office from payphones were receiving their dimes back, and that inquiries to the phone company revealed that the behaviour could be due to their line being tied in with "special equipment," making the payphone think the call was never connected. In the article, the *Freep* – as the paper was informally called – suggested the phones' malfunctions were likely the result of a tap placed on the telephone wires in an effort to spy on the newspaper staff. With the humour that marked much of the reporting by countercultural newspapers, the *Freep* encouraged its readers to call and "keep your dime ... but just be cool about what you say" (Kirby, 1970, p. 2).

Though the paper's staff at the time could not verify whether they were truly being surveilled, it is now well known that the United States government and law enforcement agencies spied on domestic activists throughout the 1960s and 1970s, using infiltration, wiretapping, and other counterintelligence tactics. In this chapter, we explore how the alternative media (Atton, 2002) of the movement – known as the underground press – addressed surveillance. To do so, we analysed articles published by two underground publications, the *Los Angeles Free Press* (or *Freep*) and the *San Francisco Good Times*, between 1968, the peak of counterculture mobilization, and 1972, the year when the FBI's surveillance practices became the subject of public debate.[1] Articles published in these two papers are accessible through the *Independent Voices* database; to our knowledge, this chapter represents the first published work that utilizes this extensive archive.

The underground press of the 1960s and 1970s provides important insights into the thoughts, priorities, and concerns of those activists who stood in opposition to United States politics and society, including widespread racism and the Vietnam War. As the first section of this chapter will explore, articles and media coverage within the *Freep* and the *Good Times*, in particular their denunciation of surveillance, was part of a general political challenge to state power that was at the heart of countercultural movements at the time. Drawing on our archival research, the next section demonstrates that activists were acutely aware of being targets of surveillance and repression, with both the *Freep* and the *Good Times* documenting and critiquing the evolution of surveillant practices and methods. Examining articles published between 1968 and 1972, we can see how state and police surveillance became more intensive, proximal, and extensive. Next, we consider the ways in which these two underground publications showcased different tactics (de Certeau, 1984) for resisting the growing ubiquity of surveillance, notably through public denunciation, humour, and legal challenges. Lastly, we suggest that what the underground press was describing and critiquing at the time draws striking parallels to current debates about surveillance. In fact, the scope and sophistication of the surveillance directed at the social movements of the 1960s and 1970s, and how they responded to it, offers significant insight into the repressive impact that contemporary regimes of surveillance can have on activism, as well as also how activists can resist them.

The *Freep*, the *Good Times*, and Surveillance of the Underground Press

The underground press that emerged in the mid-1960s corresponded to the rising New Left movements on college and university campuses and in cities nationwide, which comprised an anti-establishment rebellion against traditional cultural and political institutions, and was aided by the increased availability of affordable graphical printing provided by the mimeograph (McMillian, 2011). These "radical community newssheets" (McMillian, 2011, p. 4) also grew out of a general dissatisfaction among activists with what they perceived as the mainstream media's nonexistent or improperly framed reporting on the issues they saw as most important, particularly the Vietnam War. Articles in the underground press were both unabashedly subjective and informal in writing style, with many of the writers considering themselves primarily to be activists and secondarily writers (Peck, 1985). Accordingly, underground press papers served the activist purposes of community building, local political organizing, publicity, and, as they linked with one another via collaborative organizations like the Underground Press Syndicate, national movement building (see Armstrong, 1981; McMillian, 2011). Indeed, McMillian (2011) argues that, as the New Left's "primary means of internal communication" (p. 6), the underground press was "an attempt to build an intellectual framework for the movement's expansion" (p. 4).

With some exceptions, underground press outlets were generally unprofitable (Emery, Emery, & Roberts, 1996). Those that did achieve commercial success via advertising and paid circulation fees often found it resulted in other conflicts. Art Kunkin, the founder of the *Freep* as well as the editor until he was forced out in 1973, was regularly criticized for the pornographic or sex-related ads in the paper's classifieds section (Armstrong, 1981), while Max Scheer, the owner of the *Berkeley Barb* for its duration from 1965 to 1980, faced a staff revolt over perceptions that he was making substantial profits at the expense of an underpaid staff and writing team (Ellis, 1971). Despite these obstacles, at the height of the underground press's popularity, there were 400 underground press outlets nationwide with a circulation of 5 million copies (Armstrong, 1981).

The *Freep* was one of the earliest underground press papers and the first to publish on a regular schedule (Peck, 1985). At its peak, the *Freep* was the "most widely circulated underground paper in the nation" (McBride, 2003, p. 110) with a weekly circulation of over 100,000 copies. Unlike many of the other outlets that operated as editorial cooperatives, Kunkin ran the *Freep* in traditional management style based on his "sense of where the paper should go" (Peck, 1985, p. 187). In practice, this resulted in regular coverage of civil rights issues and various topics related to the Sunset Strip area of Los Angeles, where the paper was located. Though the paper had been operating for only a year when riots broke out in the largely Black Watts neighbourhood of South Los Angeles, it was its sympathetic coverage of the events that led to its emergence as a popular outlet for leftist views in Southern California (Rycroft, 2007). Disputes between the police and both hippies and New Left activists on the Sunset Strip, and the role played by the *Freep* in reporting on these issues and supporting the activist groups, led to both a sense of shared community and an overlap in political interests between the hippie counterculture and the New Left that was not present in other countercultural centres like San Francisco or New York (McBride, 2003). Conflicts between Kunkin and law enforcement were not limited to local police. His decision to publish an extensive list of undercover agents employed by the California State Bureau of Narcotics led to legal issues for the paper (discussed in detail below) that threatened its financial viability.

While the *Freep*'s role in Southern California was growing, underground papers were also founded in Northern California. As one of the main centres of both the New Left and the hippie movements, the San Francisco Bay Area was the home of several underground press papers. Marvin Garson, a political activist who had been jailed with other protestors at the Democratic National Convention in 1968, that same year started the *San Francisco Express Times*, a sophisticated, visually appealing newspaper focused on the political issues of the New Left. It rapidly reached a circulation of 100,000, the same as the *Freep*. Plagued by financial problems, it nevertheless managed to attract a pool of talented writers, including Todd Gitlin of Students for a Democratic Society, as well as various artists, and was considered "one of the best papers in the movement" (Glessing, 1970, p. 66). In 1969, Garson left the paper and his successors developed a new "granola-esque" (Peck, 1985, p. 227) editorial focus and renamed the paper the *San Francisco Good Times*.[2] During its run from 1968 to 1972, the FBI infiltrated the paper's staff with a paid informant, one of many tactics regularly used by intelligence agencies and law enforcement to target the New Left (Armstrong, 1981).

The use by the United States government of aggressive techniques – infiltration, wiretapping, and other counter-intelligence tactics – for spying on domestic activists throughout the 1960s and 1970s has been well documented (see Cunningham, 2004; Davis, 1997). Government and law enforcement surveillance targeted anti-war groups, student campuses, New Left groups, and movements for Black liberation, such as the Student Nonviolent Coordinating Committee and the Black Panther Party. The FBI's clandestine program, COINTELPRO, which between 1956 and 1971 aimed to monitor and disrupt domestic political activity, is perhaps the most famous of such efforts, but is hardly the only one. Local police departments and even the CIA were likewise involved in surveilling and intimidating activists throughout the United States.

As detailed by Cunningham (2004) and Davis (1997), under COINTELPRO the FBI developed a series of inventive techniques to deceive the activists of the New Left, discredit them, and even instigate them towards violent action. As was later documented by the Church Committee – the United States Senate Select Committee tasked with investigating the intelligence activities directed at American citizens – the FBI engaged in elaborate schemes to turn activists against each other, often taking advantage of visible fractures in the movement. In particular, the FBI wanted to instigate "personal conflicts or animosities between New Left leaders" and create "the impression that leaders are informants for the Bureau or other law enforcement agencies" ("Select committee to study governmental operations," 1976, p. 26). The FBI also sent out anonymous letters, for instance to inform parents of college and university students about their activities or to expose faculty members' activism to college administrations. Even more disturbing, the FBI was involved in violent actions against the New Left and its publications, from the ransacking of headquarters to the firebombing of the offices of several underground newspapers, including the *Freep* (Streitmatter, 2001).

Although the FBI's counterintelligence tactics are now the most well-known, local police departments across the United States were also engaged in surveillance and repression of various social movements; for example, departments in several cities gathered intelligence and surveilled peaceful anti-war organizations (Linfield, 1990). One particularly crucial police tactic was to infiltrate activist groups with *agents provocateurs*, undercover agents used to incite activists to commit violent acts. Undercover agents also used drug laws as a convenient method to arrest and detain movement leaders. While the counterculture undoubtedly indulged in the use of illegal substances, it is undeniable that drug laws were employed disproportionately for political repression, including the criminalization of communities of colour (McMillian, 2011; Streitmatter, 2001). The CIA was also involved in spying on domestic activists; under the CHAOS program the agency collaborated with local law enforcement departments to collect information on activists, train police agents, and offer equipment (Rips, 1981).

The most influential organizations of the movement were also the most heavily targeted by this widespread surveillance and repression. Students for a Democratic Society and the Black Panther Party, for example, were among the most surveilled and infiltrated (Cunningham, 2004), and their movement newspapers were subjected to systematic intimidation and surveillance (Rips, 1981). Indeed, repression extended to many publications connected to various countercultural movements, including student-run newspapers, underground military anti-war newspapers, feminist publications, Black liberation papers, and more (Rips, 1981).

As a way of targeting movement activists, police departments would lay obscenity charges against underground publications. While the court cases generally absolved the underground press of the charges, their length and cost effectively shut down many publications that were not financially stable even without legal expenses (Linfield, 1990; McMillian, 2011; Streitmatter, 2001). Drug charges were also disproportionately directed at underground press editors and reporters, as denounced by the Underground Press Syndicate: "the rate of arrest of underground journalists for drugs was one hundred times the general rate of narcotics arrests" (as quoted in Rips, 1981, p. 102). In addition, and as noted, underground newspapers were infiltrated by undercover agents, including the *Good Times* (Armstrong, 1981; Rips, 1981). Both the Underground Press Syndicate and the Liberation News Service, the two syndication networks of the movement, were constantly under surveillance, subjected to raids, and infiltrated by agents (McMillian, 2011; Peck, 1985). The openness of the underground press was further exploited by law enforcement to spread rumours through the movement and exacerbated existing conflicts. For instance, bait ads were planted in the newspapers (e.g., targeting communist groups), and a fictitious Liberation News Service news packet was distributed with the intent to divide the movement (Bertlet, 2012; Peck, 1985). Both the CIA and the FBI also set up their own bogus publications and fictitious news networks (Linfield, 1990; Peck, 1985).

As Kathryn Montalbano's discussion of the FBI's surveillance of the American Friends Service Committee in the previous chapter also highlights, it took a number of years, official investigations, and Freedom of Information requests to uncover the extent of the surveillance and repression that the movements of the 1960s and 1970s had to endure. It is also worth noting that it might have taken even more time to reveal the existence of COINTELPRO, if not for the actions of a group of activists – the Citizens' Commission to Investigate the FBI – who broke into an FBI field office in Media, Pennsylvania, on March 8, 1971, and stole documents detailing the agency's surveillance of political groups, which it then released to leading newspapers. The revelations were the first step in exposing the extensive political surveillance apparatus that was in place at the time; they led to the dismantling of the FBI's COINTEL programs and the official inquiries into the FBI's and the CIA's counter-intelligence activities. Even before the shocking revelations of the Citizens' Commission, however, activists could feel that they were being watched and infiltrated.

More Intensive, More Proximal, More Extensive: The Evolution of State Surveillance

As an integral part of the New Left's activist community, the underground press provided its readers with important information about the clandestine methods that government agencies were using to collect information on movements and their members. As we explore below, our examination of articles on topics related to surveillance published in the *Freep* and the *Good Times* reveals the cumulative growth of surveillance as it became more intensive, proximal, and extensive. From 1968 to 1972, coverage of surveillance in these outlets expanded from an initial focus on traditional surveillant methods, such as the use of unmarked police cars and plain-clothed officers to observe activists, to include harassment and infiltration, electronic surveillance via bugging and wiretaps, and finally, to a growing awareness of the ubiquitous nature of state surveillance, affecting not only activists but society at large. Our research found that activists writing in the *Freep* and the *Good Times* were also aware that continuous technological innovation and the growing cultural acceptance of surveillance could have long-lasting societal impacts. In many cases, their coverage raised questions and made predictions about the future of surveillance – many of these predictions draw parallels to current concerns about the relationship between government and its citizens.

Traditional Surveillance Tools

Throughout 1968 and early 1969, discussions of surveillance in the *Freep* and the *Good Times* centred on the traditional tools used, largely by local police and the FBI, to directly surveil movements. These practices included the use of unmarked cars and plain-clothed officers to stake out and spy on movement offices – including those of the underground press – and public events. Contributors to the *Freep* and the *Good Times* reported instances of police officers writing down the licence plate numbers of cars parked at these locations and canvassing people working in nearby businesses to garner information on the movements and their activities (Osborn, 1968). These publications informed readers about suspicious characters, sometimes posing as journalists or hippies, taking notes and photos during rallies. Articles also highlighted instances in which police were found to be using binoculars or telephoto lenses to observe and document demonstrations from a distance ("Trial inside and out," 1970) or to watch activities taking place indoors. One article even described the somewhat farcical scene of a prosecutor hiding in the bushes with a telephoto lens who, when approached by the students he was observing, was rescued by four police cars that immediately arrived to usher away the students (Hoffman, 1969a). An article in the *Freep* about a press conference by activist Jerry Rubin at the 1968 Democratic National Convention in Chicago and entitled "The Dick Daley Revised History of Czechago" – the sarcastic title in itself a reference to the repression activists felt at the event – provides insight into how these activists felt about the normalization of surveillance by law enforcement and the apathy of mass media. The piece argued that "police state methods have crept up on the press so stealthily that they do not yet realize what surveillance or invasion of privacy looks and sounds like" (Lipton, 1968, p. 3).

While early discussions of these traditional surveillance methods were largely focused on the targeting of movements by the state, they were also concerned with surveillance of drug users. As discussed above, drug arrests were often used by police to target and intimidate movement activists, thus reinforcing the criminalization of dissent and buttressing the narratives by which police themselves legitimated their surveillance work, as Matthew Ferguson, Justin Piché, and Kevin Walby explore in the next chapter. Articles warned of and lamented the presence of law enforcement officers who "wander through love-ins and other large public gatherings" (Rosevear, 1968, p. 30). However, our analysis of articles published close to the end of our period of investigation, that is, 1972, shows that these original concerns by activists about rudimentary surveillance were eventually supplemented by charges of increasingly invasive forms of surveillance by undercover agents and informants engaged in lengthy clandestine campaigns.

Harassment, Intimidation, and Infiltration

As time passed, writers in the *Freep* and the *Good Times* began increasingly to report on the harassment of activists, infiltration of their organizations by undercover officers and agents, and spying by informants, with an early focus on the surveillance of Black and Latino/a communities, as well as of hippies and the New Left. The publications also reported on attempts by law enforcement to infiltrate anti-racist movement organizations including the Black Panthers, the Brown Berets, and the Student Nonviolent Coordinating Committee, and in extreme cases to sabotage their activities or provoke activists to violence (Osborn, 1968; Whittaker, 1968). In one article, an *agent provocateur* reportedly admitted to working for two years infiltrating militant Chicano groups for the Bureau of Alcohol, Tobacco, and Firearms (Kunkin, 1972). In that capacity, he admitted to, among other things, jumping on a car roof during a protest, setting up purchases of weapons, and burning a house down. In another case, a man discovered his live-in girlfriend was informing the FBI of their activities with Students for a Democratic Society (Szigeti, 1969).

As mentioned, one of the major concerns expressed in the *Freep* and the *Good Times* was the increased infiltration of colleges and universities by law enforcement. A common theme we found within these articles was the dissonance between university administrators' supposed commitment to freedoms long associated with academic institutions and the complicity of those same administrators in facilitating surveillance of their students. One article posed the question of how to "reconcile an institution supposedly oriented toward free inquiry with police agents hiding behind second-story windows and on rooftops" ("Ivory tower," 1969, p. 4). Activists writing in the *Freep* and the *Good Times* also raised concerns about the use of technologies for gaining access to conversations, both in person and via telephone, that were inaccessible via infiltration and harassment.

Electronic Surveillance

Wiretapping and bugging were a significant concern for New Left activists, and examples of their use, methods for avoiding them, and the legal issues surrounding them were covered regularly in the *Freep* and the *Good Times*. This coverage only increased after the 1971 public revelations surrounding both Pentagon Papers whistleblower Daniel Ellsberg and the targeting of New Left activists via the FBI's widespread domestic surveillance program. Writers warned their fellow activists that they should presume they were being surveilled, for example saying, "assume all telephones are bugged. Avoid unnecessary mysterious comments ... If you must talk over the phone, arrange a time when someone can call you from one payphone to another" ("Pig Yoga," 1971, p. 8).

Both publications also discussed technologies and practices beyond basic wiretapping and bugging. The *Freep* and the *Good Times* criticized plans by city governments to use federal funds to expand surveillance in downtown areas via closed-circuit television (CCTV) systems and audio monitoring devices. The *Freep* reported on the installation of cameras and long-range microphones in Santa Barbara's university business district (Levitt, 1970) and the *Good Times* criticized a pilot program in a New York City suburb that utilized constantly running "low-light level cameras" ("Peep eye," 1971, p. 12) to deter criminalized activities in downtown areas. These examples of extensive public surveillance are early indicators of how underground press outlets would change their coverage of surveillance as they shifted from a focus on movements to one in which anyone could be watched at any time.

Surveillance Goes Mainstream Post-1971

After the Citizens' Commission's disclosures of 1971, the *Freep* and the *Good Times* began to discuss the pervasiveness of surveillance that reached beyond government spies targeting movement organizations. Multiple articles show instances of intra-agency spying by government institutions and surveillance of established political parties and other organizations. An extensive review in the *Good Times* of a book entitled *Who's Who in the C.I.A.* discusses how "the economic, educational and public opinion manipulation spheres in the U.S.A. have been infiltrated by the U.S. intelligence services" ("Who's who in the CIA," 1971, p. 19). The article encouraged readers to purchase the book, and to check its directory of CIA operatives for one's friends or colleagues. The *Good Times* also reported on steps taken by Congress to require banks to disclose confidential information, which would lead to increased surveillance of citizens ("Bank policies protested," 1972).

In addition, both the *Freep* and the *Good Times* reported on the increased surveillance of high school students and teachers. The *Freep*, for instance, covered instances of high school administrators tasking staff and teachers to spy on students as well as other teachers (Hoffman, 1969b), and it published articles opposing a proposed federal "Safe Schools Act," aimed at providing funding for the "expansion and training of security guards, parent patrols, surveillance and alarm systems, identification badges for students and other measures" ("Cops go to H. S.," 1972, p. 4). Concerns about civilian-on-civilian spying were not limited to schools or movement organizations. An article in the *Good Times* warned of a potential move by the mayor of Berkeley to encourage what he called "broad-based civilian activity" ("A warden on every block," 1970, p. 3) by enlisting volunteer neighbourhood block wardens to surveil their own neighbours. The article argued that this and other proposed measures would lead Berkeley to be the first fascist city in the United States. *Cop Watch*, a sporadic column ran by the *Good Times* for years, quoted a police officer telling parents it was their duty to report their own children to the police if they suspected them of recreational marijuana use (Blaine, 1969). These articles depict an increasingly distrustful society in which the government spies on itself, citizens spy on their neighbours, and family members report one another to the police. The underground papers were concerned that this culture of surveillance would become more ubiquitous over time, as changes in both political will and legal precedents would lead to new surveillance technologies and industries to produce them.

Future of Surveillance

Imagine a future where you and all your friends would be hooked up to a common cable tv system. You make your own tv tape cassette and blow it out to all the Family. This is one prospect for the revolutionary use of the cassette and cable tv. Other prospects are a bit less revolutionary and include the ripping off of the entire world in the name of profit and surveillance.

S. Silver (1971, p. 15)

This article in the *Good Times*, covering a panel discussion on the future of cinema in which presenters considered the idea of a computer-based video system, similar to today's "video on demand," foresaw the connective potential of new media, but also the potential for private corporations to track users and profit from their tastes. In this and many other instances, the underground press had a prescient understanding of how the adoption of technology could enable widespread surveillance, even through technologies like video on demand, that are not implemented with surveillance as their primary objective.

The future of surveillance was concerning to the writers of the *Freep* and the *Good Times*, both from a technological and an economical-legal point of view. Many of the technologies they talked about have since become a standard for law enforcement. For example, new prototype "transponders [that] would be linked to a computer which would monitor the wearers' locations and implement curfew and territorial restrictions" (Barkan, 1971, p. 8) are quite similar to the ankle bracelets used by law enforcement agencies to track parolees and registered sex offenders in cities nationwide.

Worries about the power of a surveillance complex were equally represented in these publications. Reporting on unprecedented support from the California law enforcement community for a bill reinstating the death penalty, the *Freep* argued that such support

> opens the way for the ominous prospect that police may use their new found political strength to get laws passed to legalize wiretapping, further limit the rights of defendants in courts, and otherwise move our society into the direction of a police controlled state. (Ridenour & Ofari, 1972, p. 5)

Concerns over the power of the surveillance community also highlighted the business interests involved. A piece in the *Good Times* foreshadowed the role that independent contractors have since come to play in both surveillance and law enforcement writ large. It describes a new Massachusetts-based firm, Universal Detective Inc., that offered to train independent narcotics agents that law enforcement agencies could rent for undercover operations ("NARC for rent," 1972). This privatization of aspects of state surveillance represents an important development not evident in the other chapters in this book, and one that has become increasingly prominent in the United States in particular.

Looking at coverage in the underground press outlets we examined, it is clear that the New Left activists who wrote for the papers believed that the surveillance practices targeting their movements were increasing in scope, as law enforcement found new ways to keep tabs on radical movements. It is important to note the compounding effect of these surveillance methods. Intelligence agents did not stop physically infiltrating movements simply because they gained the ability to wiretap them. Over time, their methods became more intensive, from initially observing activists and attending movement meetings to infiltrating movements with undercover agents for years-long stints. Indeed, surveillance became more proximal, as technology in the form of wiretapping, bugging, and the then-prototype personal electronic monitoring devices reduced the physical distance between surveillance technologies and their targets, and more extensive, as local jurisdictions adopted CCTV and encouraged citizen-on-citizen surveillance.

From Coping to Fighting Back: Tactics for Resisting Surveillance

The *Freep* and the *Good Times* are very explicit in reporting about surveillance, but in the style of the underground press, they also showcase different ways through which activists (and regular citizens) could resist surveillance. These tactics, which Kathryn Montalbano characterized as forms of "sousveillance" (see Mann, Nolan, & Wellman, 2003) in the previous chapter, can be thought of as a response of the powerless to the strategies of the powerful (de Certeau, 1984); in this case, they are attempts to resist repression that stem from the general contestation of state power that characterized the movements of this time. In particular, the articles we examined evolve over time from discussions on how to cope with surveillance to articles documenting ways in which both social movements and the general public can fight back, for instance through legal challenges. The *Freep*'s and the *Good Times*' relentless documentation of the different ways in which surveillance could and should be resisted points toward the need of going beyond the simple notion that surveillance made movements "paranoid." While paranoia might have been an important component of how activists felt, these underground papers show that it did not make them less combative. The temporal evolution from coping tactics to open challenges can be understood in the context of the progressive intensification of surveillance discussed above; however, it should also be thought of in conjunction with three key events that occurred between 1968 and 1972, namely, the publication of the Pentagon Papers, the early revelations of the Citizens' Commission, and the Watergate scandal.

The major tactics for coping with or resisting surveillance that the *Freep* and the *Good Times* showcased include public denunciation, humour, and legal challenges. Firstly, as highlighted above, the *Freep* and the *Good Times* extensively covered surveillance and denounced it publicly. This reporting is the primary way in which these underground newspapers exposed the surveillance with which they had to live. Already in 1968, the *Good Times* (still known as the *Express Times* at that point) ran an article titled "Underground Fuck" denouncing the harassment, repression, and surveillance of the underground press throughout the United States.

The article detailed the stories of several publications that were "getting a raw deal" from the "minions of law and order" (Dreyer, 1968, p. 4). Their editors and writers were subsequently hit with obscenity and libel charges, their offices busted by the police, their distributors arrested. Other less direct forms of intimidation included FBI-induced rent hikes and the unwillingness of printers to work with underground newspapers. The article also denounced the widespread surveillance of these newspapers, from the postal surveillance of the underground paper in Austin, Texas, called *The Rat* to the use of undercover officers to arrest underground press staffers, mostly on drug charges (Dreyer, 1968).

Another type of public denunciation tactic was directed at educating the readers on how to detect possible surveillance equipment. The *Freep* published a cover article titled "How to tell if your phone is bugged," written by Kunkin (1968), which explained different types of wiretapping and bugging devices (also using pictures) and how readers could detect them; it also told readers that they should "acquire some technical education about electronic surveillance" (p. 1). Kunkin instructed readers to thoroughly inspect their homes and offices if they feared they might be surveilled:

> Literally every square inch of the suspected bugged area must be searched. The counter-intelligence agent should check any visible irregularities such as a new patch of paint (particularly thin lines of silver paint which may function as a printed circuit to wire a microphone), a blob of putty, cracks in the baseboard and floorboard. All may be clues to recently planted bugs. (1968, p. 3)

But he also cautioned the readers: "It seems to me that it's essential not to panic just as it's important not to passively take that ride" (p. 1). In closing, he reflected on what surveillance might mean for activists: "it is really not possible to be either optimistic or pessimistic; it's just important that people know as much as they can and not give up" (p. 3).

The use of undercover agents, however, was the surveillance method that the underground press most ardently denounced. In fact, both the *Freep* and the *Good Times* developed a specific journalistic genre, that of the undercover agent exposé. Both papers ran articles exposing undercover officers, by disclosing their real names and even printing their pictures, or retelling their stories, especially when cops publicly disclosed the operations they were part of and tried to distance themselves from police departments (which happened relatively frequently). The newspapers further documented when former undercover agents got in trouble with their police departments. The *Good Times*' "Cop Watch," mentioned above, covered infiltration extensively.

The *Freep* repeatedly published pictures of suspected and confirmed undercover agents, complete with detailed information about pseudonyms and appearances. In fact, the *Freep* was dragged into a long legal battle with the state of California for disclosing the names of 80 undercover narcotics agents in 1969; the state filed a $10 million civil suit against the paper and charged the owner with a felony (Linfield, 1990). This did not stop the *Freep* from pursuing even more disclosures. In May 1972, it ran several pictures of undercover agents under the title "Undercover cops pose as picketers, finger activists." The *Good Times* also published pictures of presumed police spies; in October 1971, the photos of two men were annotated with the following: "these are not real people, these are counterfeits, on salary by your local police" ("These are not real people," 1971, p. 12). Public denunciation was an important tactic for the underground press, but it was clearly limited in its impact. As an article in the *Good Times* articulated: "Exposing pigs is not going to keep them from coming around, especially at a time when the agents are multiplying so quickly they've been known to inform on each other" (Phyllis, 1971, p. 13).

A second key tactic of resistance that is deployed in the underground press is that of using humour to condemn surveillance. Uncensored humour, including cartoons, was a big part of the revolutionary style of the underground press (McMillian, 2011). The *Freep* was particularly active in hosting cartoonists, such as Ron Cobb, who became crucial to the development of the underground comix movement. A cartoon by Cobb, published in the then *Express Times* in February 1968, shows two men talking on a bench under CCTV surveillance and directly observed by militarized police officers in a tank. One of the men says to the other, "Well ... at least we don't need to worry about anarchy anymore" (Cobb, 1968, p. 3). Besides making fun of mainstream tropes about surveillance – protecting good citizens from anarchy and chaos – the cartoon alerted readers about the militarization of cities and police departments. So did another cartoon in the *Freep*, which shows the globe being wiretapped; an operator, unable to find a wire still untapped, apologizes to Richard Helms, the CIA's Director of Intelligence at the time: " ... very sorry Mr Helm [*sic*] ... but there's no more holes!" (Liberation News Service, 1972b, p. 8.). The pervasiveness of surveillance was likewise highlighted by another cartoon, depicting a man sitting on the toilet surrounded by cameras and microphones and spied on by four individuals. The playful caption, "When under surveillance, always put on a good show" (Whirlwind, 1970, p. 12), points to the widespread suspicion of the existence of a concerted surveillance effort directed against the activists.

The use of undercover agents and informants was the frequent object of humour. A cartoon published by the *Freep* in 1972 pointed to the use of informants, especially on college campuses. In the cartoon, two women are seen conversing in a living room. One of them says: "Tuition has soared again this year, but we're lucky in that our Greg gets a little something from the FBI, to sort of keep an eye on his dorm floor" (Liberation News Service, 1972a, p. 4.). A satirical horoscope in the *Freep* made fun of the grooming protocols followed by undercover cops who wanted to infiltrate the movements: "August is a great vacation month. Take a vacation and try to forget your chaotic or non-existent emotional-sex life. Lay in the sun and let your hair grow real long so you can get a job spying for your local police department" (Svivananda, 1970, p. 13).

The papers also used humour to talk about their interactions with law enforcement. *Freep* writer and book author Lawrence Lipton (1969) penned an article in which he detailed humorous etiquette tips for dealing with visits from the FBI and satirized FBI tactics, like their showing up unannounced or going through people's possessions. He also poked fun at the extensive intelligence gathering conducted by the Bureau:

> Do not show too much surprise if their information about you and your views is a bit sketchy, in spite of the fact that they had a big fat secret FBI dossier on you that they could have consulted before setting out on their visit. Remember they're your guests and the rules of hospitality apply at all times, even to the uninvited fuzz. (Lipton, 1969, p. 3)

Lastly, while public denunciation and humour are present throughout the period 1968–1972, the coverage of legal battles surrounding the legality and permissibility of surveillance appears only towards the end of this period. At the beginning, in fact, legal cases that are covered in the two newspapers are predominantly those that prosecute activists. Towards the end of this intense period, in contrast, the law seems more capable of protecting activists and the courts become a viable avenue for challenging the surveillance practices of the state. Besides the three events mentioned above – the Pentagon Papers, the early revelations about COINTELPRO by the Citizens' Commission, and the Watergate scandal – one other instance that might have contributed to lending more credibility to the justice system is the landmark Supreme Court decision of June 1972, which ruled against warrantless electronic surveillance (*United States v. US District Court*, 1972).

Other examples of the *Freep* and the *Good Times* publishing articles on legal challenges against surveillance include the trial against Daniel Ellsberg and Anthony Russo for the Pentagon Papers and the lawsuit against the LA Community College District for surveilling their students. In the first instance, both papers followed closely the trial against Ellsberg and Russo, openly siding with the whistleblowers. In particular, the papers covered one of the key events in the trials: the fact that in 1972 the prosecution attempted to use wiretapped conversations of the defendants and their attorneys as evidence. A decision by the Supreme Court subsequently barred the prosecution from entering the tapes into evidence, since they were illegally obtained (*Russo v. Byrne*, 1972). A mistrial was thus declared. A new trial in 1973 resulted in the dismissal of all charges against Ellsberg and Russo. The *Freep* further reported on the use of electronic surveillance at several Los Angeles community colleges in March 1972 ("Students spied on," 1972), and closely followed the lawsuit filed by the American Federation of Teachers, together with American Civil Liberties Union and the Western Center on Law and Poverty (Ofari, 1972a). The newspaper commented on the impact of surveillance on the colleges, by highlighting how it would stifle dissent and become standard practice (Ofari, 1972a). Following up on the case in a later issue, the *Freep* detailed the instances of surveillance that had been reported by plaintiffs, and that ranged from the collection and storage of detailed information about the actions and beliefs of students to the recording of telephone conversations among students (Ofari, 1972b), which were then passed on to the LA Police Department (Donner, 1990). The newspaper also clearly articulated the significance of this lawsuit: "this will open the way for more court action by individuals and groups, particularly those politically active, to deter the use of electronic surveillance against radicals" ("Judge rules suit can be filed for wiretapping at college," 1972, p. 14).

In the space of five years, from 1968 to 1972, we thus see an evolution in the ways in which the underground press papers approached the issue of how to resist state surveillance. While public denunciation and humour were a constant feature of the *Freep*'s and the *Good Times*' coverage, it was only in 1971 and 1972 that the papers started talking about ways in which the movements could fight back against surveillance, namely, through the courts. This shift, which denoted a new optimism about the power of the law and of the court system to protect activists from surveillance and repression, was however also accompanied by the realization that surveillance had become more and more widespread, as documented above.

Conclusion

In this chapter, we focused on how two eminent alternative publications of the movements of the 1960s and 1970s, the so-called underground press, addressed the issue of surveillance. The coverage of surveillance in the *Freep* and the *Good Times* changed in two important ways in the time frame that we examined. Firstly, they documented and critiqued the transformation of surveillance itself, which over time became more intensive, more proximal, and more extensive. As the surveillance of the New Left and of the underground press intensified, the activists also became more aware of the concerted efforts of the state to monitor and repress the movements, as well as other groups, like people of colour and students. Secondly, these two publications showcased the ways in which movements employed different tactics for resisting surveillance: public denunciation, humour, and legal challenges. While the importance of public denunciation and humour is a constant feature, the possibility of a legal route to challenge surveillance emerges only in conjunction with high-profile events of 1971 and 1972 that propelled surveillance into the mainstream.

What is clear from our analysis is that the surveillance of activists in the 1960s and 1970s, as seen through its representation in the underground press, has striking parallels with today. Just like the movements of that time, the United States is still confronted with massive programs for warrantless domestic spying and with intelligence agency overreach; the scope and scale of these surveillance activities continue to be revealed through whistleblowers, be it those who broke into an FBI office in the suburbs of Philadelphia in 1971 or those who stole electronic documents from the National Security Agency. In many ways, activists and regular citizens are still attempting to cope with surveillance using similar tactics: by exposing and denouncing, through online disclosures and mainstream media coverage; by using humour, like John Oliver did when interviewing Edward Snowden in 2015; and by challenging surveillance in the courts, for instance with the lawsuits that followed the disclosure of the NSA's programs in 2013. Finally, it is striking to see how both the movements of the 1960s and 1970s and today's activists (and to a large extent, even regular citizens) had to educate themselves to be vigilant about surveillance when going about their daily lives. In 1970 this might have meant being extra careful interacting with a hippie-looking man, who could turn out to be an undercover agent, while today it might mean learning how to use encryption to protect one's electronic communications. However, unlike some of the contemporary technical and depoliticized responses to the Snowden revelations (see Gürses, Kundnani, & Van Hoboken, 2016), the contestation of surveillance by the counterculture was always placed in a general political challenge to state repression. As can be seen in the *Freep* and the *Good Times*, the activists recognized the deeply political nature of surveillance. The most important lesson to learn from the activists in the 1960s and 1970s is to continue to write, to mobilize, and to strive for a better world, while resisting and denouncing the tightening grip of state surveillance.

NOTES

1 To select our sample of articles, we searched the issues of the two newspapers published between 1968 and 1972 with multiple keywords (and their permutations) that describe practices of surveillance: surveil*, spy*, spies, infiltrat*, snoop*, wiretap*. Keywords were selected both *a priori* and through an exploratory analysis of several articles appearing in the newspapers. The keyword search yielded 184 results in the *Good Times* and 396 in the *Freep*, identifying pages in which the keywords were mentioned at least once. During the analysis, the majority of the entries were discarded because they did not address the issue of surveillance in a pertinent way, e.g., they were ads for private investigators. The final number of relevant articles is 216. Although the keyword search finds a limitation in the fact that it might not discover articles that would be relevant to the analysis, but that do contain the identified keyword, our use of multiple keywords compiled via a preliminary analysis of the papers mitigates this concern. Using thematic (Braun & Clarke, 2006) and open (Corbin & Strauss, 2008) coding, each author separately analysed half of the articles in the sample; emergent and thematic codes were then compared and reconciled by the authors.

2 In this chapter, we treat the *Good Times* proper and its predecessor, the *Express Times*, as the same newspaper.

REFERENCES

Armstrong, D. (1981). *A trumpet to arms: Alternative media in America.* Los Angeles, CA: J. P. Tarcher, Inc.

Atton, C. (2002). *Alternative media.* London, UK; Thousand Oaks, CA, and New Delhi, India: Sage Publications.

Bank policies protested. (1972, May 26). *Los Angeles Free Press,* p. 2.

Barkan, R. (1971, November 5). Government scientist proposes that computer should track citizens. *Los Angeles Free Press,* p. 8.

Bertlet, C. (2012). Muckraking gadflies buzz reality. In K. Wachsberger (Ed.), *Insider histories of the Vietnam era underground press, part 1* (pp. 267–98). East Lansing, MI: Michigan State University Press.

Blaine. (1969, September 18). Cop watch. *San Francisco Good Times*, p. 13.

Braun, V., & Clarke, V. (2006). Using thematic analysis in psychology. *Qualitative Research in Psychology, 3*(2), 77–101.

Cobb, R. (1968, February 29). [Cartoon]. *San Francisco Express Times*, p. 3.

Cops go to H. S. (1972, February 11). *San Francisco Good Times*, p. 4.

Corbin, J., & Strauss, A. (2008). *Basics of qualitative research: Techniques and procedures for developing grounded theory* (3rd ed.). Los Angeles, CA, and London, UK: Sage.

Cunningham, D. (2004). *There's something happening here: The New Left, the Klan, and FBI counterintelligence.* Berkeley, CA: University of California Press.

Davis, J. K. (1997). *Assault on the left: The FBI and the sixties antiwar movement.* Westport, CT, and London, UK: Praeger.

De Certeau, M. (1984). *The practice of everyday life* (Vol. 1). Berkeley and Los Angeles, CA: University of California Press.

Donner, F. J. (1990). *Protectors of privilege: Red squads and police repression in urban America.* Berkeley and Los Angeles, CA: University of California Press.

Dreyer, T. (1968, December 11). Underground fuck. *The San Francisco Express Times*, pp. 4–5.

Ellis, D. L. (1971). The underground press in America: 1955–1970. *The Journal of Popular Culture, V*(1), 102–24.

Emery, M. C., Emery, E., & Roberts, N. L. (1996). *The press and America: An interpretive history of the mass media* (8th ed.). Boston, MA: Allyn and Bacon.

Glessing, R. J. (1970). *The underground press in America.* Bloomington, IN: Indiana University Press.

Gürses, S., Kundnani, A., & Van Hoboken, J. (2016). Crypto and empire: The contradictions of counter-surveillance advocacy. *Media, Culture & Society, 38*(4), 576–90.

Hoffman, F. (1969a, January 10). Student breakout at Southwest College. *Los Angeles Free Press*, p. 6.

Hoffman, F. (1969b, September 19). Militant teachers at Manual, Jefferson denied teaching posts. *Los Angeles Free Press*, p. 3.

Ivory tower. (1969, May 2). *Los Angeles Free Press*, p. 4.

Judge rules suit can be filed for wiretapping at college. (1972, September 1). *Los Angeles Free Press*, p. 14.

Kirby, B. (1970, May 1). Freep phones tapped. *Los Angeles Free Press*, p. 2.

Kunkin, A. (1968, June 28). How to tell if your phone is bugged. *The Los Angeles Free Press*, pp. 1, 3.

Kunkin, A. (1972, February 4). Police spy talks. *Los Angeles Free Press*, p. 1.

Levitt, D. (1970, April 24). Isla Vista bugged. *Los Angeles Free Press*, p. 1.

Liberation News Service. (1972a, April 22). [Cartoon]. *Los Angeles Free Press*, p. 4.

Liberation News Service. (1972b, September 1). [Cartoon]. *Los Angeles Free Press*, p. 8.

Linfield, M. (1990). *Freedom under fire: US civil liberties in times of war*. Boston, MA: South End Press.

Lipton, L. (1968, September 20). The Dick Daley revised history of Czechago. *Los Angeles Free Press*, p. 3.

Lipton, L. (1969, January 24). Lawrence Lipton and his recorder entertain two FBI agents. *Los Angeles Free Press*, p. 3.

Mann, S., Nolan, J., & Wellman, B. (2003). Sousveillance: Inventing and using wearable computing devices for data collection in surveillance environments. *Surveillance & Society*, *1*(3), 331–55.

McBride, D. (2003). Death city radicals: The counterculture in Los Angeles. In J. McMillian & P. Buhle (Eds.), *The New Left revisited* (pp. 110–36). Philadelphia, PA: Temple University Press.

McMillian, J. (2011). *Smoking typewriters: The sixties underground press and the rise of alternative media in America*. New York, NY, and Oxford, UK: Oxford University Press.

NARC for rent. (1972, June 16). *San Francisco Good Times*, p. 6.

Ofari, E. (1972a, March 24). Lawsuit challenges school wiretapping. *Los Angeles Free Press*, pp. 1–4.

Ofari, E. (1972b, April 3). Spying on students and teachers continues at local colleges. *Los Angeles Free Press*, p. 28.

Osborn, J. (1968, April 12). SNCC office raided. *Los Angeles Free Press*, p. 2.

Peck, A. (1985). *Uncovering the sixties: The life and times of the underground press*. New York, NY: Pantheon Books.

Peep eye. (1971, October 29). *San Francisco Good Times*, p. 12.

Phyllis. (1971, January 29). Dallas echoes. *The San Francisco Good Times*, p. 13.

Pig Yoga. (1971, February 19). *San Francisco Good Times*, p. 8.

Ridenour, R., & Ofari, E. (1972, October 13). The death penalty. *Los Angeles Free Press*, p. 5.

Rips, G. (1981). The campaign against the underground press. In A. Janowitz & N.J. Peters (Eds.), *The campaign against the underground press*. San Francisco, CA: City Light Books.

Rosevear, J. (1968, June 14). Dope smoking – The state of the art. *Los Angeles Free Press*, p. 30.

Rycroft, S. (2007). Towards an historical geography of nonrepresentation: Making the countercultural subject in the 1960s. *Social & Cultural Geography*, *8*(4), 615–33.

Select committee to study governmental operations. (1976). *Supplementary detailed staff reports of intelligence activities and the rights of Americans* (No. 94–755) (Vol. III). Washington, DC.

Silver, S. (1971, January 29). Perceptual imperialism. *San Francisco Good Times*, p. 15.

Streitmatter, R. (2001). *Voices of revolution: The dissident press in America*. New York, NY: Columbia University Press.

Students spied on. (1972, March 10). *Los Angeles Free Press*. n.p.

Svivananda, S. (1970, January 8). 1970 the year of the goat. *San Francisco Good Times*, p. 13.

Szigeti, S. (1969, September 26). FBI informer in SDS escapes prosecution in fatal shooting. *Los Angeles Free Press*, p. 1.

These are not real people. (1971, October 29). *San Francisco Good Times*, p. 12.

Trial inside and out. (1970, August 14). *San Francisco Good Times*, p. 4.

Undercover cops pose as picketers, finger activists. (1972, May 19). *Los Angeles Free Press*, p. 6.

A warden on every block. (1970, February 27). *San Francisco Good Times*, p. 3.

Whirlwind, V. (1970, April 23). [Cartoon]. *San Francisco Good Times*, p. 12.

Whittaker, N. (1968, August 14). Panthers shot. *San Francisco Express Times*, p. 5.

Who's who in the CIA. (1971, August 20). *San Francisco Good Times*, p. 19.

Cases Cited

Russo v. Byrne, 409 US 1013 (1972).

United States v. US District Court, 407 US 297 (1972).

12 "That's Not a Conversation That Belongs to the Museum": The (In)visibility of Surveillance History at Police Museums in Ontario, Canada

MATTHEW FERGUSON, JUSTIN PICHÉ, AND KEVIN WALBY

Public police are among the many surveillance workers (G. Smith, 2012) whose jobs involve watching others (Ericson, 1994). Advances in technology provide them with increasingly sophisticated tools for collecting detailed information on the activities of the populations subject to their gaze (Lippert & Newell, 2016). Though technology has progressed, there is more continuity between the past and the present of police surveillance than one might expect. For example, Finn (2009) explains how the practice of collecting and archiving images of faces, fingerprints, and deoxyribonucleic acid (DNA) for law enforcement purposes has roots that can be traced back to the nineteenth century (also see Sekula, 1986). Little is known about the history of law enforcement surveillance as told by police information managers (Maynard, 1994) or how historical institutions, including museums, depict the past and present uses of surveillance, like the biometric practices highlighted by Finn (2009), in policing.

Our chapter is unique in the context of this book as it is centred on the present. Rather than examining a historical case, we consider how history itself is constructed, in this instance by the agents of state surveillance themselves, that is, the police. Police surveillance legitimates not only hegemonic values, but also the police's own practices. Thus, to draw on examples from this book, in South Africa police surveillance of "deviant" forms of gender expression cemented their role in constituting the racialized nation (B. Camminga, chapter 4), while in communist Romania practices of collaboration involved the performative affirmation of police authority (Cristina Plamadeala, chapter 9). In the United States, public narratives were reshaped through the policing of dissident publications (Elisabetta Ferrari & John Remensperger, chapter 11). Self-memorialization is in this sense one dimension of all police practice. In this chapter, we examine a more explicit instance. Contributing to research on police memorialization and cultural depictions of law enforcement, we demonstrate how four police museums in the Canadian province of Ontario render (in)visible the historical surveillance practices used in policing and their material impacts on the surveilled. Police museums are places where the practices of social control agents are depicted for public audiences (McNair, 2011). There are more than fifty museums in Canada where law enforcement work is represented, ranging from sections of historic forts to museums located in active police stations. Though some studies have been conducted on meaning-making in police museums, most of this research has taken place in other countries such as the United States (McNair, 2011), South Africa (Comaroff & Comaroff, 2004), Argentina (Caimari, 2012), and Mexico (Buffington, 2012).

There are seven police museums in Ontario and at least one more in the construction stage. All of these sites in the province are operated by or in partnership with police services. Drawing from field observations at four of them, we demonstrate how the idea and history of surveillance is conveyed through representations that depict these practices as innocuous instead of an invasive violation of civil liberties and constitutional rights, and as reserved only for a select group of individuals (i.e., suspects and "criminals") as opposed to the broader population. This serves as well to divorce police surveillance from its position in broader systems of often politicized state surveillance. First is the Toronto Police Museum, which is located in the atrium of the downtown headquarters of the Toronto Police Service, the largest municipal police force in the country. The site contains three levels and showcases the history of policing in the city from 1834 to the present. Second is the Ontario Provincial Police Museum, which is located in the much smaller town of Orillia at the OPP General Headquarters, a military-style complex that functions as the main base for the provincial force. The museum portrays the history of policing in Ontario since 1721, with the majority of exhibits focusing on the development of the OPP in the twentieth century. Visitors are provided with a guidebook upon entering the site containing information on the exhibit (Behind the Badge, 2014). Third is the Royal Canadian Mounted Police (RCMP) Musical Ride Centre located at the Canadian Police College in Ottawa. The Musical Ride itself is a popular travelling performance that involves RCMP officers, known as Mounties, performing elaborate horseback routines meant to show off their cavalry skills. At this federal police college, the museum provides a glimpse into the history of the Musical Ride, along with the present and future of the RCMP. These three sites were selected because they represent the three largest police services managed by their respective levels of government (municipal, provincial, and federal respectively). We have also included data from the Ontario Police College, which is located in the rural town of Aylmer and is the central training facility for police constables working in the province. In the main building, a museum sprawls out across some of the hallways and a gift shop operates during certain months of the year.

Exemplified most recently by a police museum being constructed in the new Winnipeg Police Headquarters in Manitoba, Canadian policing heritage sites are increasingly incorporated into the "front stage" of operational policing infrastructure. Inside the museums, we examine how, using exhibits that erase the political dimensions of "crime control," surveillance is portrayed as necessary in the pursuit of public safety. Blurring historical representations and current law enforcement work, surveillance is represented in two ways. The first is through depictions of historical covert investigative methods used by the police to carry out surveillance work on those deemed a community safety threat. The second highlights forensic analysis techniques used in the past and present, and the capacity for police officers to monitor and track down citizens through pieces of evidence and their bodily traces. We show that police museums often omit or neutralize controversial forms of police surveillance and malpractice, while failing to account for their racialized, gendered, heteronormative, and classist dimensions.

To provide context, we begin by reviewing relevant literature on public police, museums, and surveillance. After, we examine representations of police covert surveillance followed by a section on representations of police forensics and intelligence databases. We demonstrate how the museums present surveillance as enabling police to catch the "right" individual deemed responsible for a criminalized act. A deeper consideration about the violence of state surveillance or the relationship between "crime" and wider social issues is not encouraged at these sites. Police surveillance is portrayed as a justifiable practice used only when necessary, not indiscriminately, and in ways that prevent, rather than create, social harm. In the process, the museums obscure a longer, more troubling history of these practices and the experiences of those impacted by them. Visible and invisible historical narratives convey a desired image of surveillance in policing shaping understandings about these practices in Canada. In conclusion, we reflect on the implications of our findings with respect to the literature on representations of police, social control, and surveillance.

On Police, Museums, and Surveillance

As G. Smith (2012) argues, the surveillance work of social control agents, including public police, is often invisible, and what is known tends to be derived from cinematic representations (Lippert & Scalia, 2015; Muir, 2012) and artwork (Barnard-Wills & Barnard-Wills, 2012). Police museums are another cultural forum where depictions of this social practice are often present. Museums are places where history and heritage are staged and performed for a consuming, tourist public (L. Smith, 2006; Walby & Piché, 2015). Police museums provide visitors the opportunity to learn about history and visualize the practices of law enforcement. Many contain displays of vintage police vehicles, photographs (e.g., mug shots, missing persons, officers), and policing memorabilia such as uniforms, weaponry, badges, and communications devices (Buffington, 2012; Ferguson, Piché, & Walby, 2019; McNair, 2011). On occasion, visitors may encounter stories about police surveillance or information on the activities of those who were the object of it. For example, Caimari (2012) discusses how a room devoted to Juegos Prohibidos (Forbidden Games) in the Argentine Federal Police Museum in Buenos Aires provides insight into the practices used by those involved in illicit gambling to avoid detection by police. Representations of history in popular culture and memory institutions can serve as powerful forms of persuasion, shaping how consuming publics understand social issues and government (King & Maruna, 2006).

Museum studies scholars have noted that these representations are always limited. Curators select certain objects to display and particular narratives to frame interpretations of history (Crane, 2000). As we reveal, the visualization of police surveillance practices in police museums is partial and limited. The partiality of museum representations is amplified when curators and staff have an explicit political stance. For example, many police museums are funded, operated, and curated by police associations, unions, or foundations. These connections mean police museums are part of what Ball (2017) calls the "surveillance-industrial-entertainment complex" (p. 96), which relies on entertainment industry rationalization and normalization of police social control and its generation of fascination with surveillance practices. However, the museums we examine tend to be more driven by ideology than by a commercial logic. This fascination results in a powerful surveillant imaginary informing public views of (police) surveillance practices. As Monahan (2011) notes, more research is needed on representations of surveillance, which we address by examining historical and contemporary depictions of social monitoring in police museums.

We argue that these narrow representations can buffer museum visitors from more complete and critical accounts of the vast surveillance that Canadian police forces engage in. This is especially the case given the extent to which police forces and other institutions of state surveillance seek to throw a cloak over their past activities, suppressing access to information through a range of legislative and practical measures to keep historical documentation secret. Police already control their own history more strongly than most institutions, with museums consolidating and extending their ability to do so. Given the economic and institutional ties that many law enforcement museums have with existing police forces, we thus suggest that these forms of representation and memorialization serve as one approach of public relations management in policing aimed at mitigating ongoing public scrutiny, while co-opting efforts to establish meaningful accountability and transparency measures for the (in)actions of officers and management within these paramilitary organizations. Below, we explore two major themes that emerged from our analysis of displays at four Ontario sites. The first consisted of information about covert police investigations, while the second involved forensic analysis methods and the capacity for the police to keep track of citizens.

Representations of Covert Police Surveillance

So, I looked it up in the encyclopedia and it seemed to fit what we were doing, even wrote down the verbatim explanation in my notebook, in case I had to explain ... The expectation of an imminent cosmic cataclysm in which the supreme power destroys the ruling powers of evil and raises the righteous to a new life in a messianic kingdom. Greek origin – Apo – an uncovering of hidden things.

These are the words of retired Detective Inspector Ron Allen of the OPP, recounting the origins of the name for a 1982 wiretapping investigation dubbed Project Apocalypse. Visitors can read the story while looking at the "Undercover and Surveillance" display at the OPP Museum. The aim of Project Apocalypse was to arrest individuals suspected to be involved in criminalized activity, including a main target that had a lengthy history of smuggling prohibited drugs across the Canada-United States border. Five people were named in the investigation and eight phone lines were monitored from December 9, 1982, to February 5, 1983. As the OPP got closer to substantiating charges, the guidebook tells us, "the targets got paranoid, somehow figuring out that something was going on" and the investigators failed to obtain sufficient information to lay charges. The guidebook explains that no major arrests were made, but Revenue Canada and the banks were able to seize assets belonging to the two main targets, including "a farm, real estate holdings, and a furniture business that served as a front for criminal activity." Owing money to many people, they were never seen in Ontario again. Rumours circulated that members of a rival syndicate may have killed them for non-payment of drug debts.

Figure 12.1 Display about surveillance in the 1980s, OPP Museum.

A prominent item displayed about this case is a reproduction 1982 Project Apocalypse T-shirt (see figure 12.1). Creating T-shirts for members of
specific police investigations was a well-established practice in the 1980s.
Imprinted on it is a skeleton riding a horse with the words underneath
reading, "(maybe next year)." The guidebook notes that this image was
inspired by two topics. The first is the Four Horsemen of the Apocalypse, a
group of mythical riders representing the end times. While no additional
information is provided, the skeletal rider suggests that this is a depiction
of the fourth and final horseman, a pale horse ridden by a skeleton sym-
bolizing death. The skeleton on the shirt is bent over and holds a broken
lance, and a battle shield is visible on the ground nearby. It is clear that
members of this investigation interpreted the police as being embedded
in a conflict between good and evil, with the phrase below the image perhaps ominously hinting that an apocalypse for the targets would be coming soon, "maybe next year." It further places this historical account in a
longer narrative, one in which the police serve as a trans-historical surveillant force impossible to evade. The guidebook also notes that the image
was partly inspired by the fact that one of the targets had a history of smuggling drugs across the border inside old horses.

Previous research has highlighted a contrast in the way penal system actors and those in conflict with the law are represented to visitors in museum settings. Chen and colleagues (2016) explain how museums tend to emphasize prisoners as dangerous and cunning, while portraying staff through narratives of heroism, duty, and sacrifice. In the process, "the division between 'us' and 'them' is defined" (Chen, Fiander, Piché, & Walby, 2016, p. 36). The retelling of Project Apocalypse at the OPP Museum provides an example of the way discourses of good and evil and us versus them are demarcated in the stories shared. This narrative hints at some of the ideas that underpin our examination of representations of covert surveillance practices inside the police museums. In Ron Allen's story there is the implicit idea that the "imminent cosmic cataclysm" is a metaphor for an unavoidable clash of violence between the state (the supreme power) and the criminalized (the rulers of evil), where the avoidance of fear and death rests in the ability of the police to uncover the hidden truths leading to the neutralization of the individuals believed to be in conflict with the law. While the investigation is hampered, the story is completed with the targets being neutralized, both within the law and possibly beyond it. The lawful seizure of their private property may have led to their destruction, as Ron Allen had perhaps hoped – albeit at the hands of a more ruthless power of "evil" that rules the underground.

Covert surveillance in law enforcement has long been a concern of policing scholars (e.g., Brodeur, 1995, 1992; Skolnick, 1982). Although there is scholarship on covert investigations and surveillance tactics (e.g., O'Neill & Loftus, 2013; Wilson & McCulloch, 2012), there is little work examining representations of covert policing. In our analysis of portrayals of covert surveillance practices inside police museums we understand "covert policing" to refer to the discreet, secretive investigative work undertaken by the police that infringes on a suspect's private life, such as undercover operations and wiretapping (Loftus & Goold, 2012). Below, we demonstrate how historical representations of covert surveillance tactics validate the technological expertise of the state and provide support for the monitoring of populations deemed dangerous. The contrast with Elisabetta Ferrari and John Remensperger's examination of representations of surveillance in the underground press in the previous chapter is of course striking. Their case study and ours provide a powerful sense of the extent to which struggles over the representation and narrativization of police surveillance were and are central to the dynamics of surveillance itself.

Displays at the OPP Museum contain a number of other surveillance-related items. Exhibited is a mid-1980s briefcase surveillance camera used at the airport, the guidebook says, "when the OPP received information that a criminal 'underworld' figure was arriving or departing the province." Clicking a small red button near the briefcase handle triggered the shutter on the enclosed camera. Several black and white photos are also presented, one of them of Ron Allen working undercover as a fisherman in 1978 and another of his unmarked car parked beside a suspect's vehicle. The photo beside the T-shirt shows a Satan's Choice field meeting in 1969, which was a motorcycle group in Ontario during that period. The guidebook notes that the OPP once raided a Satan's Choice meeting using riot gear and tactics, although it "proved inadequate in the face of the Choice's aggressive opposition." The OPP were threatened by the motorcycle group and forced to retreat, leading to increased training and a rethinking of tactics used in such situations.

A mannequin of a woman placed in front of a fake street corner with graffitied bricks represents several female officers who worked OPP Intelligence and undercover operations in the 1980s. The guidebook explains the dangers faced by undercover female officers. It notes how one officer, who posed as a sex worker, claimed her favourite accessory was her cowboy boots because they were able to conceal a knife, "just in case." Another story recounts the time when a female OPP officer was a "bodyguard to a former mobster's girlfriend-turned-witness who had been threatened in 1981 and was forced into hiding." Seven years later, an author published a book about the mistress that included a personal photograph of the officer, without her permission. This put the female officer's life in danger as she was working undercover investigating a target that was "not a nice guy." The guidebook notes that nothing came of the incident. Nearby is a mannequin wearing a black leather jacket and a Harley Davidson T-shirt while holding a motorcycle helmet. This was an undercover outfit worn by an officer in the 1980s for tasks such as "drug buys, providing physical and photographic surveillance, and serving as back-up for other team members." The guidebook goes on to explain that the undercover officer wore the outfit while working as a taxi driver during project "Hack" in Kingston – an operation that resulted in the arrest of over seventy people in ten months.

One of the earliest mentions of surveillance at the OPP Museum is in a display about seized objects. The display consists of several photographs of seized material alongside three glass cases containing relics. One of them is a handgun seized by Divisional Inspector George Caldbick shortly after the formation of the OPP in 1909. He was working in Colbalt at the time, which the guidebook describes as a "rough and tumble frontier settlement" said to have a growing need for law enforcement. Caldbick gained a reputation for seizing handguns from workmen arriving by train and later using them on duty. He would have incoming train staff lock all exits except for one and then "watch" the crowd of disembarking passengers for those who appeared to be suspicious. The OPP Museum thus conveys an image of police surveillance at the beginning of the twentieth century as primarily a one-man job, often conducted by constables responsible for patrolling vast territories. Visitors learn that around this period in Orillia it was customary for a night watchman to assist the town constable by checking the locks on store doors and keeping "an eye out for vandals and fire." Nearby, a display focused on the 1920s and 1930s shares a unique example of an outside group assisting the police with surveillance. During the 1939 royal tour of Canada, the OPP were responsible for security while King George VI and Queen Elizabeth were in Ontario. On June 7, 1939, the royal couple travelled by car on the Niagara Parkway to Niagara Falls. The parkway was rural, and "given international tensions of the times" there was a possibly that "ill-wishers" could hide in the many ditches, culverts, and bushes. The OPP deputized roughly seventy Rover Scouts – young men eighteen years of age or older – for the occasion and dropped them off along the parkway to check potential hiding places and stand guard.

While touched upon at the OPP Museum, the topic of covert surveillance is non-existent at the Ontario Police College Museum and the RCMP Musical Ride Centre. At the Toronto Police Museum, covert surveillance is not a major topic either, but does materialize in a few stories about notorious criminalized acts featured on the second level. One story tells of the time in the 1980s when an undercover officer befriended a suspect named Tien Poh Su and managed to collect a sample of his blood, helping to lead to his conviction for the rape and murder of a teenage girl. Another example can be found in a nearby display case. The text inside begins with the words, "Ricky and Dwane Atkinson led one of Toronto's most sophisticated criminal gangs ... the 'Dirty Tricks' gang." The Dirty Tricks Gang was responsible for committing over fifty armed robberies in the late 1970s and early 1980s (Atkinson & Fiorito, 2017). Objects positioned in the display case include make-up, spray paint, a wig, a balaclava, a nail-studded board, and binoculars. According to the museum text, the gang's method of operation was distinctive. They wore a variety of disguises, and "carefully selected bank and credit union branches that gave them an excellent chance of escaping." The group "used state-of-the-art communications equipment," and often stole a van and loaded it with nail-studded boards to throw at pursuing police cars. It was common for them to post lookouts who "could call off a planned robbery at the first sign of police presence." In 1981, four gang members were caught and another shot dead in an attempted bank robbery on

Laird Drive in Toronto. Despite this setback, "the group was soon back in action." In August 1986, "following months of intense police surveillance, police were on the scene as the gang hit the same Laird Drive bank branch once again." The display concludes by noting that the gang members, including Ricky and Dwane Atkinson, were caught and convicted of armed robbery.

According to Loftus and Goold (2012), past state surveillance powers were directed "only at particular individuals who were deemed to be at risk or undeserving of trust" (p. 276), whereas today these powers are directed towards most of the population. Inside the police museums this narrative is prominent as well, conveyed through historical representations that suggest these practices were and remain only reserved for those deemed suspect or deviant. The effect of this vision of history is to shield visitors from encountering narratives in the museum pertaining to the vast surveillance in which Canadian police forces, as well as intelligence agencies, engage (Lyon, 2015). While many topics shared inside the police museum educate visitors about modern practices, the covert surveillance practices police carry out is often not one of them. The absence of these narratives is not surprising due to their controversial nature. The fact that the OPP Museum puts the spotlight on the covert world of policing is unique in its desire to inform visitors about an "exceptionally hidden, concealed practice" (Loftus, Goold, & Mac Giollabhui, 2016, p. 634), but there too the narrative clearly emphasizes that police surveillance was only ever directed at those who posed a threat. Modern-day surveillance is a secretive topic that could evoke unwanted criticism towards the police, and previous research has noted that police museums tend to sidestep content that does not "promote a positive public image" (Chazkel, 2012, p. 132; see also Buffington, 2012; McNair, 2011).

Representations of Forensics and Police Surveillance

Within the context of penality, forensics is the application of science and technology in a stated effort to investigate criminalized acts (Lawless, 2016). It is an often overlooked form of surveillance that is both past and future oriented; forensics can be used to look back into the past, while also collecting biological markers for future use (Hausken, 2014; Kruger, 2013). The "DNA dragnets" to which Nelkin and Andrews (1999) refer are also a dominant subject displayed at the four police museums. For example, the RCMP Musical Ride Centre shares information on modern forensic techniques used by the police, including the importance of the National DNA Databank, an investigative tool used by law enforcement in Canada containing known and unknown "criminal" DNA (Robert & Dufresne, 2008). Unlike much of what is on display inside the police museums, depictions of forensics focus on contemporary rather than historical practices. Below, we show how they are a significant means through which visitors engage with historical and cultural meanings of surveillance and penality.

Forensics occupies an important place in the exhibit at the OPP Museum. The first glimpse of a forensic relic can be seen in a display dedicated to the amalgamation of police services in Ontario. The exhibit guide explains that when a municipal police service amalgamates with the OPP, their history becomes shared with that of the OPP. Situated behind glass is a fingerprint camera that was likely used as early as 1905 (see figure 12.2), thus highlighting the early usage of biometric technologies in policing, which had its roots in colonial surveillance, as explored both by Uma Dhupelia-Mesthrie and Margaret Allen in chapter 6 and by Midori Ogasawara in chapter 7. The guidebook explains that the fingerprint camera "was used by the Petrolia Police Service to make 1:1 scale photographic copies of fingerprint cards onto glass plates." It was easy to operate because no focusing was required. Displayed in another section of the museum is an original 1977 Laser Control Unit. A large picture of a fingerprint is presented on the display case holding the relic. Visitors are taught about the way laser technology assists police in detecting substances at "crime" scenes such as fingerprints, hairs, and fibres. The display recounts how a chance discovery "changed the way we look at fingerprints" and led to the OPP becoming the first police service in the world to use a laser to detect evidence. In 1976, an OPP Senior Forensic Analyst was the neighbour of a scientist working at Xerox Research Canada, who had told him in conversation that the argon laser used in their photocopiers could make fingerprints that were invisible to the naked eye appear on paper. The laser was borrowed in a narcotics investigation later that year, where the bulky machine discovered a fingerprint on a black piece of electrical tape. The suspect was convicted and the OPP purchased its own laser soon after.

One of the largest display cases at the OPP Museum is a "mock crime scene" dedicated to the topic of bloodstain pattern analysis. Viewers are encouraged to "examine the bloodstains" and think about some of the questions they might ask if they were an analyst arriving on scene: "Was there a victim ... a nosebleed gone awry ... a fight with two assailants ... did an assault happen ... was a weapon used?" The guidebook goes into detail describing the importance of this "scientific investigation tool," noting that the first OPP officer was trained in bloodstain pattern analysis beginning in 1993. These forensic surveillance tools and

Figure 12.2 Fingerprint camera from the early 1900s, OPP Museum.

techniques constitute a form of "professional vision" (Goodwin, 1994, p. 606) legitimating the social control and monitoring efforts of police (also see Barry, 1995). Nearby, an area called the "Kidzone" provides children with a magnifying glass and encourages them to participate in forensic investigation activities, including matching the fingerprints and shoeprints of a suspect, comparing clothing fibres left at a "crime" scene, and inspecting a close-up photo of the tool marks left on a pried open door. Fostering a surveillance imaginary (Ball, 2017) for youth, one OPP Museum staff member interviewed for this study noted that the children are "right into it" and "think the science is brilliant."

Near the entrance of the Toronto Police Museum, visitors are greeted with a series of display boards that outline the history of the police service. Emphasized over two panels are the important contributions of Chief Henry James Grasett, Toronto's longest serving police chief (1886–1920) who had an "outstanding record of innovation," introducing many new practices to the service, including fingerprinting and the photograph-ing of suspects. Information on forensics at the museum is primarily conveyed through a detailed display outlining the police investigative process. Located on the second level of the museum, the arrangement consists of four boards behind glass cases, titled "*Analyzing* the Crime Scene," "*Conducting* the Investigation," "*Making* the Arrest," and "*Selected* Investigative Techniques" (original emphases). The first board discusses what police officers do when they first arrive on scene (see figure 12.3). A large vintage photograph of four investigators examining a piece of evidence appears at the centre of the board. The text underneath states: "Investigators can find traces of blood and other bodily fluids even in the most carefully cleaned crime scene. With today's technology, even a trace sample can become vital evidence for a successful prosecution." No information or dates specific to the photographs are provided. Below the board are several relics, including an antique Metro Toronto Police Homicide Forensic Kit, fingerprint dusters, a DNA swab, and a mould of a shoeprint.

The fourth board, titled "*Selected* Investigative Techniques," also shares information about forensics in police work. The board is separated into four sections – Blood Splatter, DNA, Autopsy, Angle of Entry of Bullet – and contains several black and white photos, along with text describing how each assist in police investigations. For example, the section on DNA explains that, "Since the early 1900's police have used analysis of blood type to connect a suspect with a crime. This process has now become an exact science." This is the only mention of the past in the text provided. Despite showcasing historical photos and objects, the display does not delve into the history of forensics or the ways in which its use in police investigations has changed over time, stressing in this instance a historical break. The lack of dates and other temporal markings on the antique objects sets them in a generalized and implicitly less scientific past, with the present constructed as a time when police science is wholly objective and scientific. Further into the museum, visitors can see these practices at work through several display cases that recount infamous criminalized harms that occurred in the city between the 1950s and 1990s.

On a slanted runway leading to the second level of the Toronto Police Museum, visitors can view one of the many now common practices ushered into the service at the turn of the twentieth century. Vintage mug shots are presented on two display boards, with the dates the photographs were taken not provided. A brief history on documenting those in conflict with the law is outlined at the bottom of one of the boards. It reads, "In the early 1800s, police officers often knew all of the criminals in their area. By the 1890s, the number of known criminals had become so unwieldy that detectives began to create 'mug shot' books to hold photos of criminals," an archival practice that Sekula (1986) notes helped bind "criminality" to physical bodies and bring more people

Figure 12.3 First two panels in a display about the police investigative process, Toronto Police Museum.

under state surveillance. The board goes on to discuss the rise of photo and key data for documenting "convicted criminals" on cards, which were mechanically sorted by police to find suspects matching descriptions. It concludes by noting, "With today's powerful computers, a comprehensive search can be completed within minutes."

Depictions of the history of forensics are absent at the RCMP Musical Ride Centre. The historical portion of this site is limited to the Musical Ride performance. However, depictions of forensics are present in a separate section of the museum titled "The RCMP Today and Beyond." There, the walls are lined with display boards containing detailed information about federal policing. Near the entrance are several boards on the topic of DNA analysis and the National DNA Databank that is maintained by the RCMP. One of them discusses DNA sample collection kits, which are used to collect the biological material from "convicted offenders" who are ordered to provide it (see figure 12.4). It notes that a prototype of the kit was first used in the investigation of Swissair Flight 111, which crashed in ocean waters near Peggy's Cove, Nova Scotia, in 1998, killing all 229 passengers. The contents of the kits (i.e., the blood

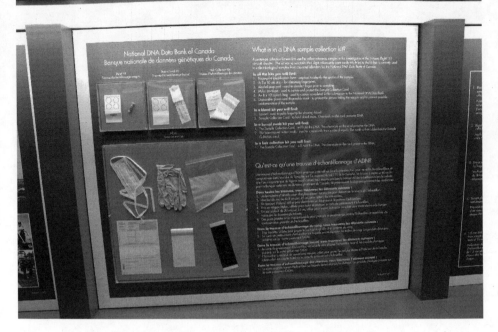

Figure 12.4 Display discussing the National DNA Databank of Canada and
DNA sample collection kits, RCMP Musical Ride Centre.

kit, buccal swab kit, and the hair collection kit) are displayed in a glass
cases attached to the board. Another board lists the benefits of having
a National DNA Databank in Canada, which include: "Linking crimes
together when there are no suspects; Helping to identify suspects; Elimi-
nating suspects when there is no match between crime scene DNA and
a profile in the NDDB [National DNA Databank]; and, Determining
whether a serial offender is involved." A series of ten steps outlines the
process from when the forensic analyst collects the DNA to when the
DNA profile is entered into the Combined DNA Index System, an electronic DNA database used by law enforcement.

Another board is dedicated to the subject of fingerprinting. It reads,
"Fingerprints help law enforcement to identify those responsible for
committing crimes," and goes on to note, "Police can search fingerprints of known criminals, and fingerprints from crime scenes. Fingerprint searches are also done for security/reliability investigations and
for civil fingerprint screening purposes." Pictured on the board are digital scans of fingerprints. A section of text reveals the RCMP holds 3.5

million sets of fingerprints in its "repository of criminal fingerprints." These retrieved biological and bodily materials become "data doubles," digital representation of ourselves that are "split off from and yet reconnectable with the individuals whose data constitutes them" (Lyon, 2001, pp. 115–16).

A similar type of display can be found at the Ontario Police College Museum outside of a door with the words "Forensic Training" written above it. An arrangement of forensic pictures is present on blue display boards, which mainly contain photographs of modern forensic practices and forensic scientists at work. Nearby is a display case dedicated to the topic of footprint evidence, which the accompanying text notes was "one of the earliest methods of placing a person at a crime scene." Inside is a footprint in a patch of dirt, a plaster cast of a footprint, a footprint comparison chart, and a black and white image of a technician. These displays operate to normalize and legitimate forensics as a pervasive form of police surveillance.

The Ontario Police College Museum also contains information about a large police database called the Canadian Police Information Centre (CPIC). Behind a glass display case are several vintage photographs and relics, such as an old telex machine. The text explains that the CPIC "was created in 1966 to provide tools to assist the police community in combating crime. It was approved by the Treasury Board in 1967 as a computerized information system to provide all Canadian law enforcement agencies with information on crimes and criminals." The placard relays how the system developed with advancements in technology such as personal computers. Text beside a miniature Canadian and American flag explains the important link between CPIC and the National Crime Information Centre, which allows Canadian and American law enforcement "access to information of mutual interest."

Chazkel (2012) notes how police museums in Latin America are shrines to modern forensic science, which applies to the four police museums examined here. The emphasis placed on forensics, along with information databases like the National DNA Databank and CPIC, is significant because it encourages and legitimizes the surveillance and control of populations deemed to be dangerous with the assistance of modern science. Inside the museums, the concept of forensics is entangled with meanings of surveillance, violence, "crime" and its control, and the general treatment of those in conflict with the law by society. Compared to other subjects at the museums, depictions of forensics lack historical depth and complexity. A focus on contemporary practices reinforces the perceived technological capabilities of Canadian police services. Writing about the National DNA Databank, Robert and Dufresne (2008) point out how DNA produces important symbolic effects for law enforcement, including strengthening the "professionalization and scientification ideal of the police force" (p. 80). However, in failing to situate these practices in historical context, the museums prevent visitors from achieving a more critical understanding of their use and impacts. These invisible or hidden histories could be learned from to better understand present and future developments in forensics and police surveillance. We illustrate this below by using the example of fingerprinting.

A dominant theme present throughout the museums is the importance of biometrics in policing, such as the use of fingerprinting, to accurately identify those who commit criminalized acts and collect their biometric information. The museums focus heavily on the importance of collecting fingerprints, while providing viewers with little historical analysis of the technique. In an attempt to justify this practice for viewers, an incomplete story is told about fingerprinting that ignores its connections to critical topics such as colonialism and racist state surveillance. Cole (2001) explains how the British system of fingerprint identification actually emerged in India, not at the famous Scotland Yard in England, "in response to administering a vast empire with a small corps of civil servants outnumbered by hostile natives" (p. 63). It was the product of a racist state agenda meant to document, monitor, and control "unruly" Indigenous populations in a way that, much like today, was believed to be scientifically accurate.

Similarly, fingerprinting also has a place in Canada's history of colonialism. In the 1920s it was suggested by white government officials that a system of identification was needed for Inuit people living in the north (D. G. Smith, 1993). It was decided that members of the RCMP would fingerprint them all, a technique that mirrored the original application of fingerprinting in India. However, many Inuit were frightened by the practice and its implied connections to criminalized activity. Fingerprinting of this kind eventually ceased, but as Alia (2006) points out, it would probably have continued if not for protests on administrative and humanitarian grounds. In the 1940s police began issuing a number on a disc that had to be worn by the Inuit at all times, an attack on their intricate naming traditions that only ended in 1971. Information about the role of fingerprinting in the progression of this racist surveillance system, which D. G. Smith (1993) describes as "violent both structurally and symbolically" (p. 66), is absent inside the police museums we examined. Its inclusion could serve as a means of developing a more critical understanding about the implications of, and those pushed to the margins through, intensive state surveillance systems (also see Fiander, Chen, Piché, & Walby, 2016).

In part through the erasure of these histories, depictions of forensics inside the police museums elicit a form of viewership where cultural fantasizing about "crime" is privileged and sustained, rather than challenged or questioned. Those under police surveillance tend to be ignored in exhibits or represented through stories about notorious criminalized acts, impeding consideration on the part of the onlooker about the everyday people who experience the violence of state surveillance and other pains associated with the penal system (see Ferguson, Piché, & Walby, 2019). This works to maintain the problematic social distance that exists between the public and those in conflict with the law, contributing to a phenomenon that Brown (2009) calls "penal spectatorship" (p. 8). It also has the effect of normalizing the surveillance creep that opens the door to an expansion of forensics within the penal system and beyond. Missing in these displays is an opportunity for viewers to make connections between individual acts of harm and the broader social conditions (e.g., patriarchy, misogyny, poverty, racism, etc.) in which a great deal of the "crime" police deal with is embedded. Without doing so, the forensics displays help bestow legitimacy to the idea that effectively dealing with criminalized acts is a process that involves finding and incapacitating the "dangerous" actors who commit them, rather than addressing conditions that gave rise to "crime." With the focus on the science of police surveillance, like other museums these sites tend not to engage with controversies related to science and technology (Delicado, 2009) that could raise ethical questions about deployment of such devices (Macdonald & Silverstone, 1992).

Conclusion

Writing about the Huntington Park International Police Museum in California, which is housed within a police station alongside other government buildings, McNair (2011) remarks upon the imposing exterior of the site: "The feeling of surveillance is heavy; the concentration of regulatory agents and governing bodies cultivates an overwhelming sense of regulation and monitoring of movement. The experience of walking the grounds feels like moving through a fortress" (p. 61). Once inside, McNair notes the open and welcoming nature of the museum, which is managed by an affable retired police officer – a sharp contrast from its surrounding structures and subject matter. A similar contrast in atmosphere exists at many police museums in Canada. The four sites examined in this study are all located on the grounds of operational police facilities. Yet, the idea and feelings of surveillance do not end at the doors of police museums. Depictions of surveillance are integral to many of the exhibitions, sometimes overtly and sometimes in more subtle ways. Several conclusions can be drawn from our findings, which are presented below.

Surveillance is represented inside the police museums as a tool used by law enforcement to catch and monitor people who are believed to have committed criminalized acts. It is depicted as being used as necessary, not indiscriminately. The use of DNA databases and the increased information gathering capabilities of police services are not presented as tools of surveillance that may be intrusive or oppressive, but rather the result of important scientific and technological progress in policing. Information on the vast expansion of police surveillance in Canada and the decisions leading to this development is absent. We are most familiar today with the expansion of surveillance since September 11, 2001, but the growth of systemic state surveillance in Canada was tied to a longer history of highly politicized campaigns targeting particular "threats," including among others the Fenians in the late nineteenth century, Indigenous people throughout Canadian history, communists in the middle part of the twentieth century, and Québec sovereignty movements in the 1960s and 1970s (Whitaker, Kealey, & Parnaby, 2012). As other chapters in this book have shown, similar developments took place around the globe. By maintaining focus on topics such as the technology available to assist police in investigations and the actions of people deemed "criminals," exhibits keep visitors at a safe distance from the feelings and experiences of those who find themselves targeted by surveillance. Suspects tend to be depicted as particularly dangerous persons, while police officers conducting surveillance are portrayed as heroic and selfless for subjecting themselves to an enhanced level of risk.

The tendency for police museums to avoid discussions that move beyond technical information about and operational successes of using surveillance, whether in the distant or recent past, shields visitors from better understanding the power of police and state surveillance in Canada and its racialized, gendered, heteronormative, and classist dimensions (Comack, 2012; Razack, 2015). Given the economic and institutional ties that many police museums have with existing police forces, these forms of representation serve as one approach of public relations management in policing that mitigates future public scrutiny, neutralizing efforts to establish meaningful accountability and transparency measures for law enforcement. In other words, it is precisely the critical perspectives and questions raised by the contributors to this book that are largely effaced in the narratives elaborated in the museums. While addressing the realities of controversial subjects such as surveillance inside police museums may be difficult (Bunch, 1992), patrons deserve to be adequately informed about the policing practices they implicitly sanction. This includes being knowledgeable about the realities of police surveillance, as well as other important topics such as the pains inflicted by police (Harkin, 2015). We conclude by sharing one example of the latter, which further highlights the significance of what is made memorable versus forgettable, visible versus invisible, within police museums.

The Ipperwash crisis was a territorial dispute in 1995 between the federal government and the Stony Point First Nation community, which the OPP monitored and in which it subsequently intervened. During the conflict, an unarmed Ojibwa man named Dudley George was shot and killed by an OPP sniper (Edwards, 2001). The OPP Museum does not have a display about this incident, mirroring the way that Alcatraz initially ignored "the memory of Native American Occupation" (Bergman & Smith, 2010, p. 184). A great deal of criticism was levelled at the OPP for their actions resulting in Dudley George's death (Edwards, 2001). The killing also resulted in an inquiry into OPP practices. In an interview, a museum staff member commented on its absence:

> Does the OPP Museum have a display about Ipperwash? No, we don't. It's not necessarily time for that yet. But what we do have is significant content from the Aboriginal Policing Bureau, and we have the backing that if someone wants to have a conversation about Ipperwash, we're certainly prepared for that. We could put you into connection with the people whose job it is to have those conservations. We're certainly not afraid of talking about our connections with Aboriginal policing and the greater political context, but that doesn't really belong, that's not a conversation that belongs to the museum. In those cases, we would facilitate conversations with people who those issues do belong to. Um, we certainly are aware of it and perhaps someday this will be an avenue to display those items, you know, what happened through the Ipperwash Inquiry, would be what we would focus on, um, the changes that happened to policing because of that. You know, the relationship with the George Family now, the fact we have Sam George's [brother of the protestor killed] tree out on the front lawn, and what really the organization learned, that's what we would focus on.

Police surveillance is about social sorting, and legitimating and/or facilitating exclusion (Walby, 2009). The inclusion of an Ipperwash display inside the OPP Museum could be a means of providing a more balanced or even critical version of history, including the role of police surveillance, which could facilitate new, empathetic understandings for visitors, while providing the provincial police force with an opportunity to assume accountability for Dudley George's death.

The imaginary display that the interviewee envisions might play a role in challenging popular representations of law enforcement targets as dangerous, violent actors that the depictions of surveillance serve to promote (also see Chen, Fiander, Piché, & Walby, 2016). It could also provoke thought about Indigenous peoples, along with the colonial processes contributing to their mass policing and incarceration. Overall, the four police museums avoid topics that may generate scrutiny towards police practices. This prevents visitors from acquiring a deeper understanding about the realities of policing in Canada, including police surveillance, leaving problematic preconceptions about "crime" and "criminal justice" unchallenged.

These issues lead to the question of what the purpose and role of police museums should be, or, drawing on the words of the interviewee, what conversations "belong" to police museums. It is often noted that museums should be safe spaces for citizens to gather and learn about the past so that they can form a clearer understanding of the present. But when it comes to museums and representations of state power, there is such a thing as playing it too safe. At the police museums we examined, an accurate understanding of the past and present often remains elusive. Much of this information remains as Project Apocalypse once was – veiled and covert.

REFERENCES

Alia, V. (2006). *Names and Nunavut: Culture and identity in Arctic Canada.* New York, NY: Berghahn Books.

Atkinson, R., & Fiorito, J. (2017). *The life crimes and hard times of Ricky Atkinson: Leader of the Dirty Tricks Gang – A True Story.* Toronto, ON: Exile Editions.

Ball, K. (2017). All consuming surveillance: Surveillance as marketplace icon. *Consumptions, Markets & Culture, 20*(2), 95–100.

Barnard-Wills, K., & Barnard-Wills, D. (2012). Invisible surveillance in visual art. *Surveillance & Society, 10*(3–4), 202–14.

Barry, A. (1995). Reporting and visualizing. In C. Jenks (Ed.), *Visual culture* (pp. 42–57). London, UK: Routledge.

Behind the badge ... The story of the Ontario Provincial Police [Gallery guide]. (2014). Orillia, ON: Queen's Printer for Ontario.

Bergman, T., & Smith, C. (2010). You were on Indian land: Alcatraz Island as recalcitrant memory space. In G. Dickinson, C. Blair, & B. Ott (Eds.), *Places of public memory: The rhetoric of museums and memorials* (pp. 160–90). Tuscaloosa, AL: University of Alabama Press.

Brodeur, J. P. (1992). Undercover policing in Canada: Wanting what is wrong. *Crime, Law and Social Change, 18*, 105–36.

Brodeur, J. P. (1995). Undercover policing in Canada: A study of its consequences. In C. Fijanut & G. Marx (Eds.), *Undercover: Police surveillance in comparative perspective* (pp. 71–102). The Hague, Netherlands: Kluwer.

Brown, M. (2009). *The culture of punishment: Prison, society, and spectacle.* New York, NY: New York University Press.

Buffington, R. M. (2012). Institutional memories: The curious genesis of the Mexican police museum. *Radical History Review, 113*, 155–69.

Bunch, L. (1992). Embracing controversy: Museum exhibitions and the politics of change. *The Public Historian, 14*(3), 63–5.

Caimari, L. (2012). Vestiges of a hidden life: A visit to the Buenos Aires Police Museum. *Radical History Review, 113*, 143–54.

Chazkel, A. (2012). Police museums in Latin America: Preface. *Radical History Review*, 113, 127–33.

Chen, A., Fiander, S., Piché, J., & Walby, K. (2016). Captive and captor representations at Canadian penal history museums. *Qualitative Sociology Review, 12*(4), 22–42.

Cole, S. (2001). *Suspect identities: A history of fingerprinting and criminal identification.* Cambridge, MA: Harvard University Press.

Comaroff, J., & Comaroff, J. (2004). Criminal obsessions, after Foucault: Postcoloniality, policing, and the metaphysics of disorder. *Critical Inquiry, 30*(4), 800–24.

Comack, E. (2012). *Racialized policing: Aboriginal people's encounters with the police.* Black Point, NS: Fernwood Publishing.

Crane, S. A. (2000). Introduction: Of museums and memory. In S. A. Crane (Ed.), *Museums and Memory* (pp. 1–13). Stanford, CA: Stanford University Press.

Delicado, A. (2009). Scientific controversies in museums: Notes from a semi-peripheral country. *Public Understanding of Science, 18*(6), 759–67.

Edwards, P. (2001). *One dead Indian: The premier, the police, and the Ipperwash Crisis.* Toronto, ON: Stoddart.

Ericson, R. (1994). The division of expert knowledge in policing and security. *British Journal of Sociology, 45*(2), 149–75.

Ferguson, M., Piché, J., & Walby, K. (2017). Representations of detention and other pains of law enforcement in police museums in Ontario, Canada. *Policing and Society, 29*(3), 318–32.

Fiander, S., Chen, A., Piché, J., & Walby, K. (2016). Critical punishment memorialization in Canada. *Critical Criminology, 24*(1), 1–18.

Finn, J. (2009). *Capturing the criminal image: From mug shot to surveillance society.* Minneapolis, MI: University of Minnesota Press.

Goodwin, C. (1994). Professional vision. *American Anthropologist, 96*(3), 606–33.

Harkin, D. (2015). The police and punishment: Understanding the pains of policing. *Theoretical Criminology, 19*(1), 43–58.

Hausken, L. (2014). Forensic fiction and the normalization of surveillance. *Nordicom Review, 1,* 3–16.

King, A., & Maruna, S. (2006). The function of fiction for a punitive public. In P. Mason (Ed.), *Captured by the media* (pp. 16–30). Cullompton, UK: Willan.

Kruger, E. (2013). Image and exposure: Envisioning genetics as a forensic-surveillance matrix. *Surveillance & Society, 11*(3), 237–51.

Lawless, C. (2016). *Forensic science: A sociological introduction.* London: Routledge.

Lippert, R. K., & Newell, B. C. (2016). Debate introduction: The privacy and surveillance implications of police body cameras. *Surveillance & Society, 14*(1), 113–16.

Lippert, R. K., & Scalia, J. (2015). Attaching Hollywood to a surveillant assemblage: Discourses of video surveillance. *Media and Communication, 3*(3), 26–38.

Loftus, B., Goold, B., & Mac Giollabhui, S. (2016). From a visible spectacle to an invisible presence: The working culture of covert policing. *British Journal of Criminology, 56*(4), 629–45.

Loftus, B., & Goold, B. (2012). Covert surveillance and the invisibilities of policing. *Criminology and Criminal Justice, 12*(3), 275–88.

Lyon, D. (2001). *Surveillance society: Monitoring everyday life.* Buckingham, UK: Open University Press.

Lyon, D. (2015). *Surveillance after Snowden.* Cambridge, UK: Polity.

Macdonald, S., & Silverstone, R. (1992). Science on display: The representation of scientific controversy in museum exhibitions. *Public Understanding of Science, 1*(1), 69–87.

Maynard, S. (1994). Through a hole in the lavatory wall: Homosexual subcultures, police surveillance and the dialectics of discovery, Toronto, 1890–1930. *Journal of the History of Sexuality, 5*(2), 207–42.

McNair, A. (2011). *The captive public: Media representations of the police and the (il)legitimacy of police power.* PhD dissertation – Department of Critical Studies. Los Angeles, CA: University of Southern California.

Monahan, T. (2011). Surveillance as cultural practice. *Sociological Quarterly, 52*(4), 495–508.

Muir, L. (2012). Control space? Cinematic representations of surveillance space between discipline and control. *Surveillance & Society, 9*(3), 263–79.

Nelkin, D., & Andrews, L. (1999). DNA identification and surveillance creep. *Sociology of Health & Illness, 21*(5), 689–706.

O'Neill, M., & Loftus, B. (2013). Policing and the surveillance of the marginal: Everyday contexts of social control. *Theoretical Criminology, 17*(4), 437–54.

Razack, S. (2015). *Dying from improvement: Inquests and inquiries into Indigenous deaths in custody.* Toronto, ON: University of Toronto Press.

Robert, D., & Dufresne, M. (2008). The social uses of DNA in the political realm or how politics constructs DNA technology in the fight against crime. *New Genetics and Society, 27*(1), 69–82.

Sekula, A. (1986). The body and the archive. *October*, *39*, 3–64.

Skolnick, J. (1982). Deception by police. *Criminal Justice Ethics*, *1*(2), 40–54.

Smith, D. G. (1993). The emergence of 'Eskimo status': An examination of the Eskimo Disc List System and its social consequences, 1925–1970. In N. Dyck & J. B Waldram (Eds.), *Anthropology, public policy, and Native peoples in Canada* (pp. 41–68). Montreal, QC, and Kingston, ON: McGill-Queen's University Press.

Smith, G. (2012). Surveillance work(ers). In K. Ball, K. Haggerty, & D. Lyon (Eds.), *Routledge handbook of surveillance studies* (pp. 107–16). London, UK: Routledge.

Smith, L. (2006). *Uses of heritage*. London, UK: Routledge.

Walby, K. (2009). Surveillance of male with male public sex in Ontario, 1983–1994. In S. Hier & J. Greenberg (Eds.), *Surveillance: Power, problems and politics* (pp. 46–58). Vancouver, BC: University of British Columbia Press.

Walby, K., & Piché, J. (2015). Staged authenticity in penal history sites across Canada. *Tourist Studies*, *15*(3), 231–47.

Whitaker, R., Kealey, G., & Parnaby, A. (2012). *Secret service: Political policing in Canada from the Fenians to fortress America*. Toronto, ON: University of Toronto Press.

Wilson, D., & McCulloch, J. (2012). (Un)controlled operations: Undercover in the security control society. In J. McCulloch & S. Pickering (Eds.), *Borders and crime: Pre-crime, mobility and serious harm in an age of globalization* (pp. 163–78). London, UK: Palgrave Macmillan.

Afterword

SIMONE BROWNE

New York City in 1966 was a time of rising crime rates, police corruption, a citywide transit strike that lasted twelve days, and the groundbreaking for the construction of the World Trade Center, while Nina Simone's *Four Women*, a "song that challenged the entangled oppressions of race and gender through the legacies of chattel slavery" (Redmond, 2013, p. 184), was banned from airplay on certain radio stations in that city. Concerned with crime and security, and tired of the slow police response to distress calls from the public in cases of emergency, Marie Van Brittan Brown and her husband, Albert L. Brown, invented what they called a "Home Security System Utilizing Television Surveillance" and filed a patent application on August 1 of that same year. Marie Van Brittan Brown was a nurse by profession and often worked the night shift, while Albert L. Brown was an electronics technician.

On December 2, 1969, the United States Patent Office issued Patent 3,482,037 to the couple for their home security system. Van Brittan Brown is listed first on the patent as its inventor and it is now referenced in dozens of other patents, including one held by Sony Corporation. The blueprint for this system details its ingenious intricacy. It includes a video scanning device at the entrance of the home, an intercom, an alarm, audio recording capabilities, and a radio-controlled lock that could allow an occupant to unbolt the door lock while lying comfortably in bed. I would like to imagine that Marie Van Brittan Brown came up with the idea for Patent 3,482,037 while talking to other nurses who worked the night shift with her. I would also like to think that they conceptualized this technology as a way outside of calling the police. Perhaps they thought of it as a strategy of self-defence, similar to what Imani Perry (2011) terms "a privacy marking gesture" in her discussion of the practice of window blinds being drawn in working-class Black neighbourhoods in the United States where she suggests that it "might be a cultural practice born in fear, or it might be a cultural practice indicating people's self-definition as rights bearing, privacy demanding citizens. Likely it is both and more" (p. 118).

There is little published work on the life of Marie Van Brittan Brown and the making of Patent 3,482,037, a precursor to contemporary remote home monitoring systems, video doorbells, and other home digital video recorder surveillance products for the consumer market. As such, this apparent absence in the official genealogy of surveillance technologies and their applications forms part of a generalized absenting of Black women's contribution when it comes to the ways that surveillance is conceptualized and historicized. Given this absence, how, if in any way, could our relationship to and knowledge of surveillance and its infrastructure be altered by knowing that there was a time when a Black woman from Queens, New York, played a pivotal role in the development of CCTV? More broadly, if our understanding of new forms of state surveillance regimes is predicated on our understanding of their analogues and continuities, to where do we turn? In their introduction to this collection, Emily van der Meulen and Robert Heynen instruct us that, by "looking at what historians do" with the archives that "carry the material traces of the surveillance work of others, most notably state actors," in order to gain insight when it comes to surveillance, we can perhaps grasp how to disrupt its methodologies of social control.

I tell the story of Marie Van Brittan Brown and raise the question of absence to point to the ways that our thinking about and questioning of surveillance and its current and most pressing tensions and concerns is necessarily sharpened when we start from a point of historicizing its technologies and the operational logics of its techniques. Collectively, the chapters in this book do precisely that work of centring what is often made absent in the official genealogies of surveillance states. From the conditions of colonial governance, to the repressive policing tactics of occupation and apartheid, to the ways that certain marginalized populations are deemed deviant and subjected to a scrutinizing bodily surveillance as part of a state's own efforts at nation-making, the theoretical approaches offered by *Making Surveillance States* provide its readers with the analytical tools for an understanding of how surveillance is conceived and how it can be challenged.

Take, for example, B Camminga's discussion of the panics surrounding perceived gender transgression and the policing of queer expressive practices. Here Camninga accounts for how the application of legislation aimed at policing queer life in public spaces makes visible the heteronormative scripts that undergirded apartheid era South Africa and that state's post-apartheid formation. In another example of how state practices do the work of rendering some subjects as marginal in the name of nation-making, Holly Caldwell's examination of public health directives concerning the Deaf in late nineteenth-century Mexico points to how the state's capacity for inclusion is about its very incapacitating of some populations as targets for disciplinary control. In that case, deafness was marked, she writes, "as a catalyst enabling the growth of eugenic surveillance and of a range of forms of moral regulation." Read together, Camminga's and Caldwell's contributions point to the sometimes bureaucratic and administrative ways that surveillance practices reveal themselves to be differently applied to marginalized individuals and groups.

Population management by way of documents, legislation, and other frameworks of control that sought to exclude many from the categories of citizen and human, and the rights said to be allotted to those who fall under its rubrics, was and still continues to be a transnational formation. Such workings are uncovered here in Uma Dhupelia-Mesthrie and Margaret Allen's examination of how Asian mobilities were managed by way of photographs, fingerprints, and paperwork in order to categorize, colonize, restrict, and exclude. This take on how power and paper files interrelate is insightfully termed "dossierveillance" by Christina Plamadeala as a way to name the workings of state policy and bureaucracy where the making of an individual's dossier is about not only producing a biographical history of one's state subjecthood, but also about the making of surveillance states.

Together, the research and analytical concepts presented in *Making Surveillance States* extend the field of surveillance studies by drawing our attention to the historical and then asking us to, as Matthew Ferguson, Justin Piché, and Kevin Walby put it in the book's final chapter, "consider how history itself is constructed." By studying surveillance in this manner, we might come away with different types of responses to the troubling questions and urgent concerns that arise from the contemporary projects of surveillance states.

REFERENCES

Perry, I. (2011). *More beautiful and more terrible: The embrace and transcendence of racial inequality in the United States.* New York, NY: New York University Press.

Redmond, S. (2013). *Anthem: Social movements and the sound of solidarity in the African diaspora.* New York, NY: New York University Press.

Contributors

Margaret Allen is a professor emerita in gender studies at the University of Adelaide, South Australia, and a member of the Fay Gale Centre. Currently she researches transnational, postcolonial, and gendered histories, especially on links between India and Australia from circa 1880 to 1940, focusing upon Indians negotiating the White Australia policy (see "I am a British subject": Indians in Australia claim rights, 1880–1940. *History Australia, 15*(3), 2018). She is also working on an Australian Research Council funded team project, "Beyond Empire: Transnational Religious Networks & Liberal Cosmopolitanisms."

Simone Browne is an associate professor in the Department of African and African Diaspora Studies at the University of Texas at Austin, United States. Her first book, *Dark Matters: On the Surveillance of Blackness* (Duke University Press, 2015), examines surveillance with a focus on transatlantic slavery, biometric technologies, branding, airports, and creative texts. Simone is also a member of Deep Lab, a feminist collaborative composed of artists, engineers, hackers, writers, and theorists.

Holly Caldwell is an assistant adjunct professor in the Department of History and Political Science at Chestnut Hill College, Pennsylvania, United States. Her research interests include deafness studies, public health policy, and social welfare reform in Latin America. Her work examines the social aspects of medicine and disabilities history, with a particular focus on the transnational connections between Mexico, Europe, and the United States. Currently, she is preparing her book manuscript on the international medicalization of deafness and deaf education reform at Mexico's *Escuela Nacional de Sordomudos*, the nation's first school for the Deaf.

B Camminga is a postdoctoral fellow at the African Centre for Migration and Society at the University of Wits, Johannesburg, South Africa. Their work considers the interrelationship between the conceptual journeying of the term "transgender" from the Global North and the physical embodied journeying of African transgender asylum seekers globally. Their first book, *Transgender Refugees and the Imagined South Africa*, was published by Palgrave Macmillan in 2018.

Uma Dhupelia-Mesthrie is a senior professor in the History Department at the University of the Western Cape, Cape Town, South Africa. Her publications reflect a sustained research interest in India-South Africa connected histories with a specific focus on transnational mobilities, travel documents, and methods of identification. She edited a special issue of *Kronos: Southern African Histories* (2014) titled "Paper Regimes" that focused on South African documentary systems.

Matthew Ferguson is a doctoral student in the Department of Criminology at the University of Ottawa, Ontario, Canada. As a member of the Carceral Cultures Research Initiative, he has conducted research on police museums and prison tourism sites in Canada. His co-authored papers have appeared in academic journals such as *Policing and Society* and the *International Journal of Tourism Research*. He is also the co-author of a chapter in *The Palgrave Handbook of Prison Tourism* (2017).

Elisabetta Ferrari is a doctoral candidate at the Annenberg School for Communication at the University of Pennsylvania, United States. She studies social movements, activists, and technologies. In particular, her current research focuses on how different social and political actors construct myths and narratives about technology and the role that such discourses play in their mobilization for social change. She investigates these phenomena in both contemporary movements and historical cases, such as the wave of 1960s–1970s activism in North America and Western Europe.

Robert Heynen is an assistant professor in the Department of Communication Studies at York University, Ontario, Canada. His areas of specialization include surveillance studies, cultural and critical theory, the politics of embodiment, media history, and digital media. He is the author of *Degeneration and Revolution: Radical Cultural Politics and the Body in Weimar Germany* (Brill, 2015, and Haymarket, 2016) and co-editor of *Expanding the Gaze: Gender and the Politics of Surveillance* (with Emily van der Meulen, University of Toronto Press, 2016).

David Lyon is the director of the Surveillance Studies Centre and a professor of Sociology and Professor of Law at Queen's University, Ontario, Canada. He is author, co-author, editor, or co-editor of 29 books that have been translated into 18 languages, the most recent of which is *The Culture of Surveillance: Watching as a Way of Life* (Polity, 2018). Lyon has received numerous awards and honours for his work, including the SSHRC Impact: Insight Award in 2015. He is currently Principal Investigator of the Big Data Surveillance project.

Kathryn Montalbano is an assistant professor in the Department of Communication Studies at Young Harris College, Georgia, United States. She specializes in communication history, media policy, and the intersection of religion, media, and the state. She is the author of *Government Surveillance of Religious Expression: Mormons, Quakers, and Muslims in the United States* (Routledge, 2018). Montalbano earned a PhD in communications from Columbia University in 2016 and a BA in English and Sociology (minor) from Haverford College in 2009.

Midori Ogasawara is a Banting Postdoctoral Fellow in the Department of Criminology at the University of Ottawa, Ontario, Canada. She holds a PhD in sociology from Queen's University, Ontario, where her research focused on identification systems and their violent consequences from postcolonial perspectives. Ogasawara also developed her investigative skills for Japan's national newspaper *Asahi Shimbun* from 1994 to 2004, and subsequently in the John S. Knight Professional Journalism Fellowships at Stanford University. In 2016, she became the first Japanese researcher and journalist to interview the NSA whistleblower Edward Snowden via video channel, which she published as *Snowden Talks About the Horrors of the Surveillance Society* (Mainichi Shimbun Publishing, 2016).

Darren Palmer is an associate professor in criminology at Deakin University, Geelong, Australia. Darren's research examines the governmentalities of policing and the multiple and often conflicting roles of surveillance in the fields of law enforcement and crime. He is co-author of *Global Criminology* (with Ian Warren, Law Book Company, 2015), which examines jurisdictional problems stemming from transnational crime, and co-editor of *National Security, Surveillance and Terror: Canada and Australia in Comparative Perspective* (with Randy K. Lippert, Kevin Walby, and Ian Warren, Palgrave Macmillan-Springer, 2016). He is also a member of the Alfred Deakin Research Institute at Deakin University.

Justin Piché is an associate professor in the Department of Criminology at the University of Ottawa, Ontario, Canada, and Co-managing Editor of the *Journal of Prisoners on Prisons*. He is also a co-investigator of the Carceral Cultures Research Initiative. He was awarded the 2012 Aurora Prize from the Social Sciences and Humanities Research Council of Canada and the 2016 Young Researcher Award from the Faculty of Social Sciences at the University of Ottawa.

Cristina Plamadeala is a doctoral student in humanities at Concordia University, Montreal, and at the École des Hautes Études en Sciences Sociales in Philosophy (cotutelle), Paris. Her SSHRC-funded research seeks to explain the reasons, motives, and methods behind the phenomena of collaboration with and resistance against the Securitate, Romania's secret police during Romania's communist era (1945–1989). Her most recent publications elaborate on the theories/concepts of dossierveillance and psuchegraphy that she had coined and developed during her doctoral studies. A graduate of Georgetown, Harvard, and Concordia Universities, Cristina currently resides in Paris, France. She is a writer and plays piano.

John Remensperger is a doctoral student at the Annenberg School for Communication at the University of Pennsylvania, United States. He researches the relationship between communication technologies and political institutions, including political parties, interest groups, and advocacy organizations. He researches how these groups use technology to organize their members, foster citizen engagement, and educate citizens to achieve policy change. His recent projects look at how organizational culture and intra-organizational dynamics impact, and are affected by, the adoption of new technologies.

Ahmad H. Sa'di is an associate professor in the Department of Politics & Government at Ben-Gurion University of the Negev, Israel. In addition to more than 40 articles and chapters, he is co-editor of *Nakba: Palestine, 1948 and the Claims of Memory* (with Lila Abu-Lughod, Columbia University Press, 1997) and author of *Thorough Surveillance: The Genesis of Israeli Policies of Population Management, Surveillance & Political Control towards the Palestinians* (Manchester University Press, 2014). His work has been published in English, Arabic, Hebrew, German, and Japanese.

Jacob Steere-Williams is an associate professor of history at the College of Charleston, South Carolina, United States. His research interests are in the social and cultural history of infectious disease and public health in nineteenth-century Britain and the British Empire. He has published on the history of epidemiology, bacteriology, chemistry, veterinary medicine, and the popularization of Victorian science and medicine, and is the author of the forthcoming monograph *The Filth Disease: Typhoid Fever and the Practices of Epidemiology in Victorian England.*

Emily van der Meulen is an associate professor of criminology at Ryerson University, Ontario, Canada. She conducts research in the areas of critical and feminist criminology, socio-legal studies, sex work studies, prison harm reduction, and surveillance studies. She has (co-)edited numerous books, including *Red Light Labour: Sex Work Regulation, Agency, and Resistance* (with Elya M. Durisin and Chris Bruckert, UBC Press, 2018) and *Expanding the Gaze: Gender and the Politics of Surveillance* (with Robert Heynen, University of Toronto Press, 2016).

Kevin Walby is an associate professor and chancellor's research chair in the Department of Criminal Justice, University of Winnipeg, Manitoba, Canada. He has authored and (co-) edited eight books, most recently *National Security, Surveillance, and Terror: Canada and Australia in Comparative Perspective* (with Randy Lippert, Ian Warren, and Darren Palmer, Palgrave, 2017).

Ian Warren is a senior lecturer in criminology at Deakin University, Geelong, Australia. His research examines contemporary and historical aspects of transnational criminal law, policing, and governance, and the emergence of geographic exclusion as an urban crime control strategy. He is co-author of *Global Criminology* (with Darren Palmer, Law Book Company, 2015), which examines jurisdictional problems stemming from transnational crime, and co-editor of *National Security, Surveillance and Terror: Canada and Australia in Comparative Perspective* (with Randy K. Lippert, Kevin Walby, and Darren Palmer, Palgrave Macmillan-Springer, 2016). He is also a member of the Alfred Deakin Research Institute at Deakin University.

Index